PRINCIPLES OF THE MAGNETIC METHODS IN GEOPHYSICS

METHODS IN GEOCHEMISTRY AND GEOPHYSICS

(Volumes 1–28 are out of print)

29. V.P. Dimri – Deconvolution and Inverse Theory – Application to Geophysical Problems
30. K.-M Strack – Exploration with Deep Transient Electromagnetics
31. M.S. Zhdanov and G.V. Keller – The Geoelectrical Methods in Geophysical Exploration
32. A.A. Kaufman and A.L. Levshin – Acoustic and Elastic Wave Fields in Geophysics, I
33. A.A. Kaufman and P.A. Eaton – The Theory of Inductive Prospecting
34. A.A. Kaufman and P. Hoekstra – Electromagnetic Soundings
35. M.S. Zhdanov and P.E. Wannamaker – Three-Dimensional Electromagnetics
36. M.S. Zhdanov – Geophysical Inverse Theory and Regularization Problems
37. A.A. Kaufman, A.L. Levshin and K.L. Larner – Acoustic and Elastic Wave Fields in Geophysics, II
38. A.A. Kaufman and Yu. A. Dashevsky – Principles of Induction Logging
39. A.A. Kaufman and A.L. Levshin – Acoustic and Elastic Wave Fields in Geophysics, III
40. V.V. Spichak – Electromagnetic Sounding of the Earth's Interior
41. A.A. Kaufman and R.O. Hansen – Principles of the Gravitational Method
42. A.A. Kaufman, R.O. Hansen and R.L.K. Kleinberg – Principles of the Magnetic Methods in Geophysics

Methods in Geochemistry and Geophysics, 42

PRINCIPLES OF THE MAGNETIC METHODS IN GEOPHYSICS

A.A. Kaufman
Emeritus Professor

R.O. Hansen†

and

Robert L. K. Kleinberg
Schlumberger-Doll Research

ELSEVIER

Amsterdam – Boston – Heidelberg – London – New York – Oxford – Paris
San Diego – San Francisco – Sydney – Tokyo

Elsevier
Radarweg 29, PO Box 211, 1000 AE Amsterdam, The Netherlands
Linacre House, Jordan Hill, Oxford OX2 8DP, UK

First edition 2009

British Library Cataloguing in Publication Data
A catalogue record for this book is available from the British Library

Library of Congress Cataloging-in-Publication Data
A catalog record for this book is available from the Library of Congress

ISBN: 978-0-444-52995-4
ISSN: 0076-6895

Printed and bound in Hungary

09 10 11 12 13 10 9 8 7 6 5 4 3 2 1

This book is dedicated to R.O. Hansen

Contents

Chapter 5

Uniqueness and the Solution of the Forward and Inverse Problems

Chapter 6

Paramagnetism, Diamagnetism, and Ferromagnetism

Chapter 7

Nuclear Magnetism Resonance and Measurements of Magnetic Field

Introduction

Magnetic methods are widely used in exploration, engineering, borehole, and global geophysics and the subjects of this monograph are the physical and mathematical principles of these methods regardless of the area of application. The first chapter is devoted entirely to the magnetic field caused solely by conduction currents. Beginning with Ampere's law we analyze the force of interaction between currents and then introduce the concept of the magnetic field and discuss its fundamental features. In order to simplify a study of magnetic field the vector potential is introduced. Special attention is paid to the system of equations of the magnetic field at regular points and at places where surface density of currents differs from zero. We also consider several examples of the field behavior because of its relationship to the application of magnetic methods in geophysics. The second chapter describes in detail the theory of the magnetic field in the presence of magnetic medium. There is a section where we study distribution of magnetization currents and association between these currents and the vectors of the inductive and remanent magnetization. The systems of equations for the magnetic field and fictitious field $\boldsymbol{H}$ are derived and we illustrate a difference between these fields considering several examples. The behavior of magnetic field in the presence of magnetic medium is described in the next chapter, where at the beginning we consider questions such as a solution of the boundary-value problems and theorem of uniqueness. Then behavior of the magnetic field and the vector of magnetization are analyzed in the presence of different magnetic bodies. In order to describe the theory of the vertical magnetometer we study several topics related to this subject, among them are the force acting on magnet, moment of rotation, interaction between two magnets, and the relationship between Ampere's and Coulomb's laws. The main magnetic field of the earth is described in Chapter 4. We begin with an introduction to the central characteristics of this field and briefly describe the history of its study. Then we consider the spherical harmonic analysis of the earth's field that is naturally preceded by information on Legendre's functions. In Chapter 5 we focus on a solution of the inverse problems in the magnetic method and describe uniqueness and nonuniqueness, ill- and well-posed problems, stable and unstable parameters, regularization, as well as different methods of solution to the forward problem. The main purpose of Chapter 6 is to describe diamagnetic, paramagnetic, and ferromagnetic substances proceeding from the atomic physics. Following the Feynman lectures, we introduce concepts of the angular momentum and magnetic moment of atom, derive an expression for the frequency of precession, and describe energy states of atomic systems. This material allows us to obtain formulas for the vector of magnetization of paramagnetic substances and investigate the relation with atomic parameters. Considering ferromagnetism we discuss the magnetization curve, spontaneous magnetization, Curie temperature and Weiss domains, as well as the principle of the fluxgate magnetometer. Finally, the last chapter is completely devoted to nuclear magnetic resonance, since this phenomenon is used for measurements of the magnetic field and also has found an application in the borehole geophysics. At the beginning we derive an equation for the vector of nuclear magnetization and describe its solution in a rotating system of coordinates. Then Bloch's equations are introduced in order to take into account the influence of a medium. Special attention is paid to measurements of relaxation processes, including topics such as the spin echoes or refocusing. As well, in this chapter we describe the principle of the proton precession magnetometer and the optically pumped magnetometer. Also, we included an appendix which describes the important role of the magnetic method in the development the plate tectonic theory.

Acknowledgments

We express our thanks to Dr. S. Akselrod, Dr. H.N. Bochman, Dr. J Fuks, Dr. A. Levshin, Dr. K. Naugolnykh, Dr. L. Osrtovsky, Dr. C. Skokan, and Dr. I. Hrvoic for their very useful comments and suggestions. Dr. M. Prouty is the coauthor of Section 7.9 of this monograph, and we would like to gratefully acknowledge his important contribution.

List of Symbols

a	major semi-axis of spheroid and ellipsoid
$\boldsymbol{a}$	acceleration
$\boldsymbol{A}$	vector potential of magnetic field
b	minor axis of spheroid and axis of ellipsoid
$\boldsymbol{B}$, $\boldsymbol{B}_e$, $\boldsymbol{B}_0$, $\boldsymbol{B}_a$, $\boldsymbol{B}_N$	magnetic field
c	axis of ellipsoid
$d\boldsymbol{M}$	the magnetic moment of an elementary volume
e	electric charge
$\boldsymbol{E}$	electric field
Ξ	electromotive force
f	frequency
$\boldsymbol{F}$	force of interaction
$\boldsymbol{F}_m$	magnetic force
f_0, f_1, f_p	precession frequency
G	Green's function
$\boldsymbol{g}$	gravitational field
h	elevation
$\boldsymbol{H}$	fictitious field
h_1, h_2, h_3	metric coefficients
$\boldsymbol{i}$, $\boldsymbol{j}$, $\boldsymbol{k}$	unit vectors in Cartesian system of coordinates
J_n	Bessel functions of n order
J'_n	derivative of the Bessel function
$j=(-1)^{1/2}$	imaginary unit
l	length of the pendulum and torsion balance
L_{qp}	distance between two points q and p
$\boldsymbol{L}$	angular momentum of electron, proton, atom
m	mass of measuring device
P_e, P_p	dipole moment of electron and proton
$\boldsymbol{p}$	magnetic moment of dipole
$\boldsymbol{P}^{av}$	average magnetic moment
$\boldsymbol{n}$	unit vector
$P_n(\mu)$, $Q_n(\mu)$	Legendre's functions of the first and second kind
$P'_n(\mu)$, $Q'_n(\mu)$	associated Legendre's functions of the first and second kind
r	radius of sphere and cylinder
S	surface, spherical harmonics
T	period of oscillations, temperature

T_c	Curie point
t	time
τ	torque
U	scalar potential of the magnetic field
$\boldsymbol{v}$	linear velocity
α	parameter, angle
γ	gyromagnetic ratio
χ	susceptibility
μ	magnetic permeability
μ_0	constant
θ	angle
φ	angle
ε, η, φ	coordinates in spheroidal system of coordinates
ω	solid angle, angular frequency

Chapter 1

Magnetic Field in a Nonmagnetic Medium

1.1. INTERACTION OF CONSTANT CURRENTS AND AMPERE'S LAW

Numerous experiments performed at the beginning of the19th century demonstrated that constant currents interact with each other; that is mechanical forces act at every element of the circuit. Certainly, this is one of the amazing phenomena of the nature and would have been very difficult to predict. In fact, it is almost impossible to expect that the motion of electrons inside of wire may cause a force on moving charges somewhere else, for instance, in another wire with current, and for this reason the phenomenon of this interaction was discovered by chance. It turns out that this force of interaction between currents in two circuits depends on the magnitude of these currents, the direction of charge movement, the shape and dimensions of circuits, as well as the their mutual position with respect to each other. The list of factors clearly shows that the mathematical formulation of the interaction of currents should be much more complicated task than that for masses or electric charges. In spite of this fact, Ampere was able to find a relatively simple expression for the force of the interaction of so-called elementary currents:

$$d\boldsymbol{F}(p) = \frac{\mu_0}{4\pi} I_1 I_2 \frac{d\boldsymbol{l}(p) \times [d\boldsymbol{l}(q) \times \boldsymbol{L}_{qp}]}{L_{qp}^3} \tag{1.1}$$

where I_1 and I_2 are magnitudes of the currents in the linear elements $dl(p)$ and $dl(q)$, respectively, and their direction coincides with that of the current density; $\boldsymbol{L}_{qp}$ the distance between these elements and is directed from the point q to the point p, which can be located at the center of these elements; finally μ_0 is a constant equal to

$$\mu_0 = 4\pi \times 10^{-7}\ \text{H/m}$$

and is often called the magnetic permeability of free space. Certainly, this is confusing definition, since free space does not have any magnetic properties. We will use the S.I. system of units where the distance is measured in meters and force in newtons. Of course, with a change of the system of units the value of μ_0 varies too. In applying Ampere's law (Equation (1.1)), it is essential that the separation between current elements must be much greater than their length; that is

$$L_{qp} \gg d\boldsymbol{l}(p) \quad \text{and} \quad L_{qp} \gg d\boldsymbol{l}(q)$$

Correspondingly, points: p and q can be located anywhere inside their elements. It is easy to see some similarity of Ampere's law and Newton's law of attraction; they describe a force between either elementary currents or elementary masses. Let us illustrate Equation (1.1) by three examples shown in Fig. 1.1. Suppose that elements $d\boldsymbol{l}(p)$ and $d\boldsymbol{l}(q)$ are in parallel with each other. Then, as follows from definition of the cross product, the force $d\boldsymbol{F}(p)$ is directed toward the element $d\boldsymbol{l}(q)$, and two current elements attract each other (Fig. 1.1(a)). If two current elements have opposite directions, the force $d\boldsymbol{F}(p)$ tries to increase the distance between elements, and therefore they repeal each other (Fig. 1.1(b)). If the elements $d\boldsymbol{l}(p)$ and $d\boldsymbol{l}(q)$ are perpendicular to each other, as is shown in Fig. 1.1(c), then in accordance

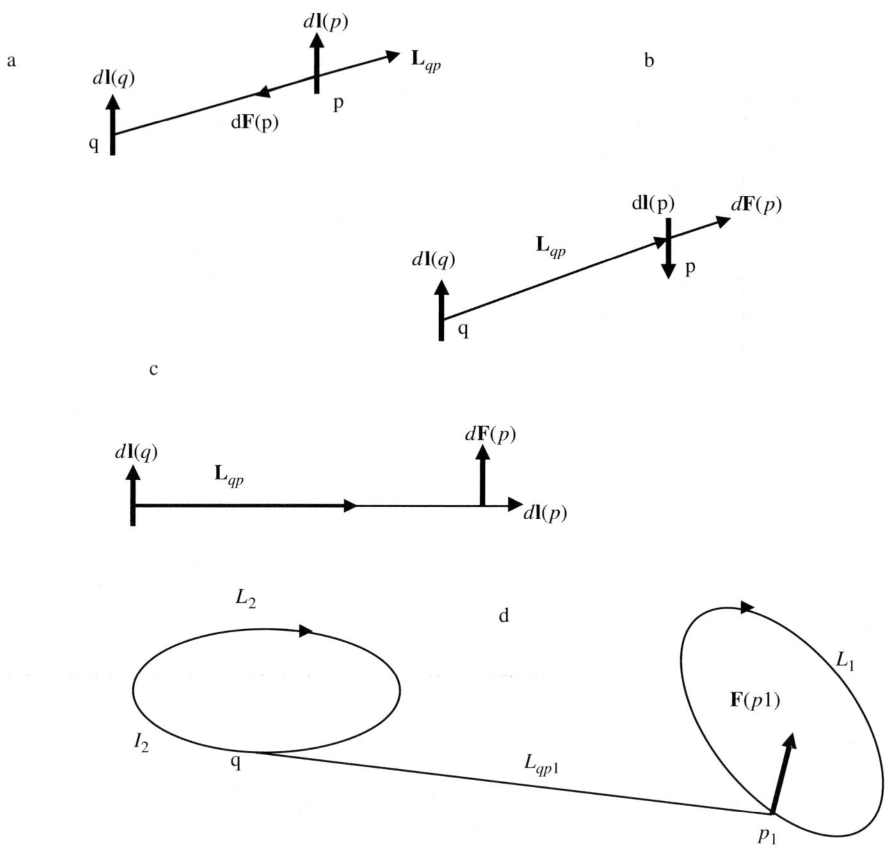

Fig. 1.1. (a) Parallel current elements. (b) Anti-parallel current elements. (c) Current elements perpendicular to each other. (d) Interaction of closed current circuits.

with Equation (1.1) the magnitude of the force acting at the element $d\boldsymbol{l}(p)$ equals

$$dF(p) = \frac{\mu_0}{4\pi} I_1 I_2 dl(p) dl(q) \frac{1}{L_{qp}^2}$$

and it is parallel to the element $d\boldsymbol{l}(q)$. At the same time, the force $d\boldsymbol{F}(q)$ at the point q is equal to zero, that is Newton's third law becomes invalid. This contradiction results from the fact that Equation (1.1) describes an interaction between current elements instead of closed current circuits. In other words, this equation is written for unrealistic case, since we cannot create a constant current in an open circuit, but as all experiments show, Equation (1.1) gives the correct result for closed current lines. For instance, applying the principle of superposition, the force of interaction between two arbitrary and closed currents (Fig. 1.1(d)) is defined as

$$\boldsymbol{F} = \frac{\mu_0}{4\pi} I_1 I_2 \oint_{L_1} \oint_{L_2} \frac{d\boldsymbol{l}(p) \times [d\boldsymbol{l}(q) \times \boldsymbol{L}_{qp}]}{L_{qp}^3} \tag{1.2}$$

The internal integral in Equation (1.2) characterizes the force acting at some point of the current line L_1, for instance, point p_1 and caused by all elements of the current line L_2. Thus, the force $\boldsymbol{F}$ represents a sum (integral) of forces applied at different points of the same circuit and, as is well known, its action causes in general a deformation, translation and a rotation of the current line L_1. It is obvious that in the case of closed circuits the interaction between them obeys Newton's third law.

The relationship between the force $\boldsymbol{F}$ and currents (Equation (1.2)) is called Ampere's law for closed circuits with constant currents, and it is impossible to overestimate its importance, since it is the theoretical foundation of many devices measuring magnetic field as well as electromotors, transforming electric energy into mechanical energy, and it has numerous applications in physics and technology. Finally, it is proper to notice the following:

(a) The force of interaction is independent of properties of the medium which surrounds the currents.

(b) The Ampere's law was formulated for currents which are independent of time. It turns out that this law allows us to calculate the force of interaction even in the case of alternating currents as long as displacement currents can be neglected.

(c) It is natural to be surprised and impressed that Ampere found Equation (1.1) since in reality he had only experimental data describing interaction for closed current circuits.

1.2. MAGNETIC FIELD OF CONSTANT CURRENTS

By analogy with the attraction field caused by masses, it is proper to assume that constant (time-invariant) currents create a field, and due to the existence of this field other current elements experience the action of the force $\boldsymbol{F}$. Such a field is called the magnetic field, and it can be introduced from Ampere's law. In fact, we can write Equation (1.1) as

$$d\boldsymbol{F}(p) = I(p)d\boldsymbol{l}(p) \times d\boldsymbol{B}(p) \tag{1.3}$$

Here

$$d\boldsymbol{B}(p) = \frac{\mu_0}{4\pi} I(q) \frac{d\boldsymbol{l}(q) \times \boldsymbol{L}_{qp}}{L_{qp}^3} \tag{1.4}$$

Equation (1.4) establishes the relationship between the elementary current and the magnetic field caused by this element, and it is called Biot–Savart law. In accordance with Equation (1.4), the magnitude of the magnetic field $d\boldsymbol{B}$ is

$$d\boldsymbol{B} = \frac{\mu_0}{4\pi} I(q) \frac{d\boldsymbol{l}}{L_{qp}^2} \sin(\boldsymbol{L}_{qp}, d\boldsymbol{l}) \tag{1.5}$$

where $(\boldsymbol{L}_{qp}, d\boldsymbol{l})$ is the angle between the vectors $\boldsymbol{L}_{qp}$ and $d\boldsymbol{l}$, and the vector $d\boldsymbol{B}$ is perpendicular to these vectors as is shown in Fig. 1.2(a). The unit vector, $\boldsymbol{b}_0$, characterizing the direction of the field, is defined by

$$\boldsymbol{b}_0 = \frac{d\boldsymbol{l} \times \boldsymbol{L}_{qp}}{|d\boldsymbol{l} \times \boldsymbol{L}_{qp}|}$$

We may say that the magnetic field exists at any point regardless of presence or absence of a current at this point. In S.I. units, the magnetic field is measured in teslas and it is related to other units such as gauss and gamma in the following way:

$$1\ \mathrm{T} = 10^9\ \mathrm{nT} = 10^4\ \mathrm{G} = 10^9\ \mathrm{gamma}$$

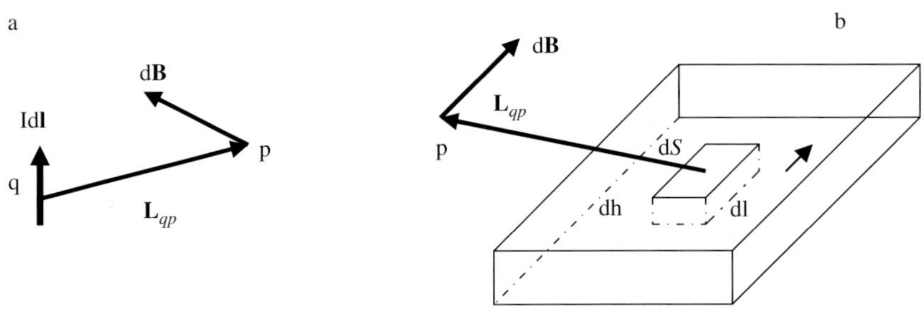

Fig. 1.2. (a) Illustration of Equation (1.4). (b) Field due to the surface currents.

1.2.1. General form of Biot–Savart law

Now we generalize Equation (1.4) assuming that along with linear currents there are also volume and surface currents. First let us represent the product $I\,d\boldsymbol{l}$ as

$$I\,d\boldsymbol{l} = \boldsymbol{j}\,dS\,dl = \boldsymbol{j}\,dV \tag{1.6}$$

since the vector of the current density $\boldsymbol{j}$ and the vector $d\boldsymbol{l}$ have the same direction and the elementary volume is equal to $dV = dS\,dl$. Thus, in place of Equation (1.4), we can write

$$d\boldsymbol{B}(p) = \frac{\mu_0}{4\pi}\frac{\boldsymbol{j}(q) \times \boldsymbol{L}_{qp}}{L_{qp}^3}\,dV \tag{1.7}$$

and this expression describes the magnetic field due to an elementary volume with the current density $\boldsymbol{j}(q)$.

If the current is concentrated in a relatively thin layer with thickness dh, which is small with respect to the distance to the observation points, it is often convenient to replace this layer by a current sheet. As is seen from Fig. 1.2(b), the product $I\,d\boldsymbol{l}$ can be modified in the following way:

$$I\,d\boldsymbol{l} = \boldsymbol{j}\,dV = \boldsymbol{j}\,dh\,dS = \boldsymbol{i}\,dS \tag{1.8}$$

Here dS is the surface element, and

$$\boldsymbol{i} = \boldsymbol{j}\,dS \tag{1.9}$$

is the surface density of currents. Correspondingly, for the magnetic field caused by the elementary surface current, we have

$$d\boldsymbol{B}(p) = \frac{\mu_0}{4\pi}\frac{\boldsymbol{i}(q) \times \boldsymbol{L}_{qp}}{L_{qp}^3} \tag{1.10}$$

Now applying the principle of superposition for all three types of currents and making use of Equations (1.4), (1.7) and (1.10), we obtain the general form of the Biot–Savart law:

$$\boldsymbol{B}(p) = \frac{\mu_0}{4\pi}\left[\int_V \frac{\boldsymbol{j}(q) \times \boldsymbol{L}_{qp}}{L_{qp}^3}\,dV + \int_S \frac{\boldsymbol{i}(q) \times \boldsymbol{L}_{qp}}{L_{qp}^3}\,dS + \sum_{n=1}^{N} I_n \oint_{L_n} \frac{d\boldsymbol{l}(q) \times \boldsymbol{L}_{qp}}{L_{qp}^3}\right] \tag{1.11}$$

In order to understand better this relationship between the magnetic field and currents (Biot–Savart law), it is appropriate to add the following:

1. Equation (1.11) allows us to calculate the magnetic field everywhere except points with linear and surface currents.

2. Unlike volume distribution of currents, linear and surface analogies are only mathematical models of real distribution of currents, which are usually introduced to simplify calculations of the field and study its behavior. For this reason, the equation

$$\boldsymbol{B}(p) = \frac{\mu_0}{4\pi} \int_V \frac{\boldsymbol{j}(q) \times \boldsymbol{L}_{qp}}{L_{qp}^3} dV \tag{1.12}$$

in essence comprises all possible cases of the current distribution and can be always used to determine the field $\boldsymbol{B}$.

3. In accordance with Biot–Savart law, the current is the sole generator of the constant magnetic field and the distribution of this generator is characterized by the magnitude and direction of the current density vector. As is well known, the vector lines of $\boldsymbol{j}(q)$ are always closed. This means that a time-invariant magnetic field is caused by generators of the vortex type and correspondingly we are dealing with a vortex field, unlike, for example, the gravitational field.

4. As was pointed out earlier, all the experiments that allowed Ampere to derive Equation (1.1) were carried out with closed circuits. At the same time, Equation (1.1), as well as Equations (1.4) and (1.7), is written for the element $d\boldsymbol{l}$, where a constant current cannot exist if this element does not constitute a part of a closed circuit. In other words, Equations (1.1) and (1.4) cannot be proved by experiment, but interaction between closed circuits takes place as if the magnetic field $\boldsymbol{B}$, caused by the current element $I\,d\boldsymbol{l}$, is described by Equation (1.4). Let us illustrate this ambiguity in the following way. Suppose that the magnetic field $d\boldsymbol{B}$ due to the elementary current $I\,d\boldsymbol{l}$ is

$$d\boldsymbol{B}(p) = \frac{\mu_0 I}{4\pi} \frac{d\boldsymbol{l}(q) \times \boldsymbol{L}_{qp}}{L_{qp}^3} + I\, dl\, \nabla\varphi$$

where φ is an arbitrary continuous function. Then, the magnetic field caused by the constant current in the closed circuit is

$$\boldsymbol{B}(p) = \frac{\mu_0 I}{4\pi} \oint_L \frac{d\boldsymbol{l} \times \boldsymbol{L}_{qp}}{L_{qp}^3} + I \oint_L \operatorname{grad}\varphi\, dl$$

As is well known from vector analysis, the circulation of a gradient is equal to zero and therefore

$$\boldsymbol{B}(p) = \frac{\mu_0 I}{4\pi} \oint_L \frac{d\boldsymbol{l}(q) \times \boldsymbol{L}_{qp}}{L_{qp}^3}$$

Thus, the ambiguity in the expression of the magnetic field due to an elementary current vanishes, when the interaction or the magnetic field of closed circuits is considered, and the magnetic field $\boldsymbol{B}$ is uniquely defined by Equations (1.11) and (1.12).

5. In accordance with Equation (1.11), the magnetic field caused by a given distribution of currents depends only on the coordinates of the observation point p; that is, it is independent of the presence of other currents. In this light, it is important to emphasize that the right-hand side of Equation (1.11) does not contain any terms that characterize physical properties of the medium where these currents are located. Therefore, the field $\boldsymbol{B}$ at the point p, generated by the given distribution of currents, remains the same if free space is replaced by a nonuniform medium. For instance, if the given current circuit is placed inside of a magnetic material like iron (Chapter 2), the field $\boldsymbol{B}$ caused by this current is the same as if it were in free space. Of course, as is well known and it will be discussed later, the presence of such medium results in a change of the total magnetic field $\boldsymbol{B}$, but this means that inside of the magnetic medium, as well on its surface, along with the given current there are other currents which also produce a magnetic field. This conclusion directly follows from Equation (1.11), which states that any change of the magnetic field $\boldsymbol{B}$ can occur only due to a change of the current distribution.
6. It is convenient to distinguish two types of currents, namely: *conduction* and *magnetization* currents (Chapter 2):

$$\boldsymbol{j} = \boldsymbol{j}_c + \boldsymbol{j}_m \tag{1.13}$$

where $\boldsymbol{j}_c$ and $\boldsymbol{j}_m$ are vectors of the current density which characterize the distribution of the conduction and magnetization currents. Thus, in place of Equation (1.12), derived for conduction currents, we can write

$$\boldsymbol{B}(p) = \frac{\mu_0}{4\pi}\int_V \frac{\boldsymbol{j}(q)\times \boldsymbol{L}_{qp}}{L_{qp}^3}dV = \frac{\mu_0}{4\pi}\int_V \frac{\boldsymbol{j}_c(q)\times \boldsymbol{L}_{qp}}{L_{qp}^3}dV + \frac{\mu_0}{4\pi}\int_V \frac{\boldsymbol{j}_m(q)\times \boldsymbol{L}_{qp}}{L_{qp}^3}dV \tag{1.14}$$

This is important generalization of Biot–Savart law, which establishes the relationship between the magnetic field and currents in any medium. Later we will take into account the influence of currents in a magnetic medium but for now it is assumed that such medium is absent and only conduction currents are considered.

7. From Ampere's and Biot–Savart laws, we have for the force with which the magnetic field acts on the elementary current $\boldsymbol{j}\,dV$:

$$d\boldsymbol{F}(p) = \boldsymbol{j}(p)\times \boldsymbol{B}(p)dV \tag{1.15}$$

As is well known from Coulomb's law, the force of the electric field acting on elementary charge with the density $\delta(p)$ is equal to

$$d\boldsymbol{F}(p) = \delta(p)\boldsymbol{E}(p)dV \tag{1.16}$$

From comparison of Equations (1.15) and (1.16), we can conclude that there is analogy between vectors $\boldsymbol{B}$ and $\boldsymbol{E}$. In fact, these two vectors determine the force acting on the corresponding generator of the field. In this sense, the vector $\boldsymbol{B}$, describing the magnetic field, is similar to the vector $\boldsymbol{E}$, which characterizes the electric field. There is another common feature of these fields, namely, each of them is caused by generators of one type only which have an obvious meaning: either charges or currents.

8. Equation (1.14) allows us to determine the magnetic field provided that the distribution of currents is known. In other words, using Biot–Savart law we can solve the forward problem. At the same time, if part of the currents is not given, Equation (1.14) becomes useless and we have to solve a boundary-value problem.
9. Earlier we emphasized that in general any constant magnetic field is caused by a combination of conduction and magnetization currents. The first one represents a motion of free charges, while magnetization current is a physical concept which allows one to take into account motion of charges within atoms. In this chapter, we focus on the field generated by the conduction currents only, but later investigate the influence of magnetization currents.
10. Biot–Savart law can be applied for calculating time-varying magnetic fields as soon as an influence of displacement currents is negligible.

1.3. THE VECTOR POTENTIAL OF THE MAGNETIC FIELD

Although calculation of the magnetic field, making use of the Biot–Savart law, is not a very complicated procedure, it is still reasonable to find a simpler way of determining the field. With this purpose in mind, by analogy with the scalar potential of the gravitational and electric fields, we introduce a new function which is more simply related to the currents than the magnetic field. Moreover, there is another reason to consider this function, namely, it allows us to derive a system of equations for the field $\boldsymbol{B}$ and simplifies the formulation of boundary-value problem, when currents can be known only if the magnetic field is already determined. Certainly, in such cases the Biot–Savart law cannot be applied and it is very useful to introduce this function. As we know, the magnetic field caused by conduction currents with density $\boldsymbol{j}$ is

$$\boldsymbol{B}(p) = \frac{\mu_0}{4\pi} \int_V \frac{\boldsymbol{j}(q) \times \boldsymbol{L}_{qp}}{L_{qp}^3} dV \tag{1.17}$$

The function $\boldsymbol{L}_{qp}/L_{qp}^3$ can be represented as

$$\frac{\boldsymbol{L}_{qp}}{L_{qp}^3} = \overset{q}{\nabla} \frac{1}{L_{qp}} = -\overset{p}{\nabla} \frac{1}{L_{qp}} \tag{1.18}$$

Here $\overset{q}{\nabla}(1/L_{qp})$ and $\overset{p}{\nabla}(1/L_{qp})$ are gradients when either point q or p changes, respectively.

Its substitution into Equation (1.17) gives

$$\boldsymbol{B}(p) = \frac{\mu_0}{4\pi}\int_V \boldsymbol{j}(q) \times \overset{q}{\nabla}\frac{1}{L_{qp}}dV = \frac{\mu_0}{4\pi}\int_V \left(\overset{p}{\nabla}\frac{1}{L_{qp}}\right) \times \boldsymbol{j}(q)dV \tag{1.19}$$

since the relative position of vectors forming the cross product is changed. Now we will make use of the equality

$$\overset{p}{\nabla} \times \frac{\boldsymbol{j}(q)}{L_{qp}} = \overset{p}{\nabla}\frac{1}{L_{qp}} \times \boldsymbol{j}(q) + \frac{\overset{p}{\nabla} \times \boldsymbol{j}(q)}{L_{qp}} \tag{1.20}$$

which follows from the vector identity

$$\nabla \times (\varphi \boldsymbol{a}) = (\nabla\varphi) \times \boldsymbol{a} + \varphi(\nabla \times \boldsymbol{a})$$

Applying Equation (1.20), we can rewrite Equation (1.19) as

$$\boldsymbol{B}(p) = \frac{\mu_0}{4\pi}\int_V \overset{p}{\nabla} \times \frac{\boldsymbol{j}}{L_{qp}}dV - \frac{\mu_0}{4\pi}\int_V \frac{\overset{p}{\nabla} \times \boldsymbol{j}}{L_{qp}}dV \tag{1.21}$$

The current density $\boldsymbol{j}$ is a function of the point q and does not depend on the location of the observation point p. Therefore, the integrand of the second integral is zero and

$$\boldsymbol{B}(p) = \frac{\mu_0}{4\pi}\int_V \overset{p}{\nabla} \times \frac{\boldsymbol{j}(q)}{L_{qp}}dV \tag{1.22}$$

Inasmuch as the integration and differentiation indicated in Equation (1.22) are carried out with respect to two different points: q and p, we can interchange the order of operations and obtain

$$\boldsymbol{B}(p) = \overset{p}{\nabla} \times \frac{\mu_0}{4\pi}\int_V \frac{\boldsymbol{j}(q)}{L_{qp}}dV \tag{1.23}$$

Let us introduce the vector $\boldsymbol{A}$:

$$\boldsymbol{A}(p) = \frac{\mu_0}{4\pi}\int_V \frac{\boldsymbol{j}(q)}{L_{qp}}dV \tag{1.24}$$

Then we obtain

$$\boldsymbol{B} = \overset{p}{\nabla} \times \boldsymbol{A} \quad \text{or} \quad \boldsymbol{B}(p) = \text{curl}\,\boldsymbol{A}(p) \tag{1.25}$$

Thus, the magnetic field $\boldsymbol{B}$, caused by constant currents, can be expressed through the function, called the vector potential $\boldsymbol{A}$ defined by Equation (1.24). Comparison of Equations (1.17) and (1.24) shows that the function $\boldsymbol{A}$ is related to currents in a much simpler way than the magnetic field is, and therefore one reason for introducing this function is already demonstrated. By definition, $\boldsymbol{A}$ is a vector unlike the potential of the gravitational and electric fields, and its magnitude and direction at every point p depend essentially on the current distribution.

Now we obtain expressions for the vector potential $\boldsymbol{A}$, caused by surface and linear currents. Making use of the equalities

$$\boldsymbol{j}\,dV = \boldsymbol{i}\,dS \quad \text{and} \quad \boldsymbol{j}\,dV = I\,\boldsymbol{dl}$$

it follows from Equation (1.24)

$$\boldsymbol{A}(p) = \frac{\mu_0}{4\pi}\int_S \frac{\boldsymbol{i}(q)}{L_{qp}}\,dS \quad \text{and} \quad \boldsymbol{A}(p) = \frac{\mu_0}{4\pi} I \oint_L \frac{\boldsymbol{dl}}{L_{qp}} \tag{1.26}$$

Applying the principle of superposition, we obtain an expression for the vector potential caused by volume, surface and line currents:

$$\boldsymbol{A}(p) = \frac{\mu_0}{4\pi}\int_V \frac{\boldsymbol{j}\,dV}{L_{qp}} + \frac{\mu_0}{4\pi}\int_S \frac{\boldsymbol{i}\,dS}{L_{qp}} + \frac{\mu_0}{4\pi}\sum_i I_i \oint_L \frac{\boldsymbol{dl}}{L_{qp}} \tag{1.27}$$

The components of the vector potential can be derived directly from this equation. For instance, in Cartesian coordinates, we have

$$\begin{aligned}
A_x(p) &= \frac{\mu_0}{4\pi}\left[\int_V \frac{j_x}{L_{qp}}\,dV + \int_S \frac{i_x}{L_{qp}}\,dS + \sum_i I_i \oint_L \frac{dl_x}{L_{qp}}\right] \\
A_y(p) &= \frac{\mu_0}{4\pi}\left[\int_V \frac{j_y}{L_{qp}}\,dV + \int_S \frac{j_y}{L_{qp}}\,dS + \sum_i I_i \oint_L \frac{dl_y}{L_{qp}}\right] \\
A_z(p) &= \frac{\mu_0}{4\pi}\left[\int_V \frac{j_z}{L_{qp}}\,dV + \int_S \frac{j_z}{L_{qp}}\,dS + \sum_i I_i \oint_L \frac{dl_z}{L_{qp}}\right]
\end{aligned} \tag{1.28}$$

Similar expressions can be written for the vector potential components in other system of coordinates. Certainly, the vector potential is related to the currents in a much simpler way than the magnetic field. For instance, as is seen from Equations (1.26), if current flows along a single straight line, the vector potential has only one component which is parallel to this line. It is also obvious that if currents are situated in a single plane, then the vector potential $\boldsymbol{A}$ at every point is parallel to this plane. Later we will consider several examples illustrating the behavior of the vector potential and the magnetic field $\boldsymbol{B}$, but now we will find two useful relations for the function $\boldsymbol{A}$, which simplify to a great extent the task of deriving the system of the magnetic field equations. First, let us determine the divergence of the vector potential $\boldsymbol{A}$. As follows from Equation (1.24), we have

$$\overset{p}{\operatorname{div}}\,\boldsymbol{A} = \overset{p}{\operatorname{div}}\frac{\mu_0}{4\pi}\int_V \frac{\boldsymbol{j}(q)}{L_{qp}}\,dV \tag{1.29}$$

Since differentiation and integration in this expression are performed with respect to different points, we can change the order of operations and obtain

$$\overset{p}{\operatorname{div}}\,\boldsymbol{A} = \frac{\mu_0}{4\pi}\int_V \overset{p}{\operatorname{div}}\frac{\boldsymbol{j}(q)}{L_{qp}}\,dV \tag{1.30}$$

The volume over which the integration is carried out includes all currents, and therefore it can be enclosed by a surface S such that outside of it currents are absent. Note that at points of the boundary S with a nonconducting medium, the normal component of the current density equals zero:

$$j_{\mathrm{n}} = 0 \tag{1.31}$$

Taking into account that the current density does not depend on the observation point ($\overset{p}{\operatorname{div}}\,\boldsymbol{j} = 0$), the integrand in Equation (1.30) can be represented as

$$\overset{p}{\nabla}\frac{\boldsymbol{j}}{L_{qp}} = \frac{\overset{p}{\nabla}\boldsymbol{j}}{L_{qp}} + \boldsymbol{j}\,\overset{p}{\nabla}\frac{1}{L_{qp}} = \boldsymbol{j}\,\overset{p}{\nabla}\frac{1}{L_{qp}}$$

Thus, we have

$$\boldsymbol{j}\,\overset{p}{\nabla}\frac{1}{L_{qp}} = -\boldsymbol{j}\,\overset{q}{\nabla}\frac{1}{L_{qp}} = -\overset{q}{\nabla}\frac{\boldsymbol{j}}{L_{qp}} + \frac{\overset{q}{\nabla}\boldsymbol{j}}{L_{qp}}$$

Applying the principle of charge conservation, $\overset{q}{\nabla}\boldsymbol{j}(q) = 0$ (current lines are closed), we obtain

$$\boldsymbol{j}\,\overset{p}{\nabla}\frac{1}{L_{qp}} = -\overset{q}{\nabla}\frac{\boldsymbol{j}}{L_{qp}} \tag{1.32}$$

Correspondingly, Equation (1.30) can be written as

$$\operatorname{div}\boldsymbol{A} = -\frac{\mu_0}{4\pi}\int_V \overset{q}{\operatorname{div}}\frac{\boldsymbol{j}(q)}{L_{qp}}\,dV \tag{1.33}$$

On the right-hand side of this equation, both integration and differentiation are performed with respect to the same point q so that we can apply Gauss' theorem. Then we have

$$\operatorname{div} \boldsymbol{A} = -\frac{\mu_0}{4\pi} \int_V \overset{q}{\operatorname{div}} \frac{\boldsymbol{j}(q)}{L_{qp}} dV = -\frac{\mu_0}{4\pi} \oint_S \frac{\boldsymbol{j}\, d\mathbf{S}}{L_{qp}} = -\frac{\mu_0}{4\pi} \oint_S \frac{j_n\, dS}{L_{qp}}$$

Taking into account the fact that the normal component of the current density j_n vanishes at the surface S, which surrounds all currents (Equation (1.31)), we obtain

$$\operatorname{div} \boldsymbol{A} = 0 \tag{1.34}$$

This is the first relation that is useful for deriving the system of field equations. The divergence is taken with respect to the observation point p. Note that this equation shows that the vector lines of the field $\boldsymbol{A}$ are closed. In this light, it is proper to point out that Equation (1.25) indicates that the vector lines of the magnetic field $\boldsymbol{B}$ are closed too, and this fact will be proved later.

The next important relation will be obtained, making use of the known expression of the potential of the attraction field caused by masses

$$U(p) = k \int_V \frac{\delta(q)}{L_{qp}} dV$$

which obeys Poisson's equation:

$$\Delta U = -4\pi k \delta$$

It is obvious that each component of the vector potential is presented in the same form as the potential $U(p)$ and this means that they obey also Poisson's equations:

$$\Delta A_x = -\mu_0 j_x, \quad \Delta A_y = -\mu_0 j_y, \quad \Delta A_z = -\mu_0 j_z$$

Multiplying each of these equations by the corresponding unit vector $\boldsymbol{i}, \boldsymbol{j}$ and $\boldsymbol{k}$ and summing, we arrive at the second useful equation for the vector potential $\boldsymbol{A}$:

$$\Delta A = -\mu_0 \boldsymbol{j} \tag{1.35}$$

1.4. MAGNETIC FIELD AND VECTOR POTENTIAL, CAUSED BY LINEAR AND SURFACE CURRENTS

To illustrate the behavior of the magnetic field, as well as its vector potential, we consider several examples; some of them will be useful in studying the field $\boldsymbol{B}$ of the earth.

1.4.1. Magnetic field of a current filament

Taking into account the axial symmetry of the problem (Fig. 1.3(a)), we will choose a cylindrical system of coordinates: r, φ, z with its origin situated on the current-carrying line. Starting from the Biot–Savart law, we can say that the magnetic field has only the component B_φ, which is independent of the coordinate φ. From the principle of superposition, it follows that the total field is the sum of fields contributed by the current elements $I\,dz$. Then we have

$$\boldsymbol{B}_\varphi = \frac{\mu_0 I}{4\pi}\int_{z_1}^{z_2}\frac{d\boldsymbol{z}\times\boldsymbol{L}_{qp}}{L_{qp}^3} \tag{1.36}$$

where $L_{qp} = (r^2+z^2)^{1/2}$ and z is the coordinate of the element dz. The coordinates of the observation point: r and $z=0$; z_1 and z_2 are coordinates of terminal points of the current line. The absolute value of the cross product is

$$|d\boldsymbol{z}\times\boldsymbol{L}_{qp}| = dz\,L_{qp}\,\sin(d\boldsymbol{z},\boldsymbol{L}_{qp}) = dz\,L_{qp}\,\sin\beta = dz\,L_{qp}\,\cos\alpha$$

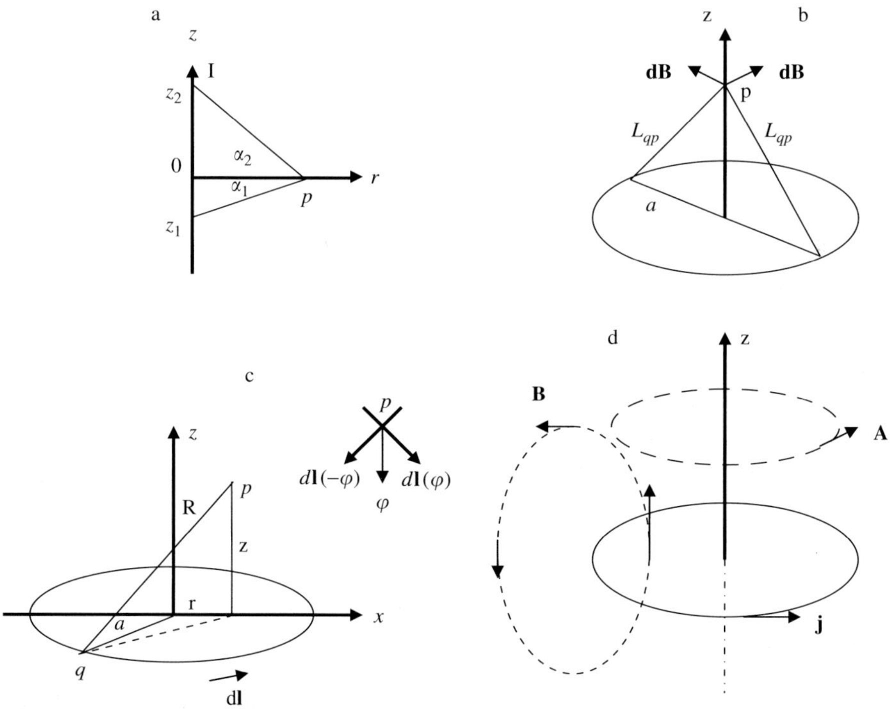

Fig. 1.3. (a) The field of current element. (b) The field at the current loop axis. (c) Illustration of Equation (1.44). (d) Geometry of the magnetic field and vector potential.

Thus

$$B_\varphi = \frac{\mu_0 I}{4\pi} \int_{z_1}^{z_2} \frac{dz}{L_{qp}^2} \cos\alpha \tag{1.37}$$

Inasmuch as $z = r\tan\alpha$, we have

$$dz = r\sec^2\alpha\, d\alpha \quad \text{and} \quad L_{qp}^2 = r^2(1+\tan^2\alpha) = r^2\sec^2\alpha$$

Substituting these expressions into Equation (1.37), we obtain

$$B_\varphi = \frac{\mu_0 I}{4\pi r} \int_{\alpha_1}^{\alpha_2} \cos\alpha\, d\alpha$$

Thus, the expression for the magnetic field caused by the current flowing along a straight line is

$$B_\varphi(p) = \frac{\mu_0 I}{4\pi r}(\sin\alpha_2 - \sin\alpha_1) \tag{1.38}$$

where α_2 and α_1 are the angles subtended by the radii from the point p to the ends of the line. Next consider two limiting cases. First, suppose that point p is far away from the line and the distance L_{qp} from the current elements to an observation point is practically the same. Then it can be taken out of the integral in Equation (1.37) and we obtain the Biot–Savart law:

$$B_\phi(p) = \frac{\mu_0 I(z_2 - z_1)}{4\pi L_{qp}^2}$$

In the opposite case of an infinitely long current line, when $\alpha_1 = -\pi/2$ and $\alpha_2 = \pi/2$, we have

$$B_\phi(p) = \frac{\mu_0 I}{2\pi r} \tag{1.39}$$

If the line is semi-infinite, $\alpha_1 = 0$ and $\alpha_2 = \pi/2$, Equation (1.38) gives

$$B_\phi(p) = \frac{\mu_0 I}{4\pi r} \tag{1.40}$$

Let us assume that $\alpha_2 = \alpha$ and $\alpha_1 = -\alpha$. Then we have

$$B_\phi(p) = \frac{\mu_0 I}{2\pi r}\sin\alpha = \frac{\mu_0 I}{2\pi r}\frac{l}{(l^2 + r^2)^{1/2}} \tag{1.41}$$

where $2l$ is the length of the current-carrying line <. If l is significantly greater than the distance r, the right-hand side of Equation (1.41) can be expanded in a series in terms of $(r/l)^2$, and we obtain

$$B_\phi(p) = \frac{\mu_0 I}{2\pi r}\left(1 + \frac{r^2}{l^2}\right)^{-1/2} \approx \frac{\mu_0 I}{2\pi r}\left(1 - \frac{1}{2}\frac{r^2}{l^2} + \frac{3}{8}\frac{r^4}{l^4} - \cdots\right)$$

We see that if the length of the current line $2l$ is four or five times larger than the separation r, the field is practically the same as that due to an infinitely long current line. Returning to Equation (1.38), it is proper to make two comments:

(a) This equation has a physical meaning when a closed circuit is considered and the line with the length $2l$ is only a part of this circuit.
(b) If r tends to zero, the field becomes infinitely large; this is understandable, since the real volume distribution of currents is replaced by a linear one, where the volume density is infinitely large. For this reason, Equation (1.38) can be only used at points located at distances greatly exceeding a diameter of a current line.

1.4.2. The vector potential A and the magnetic field B of the current in a circular loop

First, assume that the observation point p is situated on the z-axis of a loop with radius a, as is shown in Fig. 1.3(b). Then, in accordance with Equation (1.24), we have

$$\mathbf{A}(p) = \frac{\mu_0 I}{4\pi} \oint_L \frac{d\mathbf{l}}{L_{qp}}$$

Inasmuch as the distance L_{qp} is the same for all points on the loop

$$\mathbf{A}(p) = \frac{\mu_0 I}{4\pi L_{qp}} \oint_L d\mathbf{l}$$

By definition, the sum of elementary vectors $d\mathbf{l}$ along any closed path is zero. Therefore, the vector potential $\mathbf{A}$ at the z-axis of a circular current loop vanishes. Now we calculate the magnetic field on the z-axis. From the Biot–Savart law (Equation (1.4)), it can be seen that in a cylindrical system of coordinates, each current element $I\,d\mathbf{l}$ creates two components dB_z and dB_r. However, it is always possible to find two current elements $I\,d\mathbf{l}$ that contribute horizontal components with the same magnitude and opposite directions. Therefore, the magnetic field has only a vertical component along the z-axis. As can be seen from Fig. 1.3(b)

$$dB_z(p) = \frac{\mu_0 I}{4\pi}\frac{dl}{L_{qp}^2}\frac{a}{L_{qp}} = \frac{\mu_0 Ia}{4\pi}\frac{dl}{L_{qp}^3}$$

since $|\boldsymbol{dl} \times \boldsymbol{L}_{qp}| = L_{qp}\, dl$.

After integrating along the closed path of the loop, we finally obtain:

$$B_z(0,z) = \frac{\mu_0 Ia2\pi a}{4\pi(a^2+z^2)^{3/2}} = \frac{\mu_0 Ia^2}{2(a^2+z^2)^{3/2}} = \frac{\mu_0 M}{2\pi(a^2+z^2)^{3/2}} \tag{1.42}$$

where

$$M = I\pi a^2$$

with S being the area enclosed by the loop. When the distance z is much greater than the radius of the loop a, we arrive at the expression for the magnetic field, which plays an extremely important role in the theory of the magnetic field of the earth as well in a magnetic medium. Neglecting a in comparison with z, we have

$$B_z(0,z) = \frac{\mu_0 M}{2\pi z^3}, \quad \text{if } z \gg a \tag{1.43}$$

When the intensity of the field does not separately depend on the current or the loop radius, but it is defined by the product $M = IS$, we call this field that of a magnetic dipole. Thus, a relatively small current-carrying loop of radius a creates the magnetic field of a magnetic dipole having the moment $M = \pi a^2 I$ oriented along the z-axis. Later we will describe the concept of the magnetic dipole in detail. From Equation (1.42), it follows that when the distance z is at least five times greater than the radius a, the treatment of the loop as the magnetic dipole situated at the center of the loop results in an error of no more than 5%.

So far we have considered the vector potential and the magnetic field only along the z-axis. Now we will investigate a general case and first of all calculate the vector potential at any point p. Due to symmetry, the vector potential does not depend on the coordinate φ. For simplicity, we can then choose the point p in the x–z plane where $\varphi = 0$. As can be seen from Fig. 1.3(c), every pair of current-carrying elements, equally distant from point p and having coordinates φ and $-\varphi$, creates a vector potential $\boldsymbol{dA}$ located in a plane parallel to the x–y plane. Inasmuch as the whole loop can be represented as the sum of such pairs, we conclude that the vector potential $\boldsymbol{A}$ caused by the current-carrying loop has only the component A_φ. Therefore, from Equation (1.24), it follows that

$$A_\varphi = \frac{\mu_0 I}{4\pi}\oint_L \frac{dl_\varphi}{R} = \frac{\mu_0 I}{2\pi}\int_0^\pi \frac{a\cos\varphi\, d\varphi}{(a^2+r^2-2ar\cos\varphi)^{1/2}} \tag{1.44}$$

where dl_φ is the component $\boldsymbol{dl}$ along the coordinate line φ, and

$$dl_\varphi = a \cos\varphi \, d\varphi \quad \text{and} \quad R = (a^2 + r^2 - 2ar\cos\varphi)^{1/2}$$

Letting $\varphi = \pi + 2\alpha$, we have

$$d\varphi = 2\, d\alpha \quad \text{and} \quad \cos\varphi = 2\sin^2\alpha - 1$$

and therefore

$$A_\varphi = \frac{aI\mu_0}{\pi}\int_0^{\pi/2} \frac{(2\sin^2\alpha - 1)d\alpha}{[(a+r)^2 + z^2 - 4ar\sin^2\alpha]^{1/2}}$$

Introducing a new parameter

$$k^2 = \frac{4ar}{(a+r)^2 + z^2}$$

and carrying out some fairly simple algebraic operations, we obtain

$$\begin{aligned} A_\varphi &= \frac{kI\mu_0}{2\pi}\left(\frac{a}{r}\right)^{1/2}\left[\left(\frac{2}{k^2} - 1\right)\int_0^{\pi/2}\frac{d\alpha}{(1 - k^2\sin^2\alpha)^{1/2}} - \frac{2}{k^2}\int_0^{\pi/2}(1 - k^2\sin^2\alpha)^{1/2}d\alpha\right] \\ &= \frac{I\mu_0}{\pi k}\left(\frac{a}{r}\right)^{1/2}\left[\left(1 - \frac{k^2}{2}\right)K - E\right] \end{aligned} \tag{1.45}$$

where K and E are complete elliptical integrals of the first and second kind. These functions have been studied in detail and there are standard procedures for their calculation. Using the relationship between the vector potential and the magnetic field ($\boldsymbol{B} = \nabla \times \boldsymbol{A}$), we have in a cylindrical coordinate system

$$B_r = -\frac{\partial A_\varphi}{\partial z}, \quad B_\varphi = 0 \quad \text{and} \quad B_z = \frac{1}{r}\frac{\partial}{\partial r}(rA_\varphi)$$

As is known from the theory of elliptical integrals

$$\frac{\partial K}{\partial k} = \frac{E}{k(1-k^2)} - \frac{K}{k}, \quad \frac{\partial E}{\partial k} = \frac{E}{k} - \frac{K}{k}$$

and

$$\frac{\partial k}{\partial z} = -\frac{zk^3}{4ar}, \quad \frac{\partial k}{\partial r} = \frac{k}{2r} - \frac{k^3}{4r} - \frac{k^3}{4a}$$

Therefore, after differentiation, we have

$$
\begin{aligned}
B_r &= \frac{\mu_0 I}{2\pi} \frac{z}{[(a+r)^2+z^2]^{1/2}} \left[-K + \frac{a^2+r^2+z^2}{(a-r)^2+z^2} E \right] \\
B_z &= \frac{\mu_0 I}{2\pi} \frac{1}{[(a+r)^2+z^2]^{1/2}} \left[K + \frac{a^2-r^2-z^2}{(a-r)^2+z^2} E \right]
\end{aligned}
\tag{1.46}
$$

Thus, in general, the magnetic field caused by the current in a circular loop can be expressed in terms of elliptical integrals. As follows from Equation (1.45), the vector lines of the field $\boldsymbol{A}$ are circles located in the horizontal planes with centers located at the z-axis, while vector lines of the magnetic field are situated in the vertical planes (Fig. 1.3(d)). It may be proper to note that all vector lines passing through the area of the current circle appear outside.

1.4.3. The magnetic field of a magnetic dipole

Suppose that the distance from the center of the current-carrying loop to the observation point R is considerably greater than the loop radius; that is

$$
R = (r^2 + z^2)^{1/2} \gg a
$$

Then, Equation (1.44) can be simplified so that we have

$$
\begin{aligned}
A_\varphi &= \frac{\mu_0 I a}{2\pi} \int_0^\pi \frac{\cos\varphi \, d\varphi}{(R^2 - 2ar\cos\varphi)^{1/2}} = \frac{I a \mu_0}{2\pi R} \int_0^\pi \frac{\cos\varphi \, d\varphi}{\left[1 - (2ar/R^2)\cos\varphi\right]^{1/2}} \\
&\approx \frac{I a \mu_0}{2\pi R} \int_0^\pi \left(1 + \frac{ar}{R^2}\cos\varphi\right) \cos \, d\varphi \\
&= \frac{I a \mu_0}{2\pi R} \int_0^\pi \cos\varphi \, d\varphi + \frac{I a^2 r \mu_0}{2\pi R^3} \int_0^\pi \cos^2\varphi \, d\varphi
\end{aligned}
\tag{1.47}
$$

where the relation

$$
\frac{1}{(1+x)^n} \approx 1 - nx
$$

has been used assuming that $nx \ll 1$. The first integral in Equation (1.47) vanishes and we obtain

$$
A_\varphi = \frac{\mu_0 I a^2 r}{4R^3} \quad \text{or} \quad \boldsymbol{A} = A_\varphi \boldsymbol{i}_\varphi = \frac{\mu_0 I S r}{4\pi R^3} \boldsymbol{i}_\varphi
\tag{1.48}
$$

where S is the loop area. Let us introduce a spherical system of coordinates, R, θ, φ with its origin 0 at the center of the current loop and the z-axis is directed perpendicular to this loop. From this axis (as $z > 0$), the direction of the current is seen counterclockwise. Then Equation (1.48) can be rewritten as

$$\boldsymbol{A} = \frac{\mu_0 IS}{4\pi R^2} \boldsymbol{i}_\varphi \sin\theta \tag{1.49}$$

Next we will introduce the moment of the loop as a vector directed along the z-axis, whose magnitude is equal to the product of the current in the loop and its area:

$$\boldsymbol{M} = IS\boldsymbol{z}_0 = M\boldsymbol{z}_0 \tag{1.50}$$

where $M = IS$. It is proper to note that the moment $\boldsymbol{M}$ and the direction of the current flow form a right-hand system. Thus, in place of Equation (1.49), we can write

$$\boldsymbol{A} = \frac{\mu_0 M}{4\pi R^2} \boldsymbol{i}_\varphi \sin\theta \quad \text{or} \quad \boldsymbol{A} = \frac{\mu_0 \boldsymbol{M} \times \boldsymbol{R}}{4\pi R^3} \tag{1.51}$$

since

$$\boldsymbol{M} \times \boldsymbol{R} = MR\boldsymbol{i}_\varphi \sin\theta$$

Equation (1.51) will be used to account for the influence of magnetization in a magnetic medium. Now taking into account the fact that

$$\boldsymbol{B} = \text{curl}\, \boldsymbol{A} \quad \text{and} \quad A_r = A_\theta = 0$$

we obtain the following expressions for the magnetic field in a spherical system of coordinates:

$$B_R = \frac{\mu_0}{R \sin\theta} \frac{\partial (A_\varphi \sin\theta)}{\partial\theta}, \quad B_\theta = -\frac{\mu_0}{R} \frac{\partial (RA_\varphi)}{\partial R} \quad \text{and} \quad B_\varphi = 0$$

whence

$$B_R = \frac{2\mu_0 M}{4\pi R^3} \cos\theta, \quad B_\theta = \frac{\mu_0 M}{4\pi R^3} \sin\theta \quad \text{and} \quad B_\varphi = 0 \tag{1.52}$$

These equations describe the behavior of the magnetic field of a relatively small current loop; that is, its radius is much smaller than the distance from the loop center to the observation point. This is the most important condition to apply Equations (1.52), while the values of the loop radius and the distance R are not essential. We call the magnetic field, described by Equations (1.52), that of a magnetic dipole with moment $\boldsymbol{M}$. Here it is appropriate to make two comments:

1. In the case of the electric field, a "dipole" means a combination of equal charges having opposite signs, when the field is determined at distances

essentially exceeding the separation between these charges. At the same time, the notion of a "magnetic dipole" does not imply the existence of magnetic charges, but it simply describes the behavior of the magnetic field due to the current in a relatively small loop.

2. The magnetic field of any current, regardless of its shape, is equivalent to that of the magnetic dipole when the field is defined at distances much greater than loop dimensions. In other words, any current circuit creates a magnetic field such that far away from currents it coincides with the field of a magnetic dipole.

The main features of the field of the magnetic dipole directly follow from Equations (1.52) and Fig. 1.4, and they are:

(a) At points of the dipole axis z, the field has only one component B_z directed along this axis, and it decreases inversely proportional to z^3:

$$B_z = \frac{\mu_0 M}{2\pi z^3} \tag{1.53}$$

It is proper to note that this component is positive at all points of this axis.

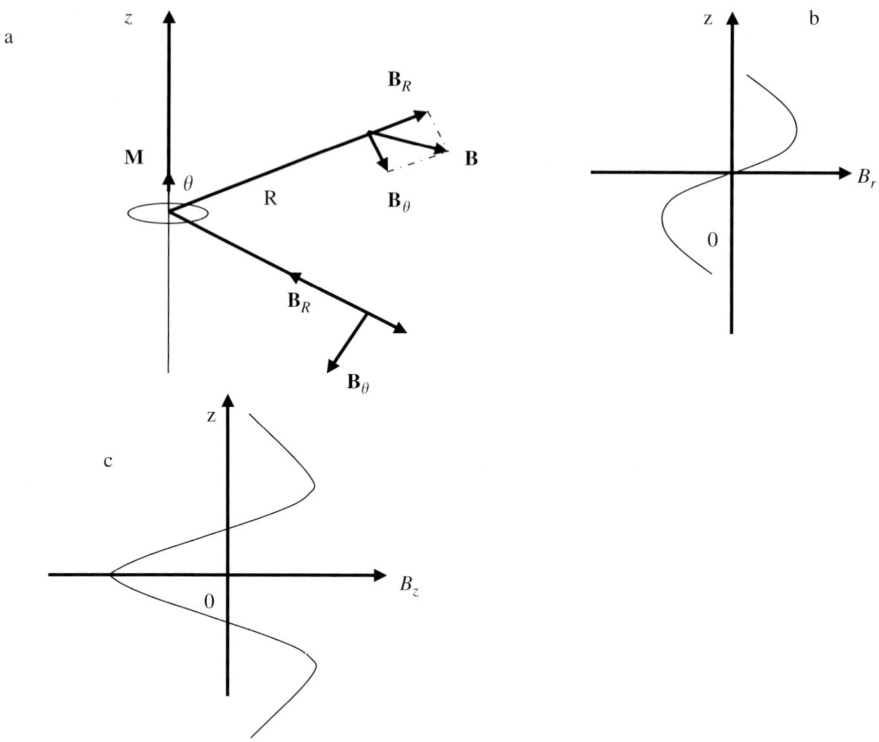

Fig. 1.4. (a) The field of a magnetic dipole. (b) The field component B_r (r = constant). (c) The field component B_z (r = constant).

(b) At the equatorial plane $\theta = \pi/2$, the radial component B_R vanishes, and the field has the direction opposite to that of the dipole moment $\boldsymbol{M}$

$$B_z = -\frac{\mu_0 M}{4\pi r^3} \tag{1.54}$$

Here r is the distance from the dipole to an observation point.

(c) Along any radius $\theta =$ constant, both components of the field, B_R and B_θ, decrease inversely proportional to R^3. At the same time, the ratio of these components, as well the orientation of the total vector $\boldsymbol{B}$ with respect to the radius R, does not change. In fact, according to Equations (1.52), we have

$$\frac{B_\theta}{B_R} = \frac{1}{2}\tan\theta \tag{1.55}$$

(d) It is useful to point out that a very simple dipole field describes the main part of the magnetic field of the earth. This fact is also useful for paleomagnetic studies.

We considered components of the magnetic field in the spherical system of coordinates.

For illustration, let us find the components of the field in the cylindrical system. As follows from Fig. 1.4(a), we have

$$B_r(r,z) = B_R \sin\theta + B_\theta \cos\theta \quad \text{and} \quad B_z(r,z) = B_R \cos\theta - B_\theta \sin\theta$$

where $R = (r^2 + z^2)^{1/2}$. Taking into account Equations (1.52), we obtain

$$\begin{gathered} B_r(r,z) = \frac{3\mu_0 M}{4\pi R^3}\sin\theta\cos\theta, \quad B_z(r,z) = \frac{\mu_0 M}{4\pi R^3}(2\cos^2\theta - \sin^2\theta) \\ \text{or} \\ B_r(r,z) = \frac{\mu_0 M r z}{4\pi(r^2+z^2)^{5/2}}, \quad B_z(r,z) = \frac{\mu_0 M}{4\pi(r^2+z^2)^{5/2}}(2z^2 - r^2) \end{gathered} \tag{1.56}$$

If we assume that r is constant, then Equation (1.56) allows us to study the behavior of the field components parallel to the dipole moment as a function of z (Fig. 1.4(b and c)). First of all, it is clear that the radial component, B_r, is an odd function of z and it changes sign in the plane of the dipole. At the same time, the vertical component is an even function of z and it changes sign at points

$$z = \pm(2)^{-1/2} r$$

1.4.4. The vector potential of a system of dipoles

Let us suppose that there is some number of relatively small current loops arbitrarily oriented with respect to each other. Each loop is characterized by its moment $\boldsymbol{M}_i$. Then, performing a summation, we have for the total moment of this system

$$\boldsymbol{M} = \sum \boldsymbol{M}_i$$

Thus, we have replaced a system of small current loops by one small loop with the moment $\boldsymbol{M}$, since it is assumed that an observation point is located far away with respect to a volume where loops are located. If there is a continuous distribution of such currents, then for the total moment we have

$$\boldsymbol{M} = \int_V \boldsymbol{P}(q)dV \tag{1.57}$$

where q is an arbitrary point of the volume and $\boldsymbol{P}$ characterizes the density of moments

$$\boldsymbol{P} = \frac{d\boldsymbol{M}}{dV} \tag{1.58}$$

In accordance with Equations (1.51) and (1.58), the vector potential $d\boldsymbol{A}$, caused by the current loops of an elementary volume, is

$$d\boldsymbol{A} = \frac{\mu_0 d\boldsymbol{M} \times \boldsymbol{L}_{qp}}{4\pi L_{qp}^3} = \frac{\mu_0 \boldsymbol{P} \times \boldsymbol{L}_{qp}}{4\pi L_{qp}^3} dV \tag{1.59}$$

where L_{qp} is the distance between any point q of the elementary volume dV and the observation point p. Now applying the principle of superposition, we obtain for the vector potential $\boldsymbol{A}$, caused by a volume distribution of current loops, the following expression:

$$\boldsymbol{A}(p) = \frac{\mu_0}{4\pi} \int_V \frac{\boldsymbol{P}(q) \times \boldsymbol{L}_{qp}}{L_{qp}^3} dV \tag{1.60}$$

which plays a very important role in the development of the theory of the magnetic field $\boldsymbol{B}$ in the presence of a magnetic medium.

1.4.5. Behavior of the of field B near surface currents

First, suppose that the current is uniformly distributed at the plane surface S and $\boldsymbol{i}(q)$ is the current density (Fig. 1.5(a)). Then, in accordance with the Biot–Savart law, the magnetic field caused by surface currents is

$$\boldsymbol{B}(p) = \frac{\mu_0}{4\pi} \int_S \frac{\boldsymbol{i}(q) \times \boldsymbol{L}_{qp}}{L_{qp}^3} dS \tag{1.61}$$

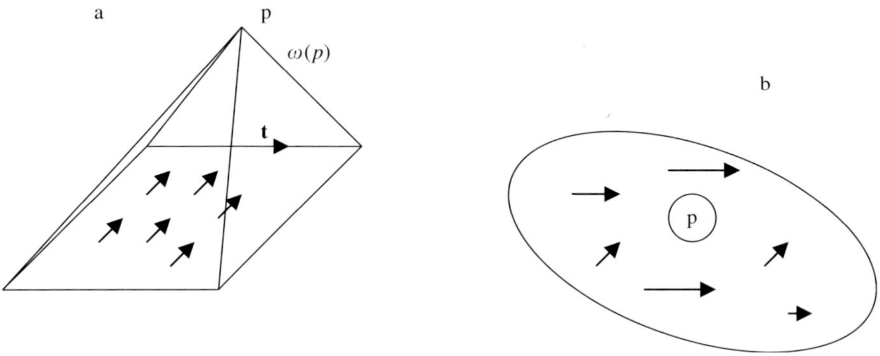

Fig. 1.5. (a) Illustration of Equation (1.65). (b) Surface current distribution.

To find the tangential component of the field, we will multiply both sides of Equation (1.61) by the unit vector $\boldsymbol{t}$, which is parallel to the surface S. This gives

$$B_t(p) = \boldsymbol{B}\cdot\boldsymbol{i} = \frac{\mu_0}{4\pi}\int_S \frac{(\boldsymbol{i}\times\boldsymbol{L}_{qp})\cdot t}{L_{qp}^3}dS$$
$$\text{or} \tag{1.62}$$
$$B_t(p) = \frac{\mu_0}{4\pi}\int_S \frac{(\boldsymbol{t}\times\boldsymbol{i})\cdot L_{qp}}{L_{qp}^3}dS$$

Inasmuch as both vectors $\boldsymbol{t}$ and $\boldsymbol{i}$ are tangential to the surface S, the cross product in Equation (1.62) can be written as

$$\boldsymbol{t}x\boldsymbol{i} = i\boldsymbol{n}\ \sin(\boldsymbol{t},\boldsymbol{i})$$

where i is the magnitude of the current density and $\boldsymbol{n}$ the unit vector perpendicular to S. Correspondingly, for the tangential component of the magnetic field, we have

$$B_t(p) = \frac{\mu_0}{4\pi}i\ \sin(\boldsymbol{i},\boldsymbol{t})\int_S \frac{\boldsymbol{L}_{qp}\cdot d\boldsymbol{S}}{L_{qp}^3}$$
$$\text{or} \tag{1.63}$$
$$B_t(p) = -\frac{\mu_0 i}{4\pi}\sin(\boldsymbol{i},\boldsymbol{t})\int_S \frac{\boldsymbol{L}_{pq}\cdot d\boldsymbol{S}}{L_{pq}^3}$$

where $d\boldsymbol{S} = dS\,\boldsymbol{n}$. As is well known, the integral is equal to the solid angle $\omega(p)$ subtended by the surface S as viewed from point p. Finally, we have

$$B_t(p) = -\frac{\mu_0 i}{4\pi}\sin(\boldsymbol{i},\boldsymbol{t})\omega(p) \tag{1.64}$$

For instance, in the direction perpendicular to the current, we obtain the total tangential component

$$B_t(p) = -\frac{\mu_0 i}{4\pi}\omega(p) \tag{1.65}$$

since $\sin(\boldsymbol{i}, \boldsymbol{t}) = 1$. The magnitude of the solid angle increases as p approaches the surface S from both the front and back sides:

$$\omega^+(p) = -2\pi \quad \text{and} \quad \omega^-(p) = 2\pi$$

respectively. Therefore, the tangential component of the field in the vicinity of the plane surface S is

$$B_t^+(p) = \frac{\mu_0 i}{2} \quad \text{and} \quad B_t^-(p) = -\frac{\mu_0 i}{2} \tag{1.66}$$

Here $B_t^+(p)$ and $B_t^-(p)$ are the total tangential components of the magnetic field at the front and back sides of S, respectively. From Equation (1.66), it follows that in general the tangential component B_t is a discontinuous function at any point of the surface S, and this discontinuity is caused by the current at this point:

$$B_t^+(p) - B_t^-(p) = \mu_0 i(p) \tag{1.67}$$

where $p \to q$ and $\boldsymbol{t}$ and $\boldsymbol{i}$ are perpendicular to each other and they are tangential to the plane S. In particular, if the surface S is an infinite plane, the magnitude of the solid angle at any point p is equal to 2π and, correspondingly, the tangential component B_t from both sides of the plane does not change and equals

$$B_t = \pm\frac{\mu_0 i}{2} \tag{1.68}$$

regardless of the position of the observation point. At the same time, the normal component B_n vanishes and it follows from symmetry. Now we will study the behavior of the tangential component B_t near an arbitrary surface S when the current density $\boldsymbol{i}$ is some function of the point q (Fig. 1.5(b)). It is clear that the field $B_t(p)$ can be represented as a sum of two fields:

$$B_t(p) = B_t^q(p) + B_t^{S-q}(p) \tag{1.69}$$

where B_t^q and B_t^{S-q} are tangential components of field, generated by the current element $\boldsymbol{i}\,dS(q)$ and the remainder of the currents. Considering the behavior of the field near the point q, we can say that the field B_t^{S-q} is a continuous function, since its generators are located at some distance from this point. At the same time, when p approaches the surface $p \to q$, the solid angle subtended by the element $dS(q)$ tends to $\pm 2\pi$. Therefore, we can write

$$B_t^+(p) = \frac{\mu_0 i(p)}{2} + B_t^{S-p}(p), \quad B_t^-(p) = -\frac{\mu_0 i(p)}{2} + B_t^{S-p}(p) \tag{1.70}$$

The latter shows the discontinuity of the tangential component at any point of the current surface is always defined by the current density in this point only, and it is equal to

$$B_t^+(p) - B_t^-(p) = \mu_0 i(p) \tag{1.71}$$

This equation is often called the surface analogy of the first field equation, and it can be written as

$$\text{curl}\, \boldsymbol{B} = \mu_0 \boldsymbol{i} \quad \text{or} \quad \boldsymbol{n} \times (\boldsymbol{B}^+ - \boldsymbol{B}^-) = \mu_0 \boldsymbol{i} \tag{1.72}$$

where $\boldsymbol{B}^+$ and $\boldsymbol{B}^-$ are the magnetic fields at the front and back sides of the current surface, respectively. It is proper to notice that Equation (1.72) also remains valid for a wide range of electromagnetic fields applied in geophysics. At the same time, the normal component of the field is a continuous function, since at the vicinity of the point q it is caused only by currents around the element $dS(q)$. It is also clear that in the vicinity of surface currents, the magnetic field does not tend to infinity.

1.5. SYSTEM OF EQUATIONS OF THE MAGNETIC FIELD *B* CAUSED BY CONDUCTION CURRENTS

In principle, the Biot–Savart law allows us to determine the magnetic field if the currents are known. However, in many cases of a current distribution in a nonuniform conducting medium and in the presence of magnetic medium, it is impossible to specify some of the currents, if the field $\boldsymbol{B}$ is unknown. In other words, the Biot–Savart law becomes useless, and we have to formulate a system of field equations and boundary-value problems. To solve this task, we start from Equation (1.25) which shows that divergence of the field $\boldsymbol{B}$ vanishes. In fact, we have

$$\text{div}\, \boldsymbol{B} = \text{div curl}\, \boldsymbol{A} \tag{1.73}$$

As follows from the vector analysis, the right-hand side of Equation (1.73) is identically zero. Therefore

$$\text{div}\, \boldsymbol{B} = 0 \tag{1.74}$$

This means that the magnetic field does not have sources and, correspondingly, the vector lines of the magnetic field $\boldsymbol{B}$ are closed. Next, applying Gauss' theorem, we obtain the integral form of this equation

$$\oint_S \boldsymbol{B} \cdot d\boldsymbol{S} = 0 \tag{1.75}$$

That is, the total flux of the field $\boldsymbol{B}$ through any closed surface is always equal to zero. Certainly, this is a fundamental feature of a magnetic field; we can imagine an unlimited number of different closed surfaces as well as currents, but for all of them Equation (1.75) is valid, and this happens because magnetic charges are absent. Now we will derive the surface analogy of Equation (1.74), and with this purpose in mind consider a very thin layer with current density $\boldsymbol{j}$. For the flux of the field through an elementary cylindrical surface, as is shown in Fig. 1.6(a), we have

$$\boldsymbol{B}^{(2)} \cdot d\boldsymbol{S}_2 + \boldsymbol{B}^{(1)} \cdot d\boldsymbol{S}_1 + \boldsymbol{B} \cdot d\boldsymbol{S}_* = 0 \tag{1.76}$$

where

$$d\boldsymbol{S}_2 = dS\boldsymbol{n} \quad \text{and} \quad d\boldsymbol{S}_1 = -dS\boldsymbol{n}$$

and $d\boldsymbol{S}_*$ is the lateral surface of the cylinder. Then reducing the layer thickness h to zero, Equation (1.76) becomes

$$B_{\rm n}^{(2)}\, dS - B_{\rm n}^{(1)}\, dS = 0 \quad \text{or} \quad B_{\rm n}^{(2)} = B_{\rm n}^{(1)} \tag{1.77}$$

Thus, the normal component of the magnetic field $\boldsymbol{B}$ is always a continuous function of spatial variables; otherwise we would have magnetic charges. We have derived three forms of one of the equations which show that the magnetic field is caused by the currents only:

$$\oint_S \boldsymbol{B} \cdot d\boldsymbol{S} = 0, \quad \text{div}\, \boldsymbol{B} = 0, \quad B_{\rm n}^{(2)} - B_{\rm n}^{(1)} = 0 \tag{1.78}$$

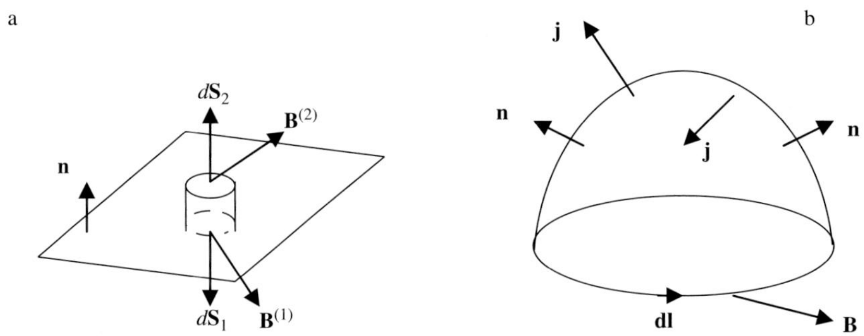

Fig. 1.6. Illustration of (a) Equation (1.76) and (b) Equation (1.81).

Each of them expresses the same fact, namely the absence of magnetic charges. Let us make two comments:

(a) Equations (1.78) have been derived assuming that the field $\boldsymbol{B}$ is caused by conduction currents. However, they remain valid in the presence of magnetic materials, when the field is also generated by magnetic dipoles of atoms.

(b) We obtained these equations from the Biot–Savart law for time-invariant currents, but actually they are still valid for alternating magnetic fields and in effect represent the fourth equation of the Maxwell' s system of equations describing electromagnetic fields.

Next we will develop the second equation of the magnetic field, making use again of the equation

$$\boldsymbol{B} = \text{curl}\,\boldsymbol{A}$$

and the identity

$$\text{curl curl}\,\boldsymbol{A} = \text{grad div}\,\boldsymbol{A} - \Delta\boldsymbol{A}$$

Considering the fact that

$$\text{div}\,\boldsymbol{A} = 0$$

and taking into account Equation (1.35), we obtain

$$\text{curl}\,\boldsymbol{B} = \mu_0 \boldsymbol{j} \tag{1.79}$$

This equation of the magnetic field in this differential form shows that at any regular point the curl of the magnetic field characterizes the volume density of the current at the same point. In particular, if we consider an elementary volume where this density is absent, then

$$\text{curl}\,\boldsymbol{B} = 0 \tag{1.80}$$

Equation (1.79) expresses the fact that currents are generators of the vortex type which create the magnetic field. Applying Stokes' theorem, we obtain the integral form of the first equation of the field

$$\oint\limits_L \boldsymbol{B}\cdot d\boldsymbol{l} = \int_S \text{curl}\,\boldsymbol{B}\cdot d\boldsymbol{S} = \mu_0 \int_S \boldsymbol{j}\cdot d\boldsymbol{S}$$

$$\text{or} \tag{1.81}$$

$$\oint\limits_L \boldsymbol{B}\cdot d\boldsymbol{l} = \mu_0 I$$

where I is the current flowing through the surface S, bounded by the path L (Fig. 1.6(b)). It is proper to notice that the mutual orientation of vectors $d\boldsymbol{l}$ and $d\boldsymbol{S}$ is not arbitrary but it is defined by the right-hand rule. Thus, circulation of the magnetic is defined by the flux of the current density, that is, the current I piercing the surface S surrounded by the contour L, and it does not depend on currents located outside of the perimeter of this area. Sometimes Equation (1.81) is called Ampere's law. It should be obvious that if the circulation is zero, it does not follow that the magnetic field is also zero at every point along L or charges do not move through the surface S. Of course, the path L can pass through media with different physical properties. Earlier we demonstrated that in the presence of surface currents, the tangential component of the magnetic field is a discontinuous function:

$$\boldsymbol{n} \times (\boldsymbol{B}^{(2)} - \boldsymbol{B}^{(1)}) = \mu_0 \boldsymbol{i} \tag{1.82}$$

and this represents the third form of the first equation. In particular, in real conditions when $\boldsymbol{i}=0$, the tangential component of $\boldsymbol{B}$ is a continuous function. Thus, we have derived three forms of Ampere's law, which show that the circulation of the magnetic field is defined by the current flux through any surface bounded by the path of integration, and currents are vortices of the magnetic field. These forms are

$$\oint_L \boldsymbol{B}\, d\boldsymbol{l} = \mu_0 I, \quad \text{curl}\, \boldsymbol{B} = \mu_0 \boldsymbol{j}, \quad B_t^{(2)} = B_t^{(1)} \tag{1.83}$$

It is interesting to notice that the last of these equations is valid for any alternating field, and it is usually regarded as the surface analogy of Maxwell's third equation. On occasion it is convenient to replace this equation by Equation (1.82), when we introduce the surface current density. Although the first two equations of the set (1.83) were derived from the expression for the magnetic field caused by constant currents, they remain valid for so-called quasi-stationary fields, which are widely used in geophysics. Now let us summarize these results in the form shown as follows:

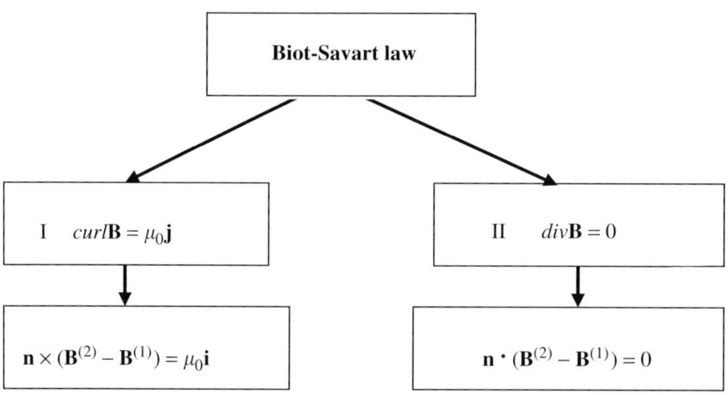

(1.84)

It is proper here to make several comments concerning Equations (1.84):

1. The system (1.84) together with boundary conditions contains the same information about the magnetic field as the Biot–Savart law, and this field is a classical example of a vortex field. Its generators are currents characterized by the current density field $\boldsymbol{j}$.
2. At surfaces where the current density $\boldsymbol{j}$ equals zero, both the normal and tangential components of the magnetic field are continuous functions.
3. The system (1.84) characterizes the behavior of the field in both a conducting and a nonconducting medium. Moreover, it is valid even in the presence of a medium that has an influence on the field (magnetic material), provided that the right-hand side of the first equation:

$$\text{curl}\,\boldsymbol{B} = \mu_0 \boldsymbol{j}$$

includes also the magnetization currents inside of the magnetic medium, that is

$$\boldsymbol{j} = \boldsymbol{j}_c + \boldsymbol{j}_m \tag{1.85}$$

where $\boldsymbol{j}_c$ and $\boldsymbol{j}_m$ are the densities of the conduction and magnetization currents, respectively. They are sole generators of the constant magnetic field. By analogy, in the case of the surface currents, we have

$$\boldsymbol{i} = \boldsymbol{i}_c + \boldsymbol{i}_m \tag{1.86}$$

Now let us consider three examples which illustrate an application of equations for the field $\boldsymbol{B}$ and its vector potential $\boldsymbol{A}$.

1.5.1. Example one: Magnetic field due to a current in a cylindrical conductor

Consider an infinitely long and homogeneous cylindrical conductor (Fig. 1.7(a)), with the radius a and current I. In such a case, the current density $\boldsymbol{j}$ is uniformly distributed over the cross-section S and everywhere inside has only a z-component, which is constant:

$$j = j_z = \text{constant} \tag{1.87}$$

In the cylindrical system of coordinates r, φ, z where the z-axis is directed along the conductor, the magnetic field can be characterized by three components B_r, B_φ and B_z. However, it turns out that two components are equal to zero. As follows from the Biot–Savart law, the magnetic field caused by the current element is perpendicular to the current density $\boldsymbol{j}$ and therefore the vertical component B_z equals zero. Next, consider two current elements located symmetrically with respect to the plane $\varphi = \text{constant}$ which passes through an observation point p (Fig. 1.7(b)). It is clear that the sum of radial components is zero. Since the current field can be represented as sum of such pairs, we can say the total magnetic

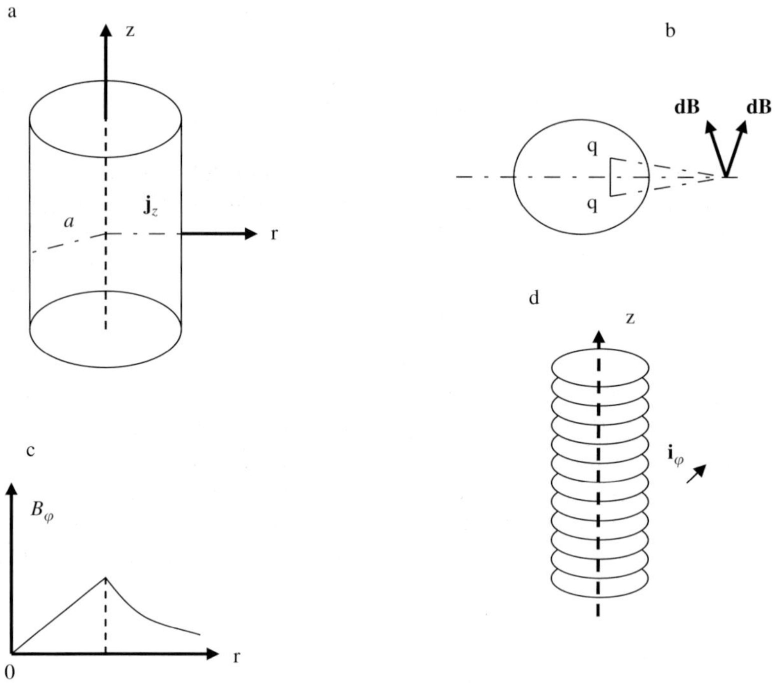

Fig. 1.7. (a) Cylindrical conductor. (b) Radial component due to current elements. (c) Behavior of the magnetic field inside and outside the cylinder. (d) Infinitely long solenoid.

field does not have a radial component, $B_r = 0$. Thus, we demonstrated that

$$\boldsymbol{B} = (0, B_\varphi, 0)$$

Taking into account the symmetry of the distribution of currents, we see that the vector lines of the magnetic field are circles located in horizontal planes and their centers are situated on the z-axis. In order to determine the component B_φ, we take one such line and apply the first equation in the integral form. This gives

$$\oint\limits_L \boldsymbol{B} \cdot d\boldsymbol{l} = \oint\limits_L B_\varphi \, dl = B_\varphi \oint\limits_L dl = 2\pi r B_\varphi = \mu_0 I_s$$

Here I_s is the current passing through any area bounded by the current line. In deriving this equality, we took into account the fact that the magnitude of the field does not vary along this circle and both vectors: $\boldsymbol{B}$ and $d\boldsymbol{l}$, are parallel to each other. Thus, the field outside and inside of the current is

$$B^e_\varphi = \frac{\mu_0 I}{2\pi r}, \quad \text{if } r \geq a \tag{1.88}$$

and

$$B_{\varphi}^{i} = \frac{\mu_0 j}{2} r, \quad \text{if } r \le a \tag{1.89}$$

since $I_s = \pi r^2 j$. In accordance with Equations (1.88) and (1.89), the magnetic field is equal to zero at the z-axis and increases linearly inside. At the surface of the conductor, it reaches a maximum, equal to

$$B_{\varphi}(a) = \frac{\mu_0 j}{2} a \tag{1.90}$$

and then the field decreases inversely proportional to the distance r (Fig. 1.7(c)). In this light, let us notice the following. Considering the magnetic field of the linear current, we found out that the field tends to infinity when an observation point approaches the surface of the current line. As was pointed out earlier, it is a result of a replacement of real distribution of currents by its fictitious model. As is seen from Equation (1.90), at the surface of a conductor, the field has a finite value which is usually rather small.

1.5.2. Example two: Magnetic field of an infinitely long solenoid

Suppose that at each point of the cylindrical surface S, a distribution of currents is characterized by the density $\boldsymbol{i}_{\phi}$ and it has everywhere the same magnitude (Fig. 1.7(d)). Inasmuch as the current has a component in the φ-direction, we have: $B_{\varphi} = 0$. It is a simple matter to show that the radial component also vanishes. In fact, consider two elementary current circuits, located symmetrically with respect to plane where an observation point is located (Fig. 1.8(a)). We can see that the sum of radial components is equal to zero. Taking into account the fact that the solenoid is infinitely long, we can always find such a pair of current loops and therefore the resultant radial component of the solenoid is also equal to zero. Thus, the total field

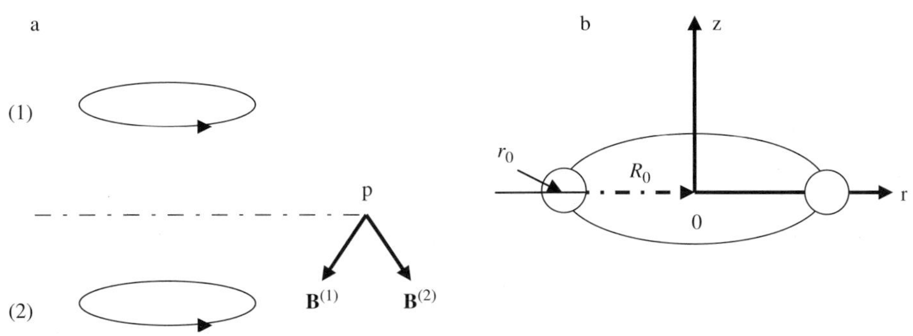

Fig. 1.8. (a) Radial component due to symmetrical current loops. (b) Toroid.

can have only a z-component:

$$\boldsymbol{B} = (0, 0, B_z) \tag{1.91}$$

This result greatly simplifies the algebra, because we have to focus on one component only. In principle, it can be evaluated by an integration of the fields caused by elementary current circles with the same radius a, but this is rather cumbersome. For this reason, we will make use of a different approach, based on Poisson's equation for the vector potential

$$\Delta \boldsymbol{A} = -\mu_0 \boldsymbol{j} \tag{1.92}$$

Taking into account symmetry and the fact that the vector potential has the same component as the current density, we have

$$\boldsymbol{A} = A_\varphi(r)\boldsymbol{i}_\varphi \tag{1.93}$$

Outside of the currents, this function obeys Laplace's equation:

$$\Delta \boldsymbol{A} = \Delta(A_\varphi \boldsymbol{i}_\varphi) = \boldsymbol{i}_\varphi \Delta A_\varphi + A_\varphi \Delta \boldsymbol{i}_\varphi = 0 \tag{1.94}$$

In a cylindrical system of coordinates, the operator Δ is

$$\Delta = \frac{1}{r}\frac{\partial}{\partial r}\left(r\frac{\partial}{\partial r}\right) + \frac{1}{r^2}\frac{\partial^2}{\partial \varphi^2} + \frac{\partial^2}{\partial z^2} \tag{1.95}$$

and

$$\boldsymbol{i}_\varphi = -\boldsymbol{i}_x \sin\varphi + \boldsymbol{i}_y \cos\varphi$$

where $\boldsymbol{i}_x$ and $\boldsymbol{i}_y$ are unit vectors in Cartesian system of coordinates and are independent of the coordinates of a point. First, we will find an expression of $\Delta \boldsymbol{i}_\varphi$. It is clear that the derivatives with respect to r and z are equal to zero and

$$\frac{\partial}{\partial \varphi}\boldsymbol{i}_\varphi = -\boldsymbol{i}_x \cos\varphi - \boldsymbol{i}_y \sin\varphi$$

Thus

$$\frac{\partial^2}{\partial \varphi^2}\boldsymbol{i}_\varphi = \boldsymbol{i}_x \sin\varphi - \boldsymbol{i}_y \cos\varphi = -\boldsymbol{i}_\varphi$$

Substitution of the latter into Equation (1.94) gives Laplace's equation with respect to a scalar component A_φ that greatly simplifies our task:

$$\frac{d}{dr}\left[r\frac{dA_\varphi(r)}{dr}\right] - \frac{A_\varphi(r)}{r} = 0 \tag{1.96}$$

This is an ordinary differential equation of the second order and its solution is

$$A_\varphi(r) = Cr + Dr^{-1} \tag{1.97}$$

Taking into account the fact that the magnetic field has to have a finite value and tends to zero at infinity, we represent the potential inside and outside of the solenoid as

$$A_\varphi^{(i)} = Cr \quad \text{and} \quad A_\varphi^{(e)} = Dr^{-1} \tag{1.98}$$

where C and D are unknown coefficients. By definition

$$\boldsymbol{B} = \text{curl}\,\boldsymbol{A}$$

or in the cylindrical system of coordinates

$$\boldsymbol{B} = \frac{1}{r}\begin{vmatrix} \boldsymbol{i}_r & r\boldsymbol{i}_\varphi & \boldsymbol{i}_z \\ \dfrac{\partial}{\partial r} & \dfrac{\partial}{\partial \varphi} & \dfrac{\partial}{\partial z} \\ 0 & rA_\varphi & 0 \end{vmatrix}$$

whence

$$B_r = 0, \quad B_\varphi = 0 \quad \text{and} \quad B_z = \frac{1}{r}\frac{\partial}{\partial r}(rA_\varphi) \tag{1.99}$$

Substitution of $A_\varphi^{(e)}$ into Equation (1.99) yields

$$B_z^{(e)} = 0, \quad \text{if } r > a \tag{1.100}$$

and we have proved that surface currents of the solenoid do not create a magnetic field outside the solenoid. In the same manner, for the field inside of the solenoid, we obtain

$$B_z^{(i)} = 2C, \quad \text{if } r < a$$

that is, this magnetic field is constant. In order to determine C, we recall that the difference of tangential components at both sides of the solenoid is

$$2C = \mu_0 i \quad \text{or} \quad B_z^{(i)} = \mu_0 i$$

Thus, for the field $\boldsymbol{B}$, caused by currents in the solenoid, we have

$$B_z^{(i)} = \mu_0 i_\varphi, \ \text{if } r < a \quad \text{and} \quad B_z^{(e)} = 0, \ \text{if } r > a \tag{1.101}$$

We may say that the magnetic field of the solenoid is concentrated only inside of it. Certainly, this is a very simple behavior, but such result is hardly obvious. First, it is difficult to predict that the field inside, $B_z^{(i)}$, is uniform over the cross-section, since the field due to a single current loop varies greatly. Also it is not obvious before calculations that the field outside of the solenoid is zero; that is, the sum of fields caused by all elementary current loops compensates each other. Consider a plane $z = \text{constant}$ where an observation point p is situated. Current circuits located relatively close to this plane generate a negative component $dB_z^{(e)}$ at the point p, while current loops situated far away give a positive contribution. Correspondingly, the field outside is a result of subtraction of elementary fields, and it turns out that in the case of an infinitely long solenoid this difference equals zero. Note that inside the solenoid all terms of this sum are positive. Of course, if a solenoid has a finite extension along the z-axis, the field outside is not zero and it is characterized everywhere by both components, B_r and B_z, and the latter prevails at its central plane.

1.5.3. Example three: Magnetic field of a current toroid

Consider a toroid with the current density $\boldsymbol{i}$ shown in Fig. 1.8(b) and introduce a cylindrical system of coordinates with the z-axis perpendicular to the toroid. Taking into account the axial symmetry, we see that the vector potential and magnetic field are independent of the coordinate φ. Also imagine two current loops of the toroid located symmetrically with respect to the vertical plane where a point of observation is located. It is clear that the sum of vector potentials, due to these elementary currents, does not give the φ-components. Thus, for the vector potential, we have

$$\boldsymbol{A} = (A_r, 0, A_z) \tag{1.102}$$

Taking into account the fact that

$$\boldsymbol{B} = \frac{1}{r}\begin{vmatrix} \boldsymbol{i}_r & r\boldsymbol{i}_\varphi & \boldsymbol{i}_z \\ \dfrac{\partial}{\partial r} & \dfrac{\partial}{\partial \varphi} & \dfrac{\partial}{\partial z} \\ A_r & 0 & A_z \end{vmatrix}$$

we obtain

$$B_r = \frac{1}{r}\frac{\partial A_z}{\partial \varphi} = 0, \quad B_\varphi = \left(\frac{\partial A_r}{\partial z} - \frac{\partial A_z}{\partial r}\right), \quad B_z = \frac{1}{r}\frac{\partial A_r}{\partial \varphi} = 0 \tag{1.103}$$

Thus, the magnetic field has only one component B_φ and it cannot be calculated from Equation (1.103), since neither component of the vector potential is known. However, this task can be easily solved by using the first equation of the field in the integral form:

$$\oint_L \boldsymbol{B}\, d\boldsymbol{l} = \mu_0 I_S$$

Taking into account the axial symmetry and the fact that $\boldsymbol{B}$ and $d\boldsymbol{l}$ have the same direction, this equality is greatly simplified and gives

$$B_\varphi 2\pi r = \mu_0 I_S \tag{1.104}$$

where L is a circular path of radius r, located in the horizontal plane with the center situated at the toroid axis, and I_S the current passing through a surface S surrounded by this path L. First, consider a point p, located outside a toroid. In such a case, the current either does not intersect the surface S or its value is equal to zero. This means that $B_\varphi = 0$ and, therefore, the magnetic field is absent outside the toroid, as in the case of the solenoid

$$\boldsymbol{B}_\varphi^{(e)} = 0 \tag{1.105}$$

Next, consider the magnetic field inside of the toroid. As follows from Equation (1.104), the field $B_\varphi^{(i)}$ is not uniform and equals

$$B_\varphi^{(i)} = \frac{\mu_0 I_S}{2\pi r} \tag{1.106}$$

In this case, the path of integration is inside the toroid. Suppose that it is located in the plane $z=0$ and a change of its radius does not change the flux of the current density. Therefore, within the range:

$$R_0 - r_0 < r < R_0 + r_0$$

an increase of r results in a decrease of the field inversely proportional to r. If we consider circular paths in planes with $z \neq 0$ ($z \leq r_0$), then the current I_s becomes smaller with increase of z. Thus, we observe a nonuniform magnetic field inside the toroid. It is natural to expect that with an increase of the ratio of the toroid radius R_0 to that of its cross-section r_0, the field inside becomes more uniform. It may be proper to note that if the toroid has an arbitrary but constant cross-section and the current density is independent on the coordinate φ, we can still apply Equation (1.104) and conclude that the field $\boldsymbol{B}$ is equal to zero outside the toroid. Of course, if a current density is not constant in the last two examples, the magnetic field appears outside, $\boldsymbol{B}^{(e)} \neq 0$.

1.6. THE SYSTEM OF EQUATIONS FOR VECTOR POTENTIAL A

Earlier we have derived the system of equations of the magnetic field caused by conduction currents. It is also useful to derive a system of equations for the vector potential $\boldsymbol{A}$. As we know

$$\boldsymbol{B} = \nabla \times \boldsymbol{A}$$

and

$$\boldsymbol{A}(p) = \frac{\mu_0}{4\pi}\int_V \frac{\boldsymbol{j}(q)}{L_{qp}}\,dV, \quad \Delta \boldsymbol{A} = -\mu_0 \boldsymbol{j}, \quad \nabla \boldsymbol{A} = 0$$

Also if there is an interface where surface currents are introduced, the tangential and normal components of the magnetic field behave as

$$\boldsymbol{n} \times (\boldsymbol{B}^{(2)} - \boldsymbol{B}^{(1)}) = \mu_0 i \quad \text{and} \quad \boldsymbol{n} \cdot \boldsymbol{B}^{(2)} = \boldsymbol{n} \cdot \boldsymbol{B}^{(1)}$$

or in terms of the vector potential:

$$\boldsymbol{n} \times (\nabla \times \boldsymbol{A}^{(2)}) - \boldsymbol{n}x(\nabla \times \boldsymbol{A}^{(1)}) = \mu_0 \boldsymbol{i} \quad \text{and} \quad (\nabla \times \boldsymbol{A}^{(2)})_{\mathrm{n}} = (\nabla \times \boldsymbol{A}^{(1)})_{\mathrm{n}}$$

Inasmuch as the normal component $(\nabla \times \boldsymbol{A})_{\mathrm{n}}$ includes only derivatives in directions tangential to the interface, the last equality remains valid if we require continuity of the vector potential:

$$\boldsymbol{A}^{(1)} = \boldsymbol{A}^{(2)}$$

Correspondingly, the system of equations for the vector potential is given in Equation (1.1.07). Now the following question arises. Why do we need Equations (1.1.07) if we know the expression for the vector potential in terms of the current density (Equation (1.24))? Certainly it is much easier to find the vector potential from Equation (1.24) than perform a solution of the system (1.1.07), if all conduction currents are known. However, a completely different situation takes place if currents are given only within some volume surrounded by a surface S.

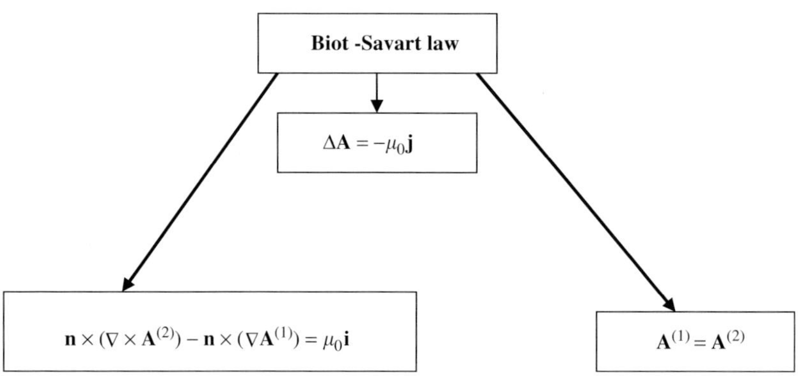

(1.107)

In such a case, Equation (1.24) allows us to calculate only the function $\mathbf{A}$ caused by currents inside the volume V, but the vector potential generated by currents located outside the volume remains unknown. In order to solve this task, we have to have additional information about the vector potential on the surface S. This leads us to formulation of boundary-value problems and theorems of uniqueness. We will discuss this subject in detail in the next chapter, but now let us make one comment. There are two systems of equations, namely, with respect to the magnetic field and the vector potential. Often it is more convenient to apply Equations (1.1.07) because in many important cases the field $\mathbf{A}$ has one or at maximum two components, which may greatly facilitate the solution of the boundary-value problems.

Chapter 2

Magnetic Field Caused by Magnetization Currents

2.1. MAGNETIZATION CURRENTS AND MAGNETIZATION: BIOT–SAVART LAW

As is well known, some substances, for instance iron after being placed in a magnetic field $\boldsymbol{B}$, produce a noticeable change in this field, while other materials have an extremely small influence. This happens due to a magnetization which is displayed in varying degrees by all materials. Moreover, there is such a group of magnetic materials whose magnetization remains even if the external field $\boldsymbol{B}$ disappears. The existence of these materials, for example, ferromagnetic, is vital for magnetic methods in geophysics as well as in numerous applications in other areas. Taking into account our purpose a magnetization can be described in the following way. First, suppose for simplicity that a magnetic material is an insulator and consequently conduction currents are absent. In spite of this fact, within every atom different types of motions of charges occur that can be approximately visualized as an elementary current (Chapter 6). Therefore, every small volume contains practically an unlimited number of such currents. If the external magnetic field is absent now as well as in the past, then these currents are randomly distributed and their magnetic field vanishes inside and outside of a magnetic material. In the presence of the external magnetic field $\boldsymbol{B}_0$ we observe a completely different picture. As will be shown in Chapter 3, any small current loop is subjected to a rotation and the moment of rotation is equal to

$$\boldsymbol{M}_{\mathrm{r}}(p) = \boldsymbol{M}(p) \times \boldsymbol{B}(p) \tag{2.1}$$

Here, $\boldsymbol{M}$ is the moment of the current loop, $(IS\mathbf{n})$, and $\boldsymbol{B}$, the magnetic field, while $\boldsymbol{n}$ is the unit vector, normal to the surface S, bounded by the elementary current. Let us note that the field $\boldsymbol{B}(p)$ is caused by all currents except that in the vicinity of the point p. We see that due to the magnetic field an elementary current tends to rotate until both vectors $\boldsymbol{M}$ and $\boldsymbol{B}$ become parallel each other, that is, they have the same or opposite directions. In such case, a motion stops and equilibrium is observed. This process is called magnetization and elementary currents are mainly oriented orderly. For this reason they create a magnetic field $\boldsymbol{B}_{\mathrm{m}}$ inside and outside of magnetic materials and at each point the resultant field $\boldsymbol{B}$ is a sum:

$$\boldsymbol{B} = \boldsymbol{B}_0 + \boldsymbol{B}_{\mathrm{m}} \tag{2.2}$$

If a medium also possesses conductivity, we will distinguish two types of currents, namely conduction currents which describe a motion of free charges through a medium, and magnetization ones, which are closed within an elementary volume. To calculate the magnetic field $\boldsymbol{B}_{\mathrm{m}}$, caused by the latter, we will perform a transformation from the micro to macro scale. This means that a system of atomic currents in such a volume is replaced by a single macroscopic current with the density $\boldsymbol{j}_{\mathrm{m}}$ and it is called the density of magnetization currents. To some extent a similar procedure is also performed for the molecular electric dipoles in a dielectric medium that leads to appearance of bounded charges. It may be proper to emphasize that the notion "magnetization currents" means a macroscopic quantity which gives the same result as a system of atomic currents. Then, for the total density of the volume and surface currents we have

$$\boldsymbol{j} = \boldsymbol{j}_{\mathrm{c}} + \boldsymbol{j}_{\mathrm{m}} \quad \text{and} \quad \boldsymbol{i} = \boldsymbol{i}_{\mathrm{c}} + \boldsymbol{i}_{\mathrm{m}} \tag{2.3}$$

Consequently, the Biot–Savart law is written as

$$\boldsymbol{B}(p) = \frac{\mu_0}{4\pi}\left[\int_V \frac{(\boldsymbol{j}_{\mathrm{c}} + \boldsymbol{j}_{\mathrm{m}}) \times \boldsymbol{L}_{qp}}{L_{qp}^3}\, dV + \int_S \frac{(\boldsymbol{i}_{\mathrm{c}} + \boldsymbol{i}_{\mathrm{m}}) \times \boldsymbol{L}_{qp}}{L_{qp}^3}\, dS\right] \tag{2.4}$$

and it describes the magnetic field $\boldsymbol{B}$ at every point inside and outside the magnetic substance. By analogy with the case of a nonmagnetic medium, the expression for the vector potential $\boldsymbol{A}$, ($\boldsymbol{B} = \mathrm{curl}\, \boldsymbol{A}$) is

$$\boldsymbol{A}(p) = \frac{\mu_0}{4\pi}\left[\int_V \frac{\boldsymbol{j}_{\mathrm{c}} + \boldsymbol{j}_{\mathrm{m}}}{L_{qp}}\, dV + \int_S \frac{\boldsymbol{i}_{\mathrm{c}} + \boldsymbol{i}_{\mathrm{m}}}{L_{qp}}\, dS\right] \tag{2.5}$$

which directly follows from Equation (2.4). It is appropriate to emphasize that

(a) The Biot–Savart law defines the macroscopic field $\boldsymbol{B}$ which is the mean microscopic field within every elementary volume or a surface.

(b) The coefficient on the right-hand side of Equation (2.4)

$$\frac{\mu_0}{4\pi}$$

is independent of magnetic materials. In other words, the Biot–Savart law correctly describes the magnetic field $\boldsymbol{B}$ in any magnetic medium provided that all currents are taken into account.

(c) Inasmuch as the distribution of magnetization currents is usually unknown, the Biot–Savart law cannot be used for field calculation in the presence of magnetic materials, and therefore in general we have to refer to a system of field equations, which will be described in the next and following sections.

(d) Both densities $\boldsymbol{j}_{\mathrm{m}}$ and $\boldsymbol{i}_{\mathrm{m}}$ are macroscopic quantities characterizing a distribution of currents within an elementary volume and elementary surface, which are in many orders larger than dimensions of atoms.

2.2. SYSTEM OF EQUATIONS OF THE FIELD *B* IN THE PRESENCE OF A MAGNETIC MEDIUM

The system of field equations in a nonmagnetic medium, (Equation (1.84)) has been derived from the Biot–Savart law:

$$\boldsymbol{B}(p) = \frac{\mu_0}{4\pi}\int_V \frac{\boldsymbol{j}_c \times \boldsymbol{L}_{qp}}{L_{qp}^3}\, dV$$

Comparing the latter with Equation (2.4) and taking into account Equation (1.84) we come to the conclusion that the system of field equations in differential form in the presence of magnetic materials at regular points is

$$\text{curl}\, \boldsymbol{B} = \mu_0(\boldsymbol{j}_c + \boldsymbol{j}_m) \quad \text{div}\, \boldsymbol{B} = 0 \tag{2.6}$$

and at interfaces

$$\text{Curl}\, \boldsymbol{B} = \mu_0(\boldsymbol{i}_c + \boldsymbol{i}_m) \quad \text{Div}\, \boldsymbol{B} = 0 \tag{2.7}$$

Here

$$\text{Curl}\, \boldsymbol{B} = \boldsymbol{n} \times (\boldsymbol{B}^{(2)} - \boldsymbol{B}^{(1)}) \quad \text{and} \quad \text{Div}\, \boldsymbol{B} = \boldsymbol{n} \cdot (\boldsymbol{B}^{(2)} - \boldsymbol{B}^{(1)})$$

Correspondingly, the integral form of these equations is

$$\oint \boldsymbol{B} \cdot d\boldsymbol{l} = \mu_0(I_c + I_m) \quad \text{and} \quad \oint \boldsymbol{B} \cdot d\boldsymbol{S} = 0$$

where I_c and I_m are the conduction and magnetization currents passing through an area surrounded by the path of integration. From a theoretical point of view this system does not differ from Equations (1.84). In fact, both of them describe a vortex field. However, there is one essential difference: namely, the right-hand side of the first equation in sets (2.6)–(2.7) contains the unknown density of magnetization currents. In the case when the field is caused by conduction currents (Equations (1.84)) the latter can be either specified everywhere or an absence of knowledge about them may be replaced by boundary conditions. A completely different situation takes place in the presence of magnetization currents which depend on unknown magnetic field. Thus, in order to determine field $\boldsymbol{B}$ we have to know the density of magnetization currents, but in principle it can be evaluated if the field is already calculated. This is a classical example of so-called closed circle problem. In order to overcome this formidable obstacle we have to bring some information about a magnetic medium and recall that every elementary volume may have a dipole moment $d\boldsymbol{M}$ and its magnetization is characterized by a vector of magnetization $\boldsymbol{P}$.

2.3. RELATION BETWEEN MAGNETIZATION CURRENTS AND MAGNETIZATION

With this purpose in mind we demonstrate that the density of magnetization currents, which differs from zero at certain places of a magnetic medium, is related to the dipole moment of an elementary volume. In other words, we will proceed from the fact that the closed currents within an elementary volume create a field $\boldsymbol{B}$ that coincides with the field of a magnetic dipole with some moment $d\boldsymbol{M}$. Now we will show again the importance of the concept of the magnetic dipole. Taking into account Equations (1.58)–(1.60), the vector potential caused by magnetization currents is

$$\boldsymbol{A}_{\text{m}}(p) = \frac{\mu_0}{4\pi} \int_V \frac{\boldsymbol{P}(q) \times \boldsymbol{L}_{qp}}{L_{qp}^3} \, dV \tag{2.8}$$

where

$$\boldsymbol{P}(q) = \frac{d\boldsymbol{M}(q)}{dV}$$

The vector $\boldsymbol{P}(q)$ is called the vector of magnetization, and it characterizes the magnitude and orientation of the dipole moment $d\boldsymbol{M}(q)$. It is also clear that the direction of the vector $\boldsymbol{P}(q)$ is perpendicular to the plane where the current loop is located (Fig. 2.1(a)). As follows from the definition of the dipole moment, the unit of measurement of the vector $\boldsymbol{P}$ is amperes per meter

$$[\boldsymbol{P}] = \text{A/m}$$

Taking into account the fact that

$$\frac{\boldsymbol{L}_{qp}}{L_{qp}^3} = \overset{q}{\nabla} \frac{1}{L_{qp}}$$

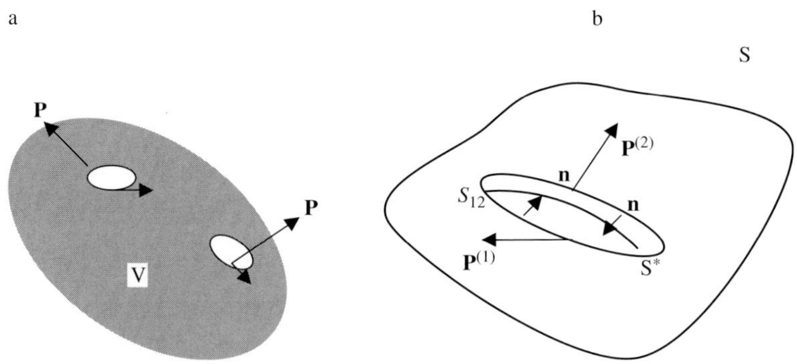

Fig. 2.1. (a) Magnetization currents and the vector of magnetization. (b) Illustration of Equation (2.16).

we represent Equation (2.8) as

$$A_{\mathrm{m}}(p) = \frac{\mu_0}{4\pi} \int_V \boldsymbol{P}(q) \times \overset{q}{\nabla} \frac{1}{L_{qp}} dV \tag{2.9}$$

Our next transformations imply that all points of a medium are regular, that is, the vector of magnetization $\boldsymbol{P}(q)$ has the first derivatives. Later we will consider a more general case with interfaces inside a medium where the vector $\boldsymbol{P}$ is a discontinuous function.

Then, making use of equality

$$\text{curl}\,(\varphi\mathbf{a}) = \varphi\text{curl}\,\mathbf{a} - \mathbf{a} \times \text{grad}\,\varphi$$

and letting $\boldsymbol{a} = \boldsymbol{P}$, $\varphi = 1/L_{qp}$, we have

$$A_{\mathrm{m}}(p) = \frac{\mu_0}{4\pi} \int_V \frac{\overset{q}{\text{curl}}\,\boldsymbol{P}}{L_{qp}} dV - \int_V \overset{q}{\text{curl}} \frac{\boldsymbol{P}}{L_{qp}} dV \tag{2.10}$$

It is essential that the integration and differentiation are performed with respect to the same point q, where the magnetization currents are situated. Now we demonstrate that the second integral becomes equal to zero. To prove this fact we use the equality

$$\int_V \text{curl}\,\boldsymbol{a} dV = \oint_S \boldsymbol{n} \times \boldsymbol{a} dS \tag{2.11}$$

where S is the surface surrounding the volume and $\boldsymbol{n}$ is the unit vector directed outside of this surface. Therefore, we obtain

$$A_{\mathrm{m}}(p) = \frac{\mu_0}{4\pi} \int_V \frac{\overset{q}{\text{curl}}\,\boldsymbol{P}}{L_{qp}} dV - \frac{\mu_0}{4\pi} \oint_S \frac{\boldsymbol{n} x \boldsymbol{P}}{L_{qp}} dS \tag{2.12}$$

The vector potential $A_{\mathrm{m}}(p)$ is caused by all magnetization currents including those located far away from observation points, that is, at points of the surface S which are at infinity and where a magnetic medium is absent and therefore $\boldsymbol{P}(q) = 0$. Consequently, the surface integral vanishes and we have

$$A_{\mathrm{m}}(p) = \frac{\mu_0}{4\pi} \int_V \frac{\text{curl}\,\boldsymbol{P}}{L_{qp}} dV \tag{2.13}$$

As follows from Equation (2.5) the vector potential due to magnetization currents can be also written as

$$A_{\mathrm{m}}(p) = \frac{\mu_0}{4\pi} \int_V \frac{\boldsymbol{j}_{\mathrm{m}}}{L_{qp}} dV \tag{2.14}$$

Comparing these last two equations we arrive at a relationship between the volume density of magnetization currents $\boldsymbol{j}_{\mathrm{m}}$ and the vector of magnetization $\boldsymbol{P}$:

$$\boldsymbol{j}_{\mathrm{m}} = \operatorname{curl} \boldsymbol{P} \tag{2.15}$$

Thus, we have shown that in place of dipoles we can imagine a distribution of magnetization currents with a help of the volume density $\boldsymbol{j}_{\mathrm{m}}$. As will be demonstrated later, this algebra is designed to simplify the system of equations of the magnetic field $\boldsymbol{B}$ and it also proves that Equation (2.5) is justified. In accordance with Equation (2.15) in the curvilinear orthogonal system of coordinates x_1, x_2, and x_3 with metric coefficients h_1, h_2, h_3:

$$\boldsymbol{j}_{\mathrm{m}} = \frac{1}{h_1 h_2 h_3} \begin{vmatrix} h_1 \boldsymbol{i}_1 & h_2 \boldsymbol{i}_2 & h_3 \boldsymbol{i}_3 \\ \dfrac{\partial}{\partial x_1} & \dfrac{\partial}{\partial x_2} & \dfrac{\partial}{\partial x_3} \\ h_1 P_1 & h_2 P_2 & h_3 P_3 \end{vmatrix}$$

and it clearly shows that the current density $\boldsymbol{j}_{\mathrm{m}}$ arises only in those places where a certain combination of derivatives from the vector of magnetization differs from zero. For instance, if in the vicinity of some point the vector $\boldsymbol{P}$ is constant, that is, the dipole moment remains the same, the magnetization current density is equal to zero. In other words, the appearance of this current is caused by a change of the dipole moment.

Next let us find a relationship between the vector of magnetization and surface magnetization currents. In deriving Equation (2.12) we assumed that the vector $\boldsymbol{P}$ is a continuous function inside of the medium; otherwise the equality (2.11) is invalid. Now we consider a more complicated model of a magnetic medium with some interface S_{12}, where the vector of magnetization $\boldsymbol{P}$ is a discontinuous function (Fig. 2.1(b)). In this case we will perform the same transformation with Equation (2.9) as before, but preliminarily it is necessary to enclose the surface S_{12} by another surface S^* and then apply Equation (2.11) in the volume V surrounded by the surface S and S^*. The necessity of this procedure is related to the fact that curl $\boldsymbol{P}$ does not have meaning at the surface S_{12}. Thus, instead of Equation (2.12), we have

$$\boldsymbol{A}_{\mathrm{m}}(p) = \frac{\mu_0}{4\pi} \int_V \frac{\operatorname{curl} \boldsymbol{P}}{L_{qp}} dV - \frac{\mu_0}{4\pi} \oint_{S^*} \frac{\boldsymbol{n}^* \times \boldsymbol{P}}{L_{qp}} dS - \frac{\mu_0}{4\pi} \oint_{S} \frac{\boldsymbol{n} \times \boldsymbol{P}}{L_{qp}} dS \tag{2.16}$$

where $\boldsymbol{n}^*$ is the unit vector perpendicular to the surface S^* and directed outside of volume V. As S^* approaches S_{12}, and neglecting the last integral since S is located at infinity, we have

$$\boldsymbol{A}_{\mathrm{m}}(p) = \frac{\mu_0}{4\pi} \int_V \frac{\operatorname{curl} \boldsymbol{P}}{L_{qp}} dV - \frac{\mu_0}{4\pi} \oint_{S^*} \frac{\boldsymbol{n}^* \times \boldsymbol{P}}{L_{qp}} dS$$

Taking into account the fact that integration over the surface S^* consists of integration over the back and front sides of the interface S_{12}, we obtain

$$\oint_{S^*} \frac{\boldsymbol{n}^* \times \boldsymbol{P}}{L_{qp}} dS = \int_{S_{12}} \frac{(\boldsymbol{n}^* \times \boldsymbol{P})_1 + (\boldsymbol{n}^* \times \boldsymbol{P})_2}{L_{qp}} dS$$

where the indexes "1" and "2" indicate the back and front sides of the interface S_{12}, respectively. As is seen from Fig. 2.1(b)

$$\boldsymbol{n}_1^* = \boldsymbol{n} \quad \text{and} \quad \boldsymbol{n}_2^* = -\boldsymbol{n}$$

Here $\boldsymbol{n}$ is the unit vector, normal to the surface S_{12}. Consequently, we arrive at the following expression for the vector potential caused by volume and surface magnetization currents:

$$\boldsymbol{A}_{\mathrm{m}}(p) = \frac{\mu_0}{4\pi} \int_V \frac{\operatorname{curl} \boldsymbol{P}}{L_{qp}} dV + \frac{\mu_0}{4\pi} \int_S \frac{\boldsymbol{n} \times (\boldsymbol{P}_2 - \boldsymbol{P}_1)}{L_{qp}} dS \qquad (2.17)$$

where $\boldsymbol{P}_1$ and $\boldsymbol{P}_2$ are the vectors of magnetization at the back and front sides of the interface, respectively. Comparing Equations (2.5) and (2.17) we see that

$$\boldsymbol{i}_{\mathrm{m}} = \boldsymbol{n} \times (\boldsymbol{P}_2 - \boldsymbol{P}_1) = \operatorname{Curl} \boldsymbol{P} \qquad (2.18)$$

That is, the difference of tangential components of the magnetization vector at both sides of the interface defines the density of magnetization currents $\boldsymbol{i}_{\mathrm{m}}$. Thus, we have established relationships between the volume and surface density of magnetization currents and the vector $\boldsymbol{P}$:

$$\boldsymbol{j}_{\mathrm{m}}(q) = \operatorname{curl} \boldsymbol{P}(q) \quad \text{and} \quad \boldsymbol{i}_{\mathrm{m}}(q) = \operatorname{Curl} \boldsymbol{P}(q) \qquad (2.19)$$

These formulas are very important because they allow us to obtain a system of the field equations, where the right-hand side is known. Besides, they are very useful to determine places of a magnetic medium where generators of the magnetic field; that is, magnetization currents arise. It may be proper to note again that these macroscopic currents produce the same effect as a real distribution of magnetic dipoles of atoms.

2.4. SYSTEM OF EQUATIONS WITH RESPECT TO THE MAGNETIC FIELD *B*

Now we are prepared to make some important changes in the sets (2.6)–(2.7). Substitution of Equation (2.19) into these sets gives

$$\operatorname{curl} \boldsymbol{B} = \mu_0(\boldsymbol{j}_{\mathrm{c}} + \operatorname{curl} \boldsymbol{P}) \quad \operatorname{div} \boldsymbol{B} = 0$$
$$\operatorname{Curl} \boldsymbol{B} = \mu_0(\boldsymbol{i}_{\mathrm{c}} + \operatorname{Curl} \boldsymbol{P}) \quad \operatorname{Div} \boldsymbol{B} = 0$$

or

$$\begin{aligned} \text{curl}\,(\boldsymbol{B} - \mu_0 \boldsymbol{P}) &= \mu_0 \boldsymbol{j}_c \quad \text{div}\,\boldsymbol{B} = 0 \\ \text{Curl}\,(\boldsymbol{B} - \mu_0 \boldsymbol{P}) &= \mu_0 \boldsymbol{i}_c \quad \text{Div}\,\boldsymbol{B} = 0 \end{aligned} \tag{2.20}$$

where the right-hand side of the first equation contains only the density of conduction currents which can be usually specified. It may be proper to emphasize that we were able to represent the set (2.20) in this form because the magnetization currents are expressed in terms of curl of the vector $\boldsymbol{P}$. Otherwise, we would not be able to combine this term with curl $\boldsymbol{B}$. However, there is still one very serious obstacle to overcome since the system has two unknown vectors $\boldsymbol{B}$ and $\boldsymbol{P}$. As we will see in order to solve this problem it is necessary to know the magnetic properties of a medium, which are obtained by experiments or analytically. A similar situation occurs in dielectric insulators and conducting media.

2.5. FIELD *H* AND RELATIONSHIP BETWEEN VECTORS *B*, *P*, AND *H*

First, we introduce a new vector field $\boldsymbol{H}$ as

$$\mu_0 \boldsymbol{H} = \boldsymbol{B} - \mu_0 \boldsymbol{P} \tag{2.21}$$

and rewrite Equation (2.20) as a system of equations with respect two vectors, $\boldsymbol{B}$ and $\boldsymbol{H}$:

$$\begin{aligned} \text{curl}\,\boldsymbol{H} &= \boldsymbol{j}_c \quad \text{div}\,\boldsymbol{B} = 0 \\ \text{Curl}\,\boldsymbol{B} &= \boldsymbol{i}_c \quad \text{Div}\,\boldsymbol{B} = 0 \end{aligned} \tag{2.22}$$

Certainly, we have advanced in deriving the system of field equations, but as was pointed out it still contains two unknown fields, and in this light it is proper to make several comments which emphasize this fact.

1. In accordance with Equation (2.21), we have

$$\boldsymbol{H} = \frac{1}{\mu_0}\boldsymbol{B} - \boldsymbol{P} \tag{2.23}$$

   that is, $\boldsymbol{H}$ is the difference of two fields with completely different physical meaning. Indeed, one of them up to a constant μ_0 describes the magnetic field; in other words, the real force acting on an elementary current. The other characterizes the density of the dipole moments; that is, a distribution of magnetization currents. Such combination can hardly be explained from a physical point of view. In other words, the function $\boldsymbol{H}$ is a pure mathematical concept. Later we will demonstrate that, in general, fictitious sources along with conduction currents "create" this fictitious field. This shows once more that $\boldsymbol{H}$ is an auxiliary field, which only allows us to derive a system of field equations where the right-hand side is usually given.

2. In the vicinity of points where the magnetization vector $\boldsymbol{P}$ is absent, in particular, in free space, the fields $\boldsymbol{B}$ and $\boldsymbol{H}$ differ by the constant μ_0 only:

$$\boldsymbol{B} = \mu_0 \boldsymbol{H} \tag{2.24}$$

but this does not change the fact that they are fundamentally different from each other.

3. The field $\boldsymbol{H}$ is often called the magnetic field, and from an historical point of view such terminology can be easily justified. However, here it will be called only the field $\boldsymbol{H}$.
4. This field, as well as the magnetization vector $\boldsymbol{P}$, is measured in amperes per meter; and this unit is related to that in Gauss' system by

$$1\ \mathrm{A/m} = 4\pi \times 10^{-3}\ \text{oersted} \quad \text{or} \quad 1\ \text{oersted} = 79.6\ \mathrm{A/m}$$

2.6. THREE TYPES OF MAGNETIC MEDIA AND THEIR MAGNETIC PARAMETERS

2.6.1. *Inductive and residual magnetization*

As was pointed out earlier in order to use the system of Equation (2.22) we have to establish a relation between the vector of magnetization $\boldsymbol{P}$ and either the field $\boldsymbol{B}$ or $\boldsymbol{H}$. From the physical point of view it is natural to deal with the function $\boldsymbol{P} = \boldsymbol{P}(\boldsymbol{B})$ because the magnetic field produces the magnetization, but in the past and sometimes now the function $\boldsymbol{P} = \boldsymbol{P}(\boldsymbol{H})$ is used. Experimental studies show that this linkage is rather complicated and has a form

$$\boldsymbol{P} = \chi(\boldsymbol{H})\boldsymbol{H} + \boldsymbol{P}_{\mathrm{r}} \tag{2.25}$$

where $\chi(\boldsymbol{H})$ is a function that usually depends on the field strength and the past history of the magnetic material; and $\boldsymbol{P}_{\mathrm{r}}$ is the residual magnetization, which remains even when the magnetic field vanishes, (Chapter 6). In many cases, however, it is possible to use the approximate relation

$$\boldsymbol{P} = \chi \boldsymbol{H} + \boldsymbol{P}_{\mathrm{r}} \tag{2.26}$$

where χ is a dimensionless constant of the magnetic medium, and it is called the magnetic susceptibility. It is clear that the parameter χ is dimensionless. In accordance with Equation (2.26) the vector of magnetization $\boldsymbol{P}$ is a sum of two vectors. One of them

$$\boldsymbol{P}_{\mathrm{in}} = \chi \boldsymbol{H} \tag{2.27}$$

is called the inductive magnetization which is caused by the magnetic field $\boldsymbol{B}$ existing at the moment of a field measurements. In contrary, the residual magnetization $\boldsymbol{P}_r$ appears as a result of the action of the magnetic field in the past. Certainly, both vectors, $\boldsymbol{P}_{in}$ and $\boldsymbol{P}_r$ have the same inductive origin, since they appear due to the magnetic fields $\boldsymbol{B}$.

2.6.2. Types of magnetic medium

There are several groups of magnetic materials, where a behavior of the parameter χ is completely different. Here, we briefly characterize three main groups: (a) *diamagnetic*; (b) *paramagnetic*; and (c) *ferromagnetic*, but in Chapter 6 these groups and others will be described in some detail.

In diamagnetic substances χ is extremely small ($\approx 10^{-6}$) and negative, so that the magnetization is very small. Correspondingly, the vectors $\boldsymbol{P}$ and $\boldsymbol{H}$ have opposite directions. The susceptibility of paramagnetic materials is positive and it is about 10^{-4}, that is, it is also very small. Unlike the previous case the vector of magnetization and vector $\boldsymbol{H}$ have the same direction. It is essential that the residual magnetization is absent in both these groups of magnetic materials and instead of Equation (2.26), we have

$$\boldsymbol{P} = \chi \boldsymbol{H} \tag{2.28}$$

Ferromagnetic is usually characterized by large and positive values of susceptibilities, and it is able to sustain magnetization in the absence of an external magnetic field. The susceptibility of rocks is mainly defined by the presence of ferromagnetic, such as magnetite. Table 2.1 demonstrates values of susceptibility of rocks. Equation (2.26) applies with a sufficient accuracy for one type of ferromagnetic, called soft magnetic materials, provided that magnetic field strength changes within a certain range.

Also it is proper to emphasize that there is a temperature called the Curie point, above which ferromagnetic properties vanish, that is, due to the high energy of

Table 2.1. Values of susceptibility of rocks.

	$\chi \times 10^6$		$\chi \times 10^6$
Graphite	−100	Gabbro	3,800–90,000
Quartz	−15.1	Dolomite (impure)	20,000
Anhydrite	−14.1	Pyrite (pure)	35–60
Rock salt	−10.3	Pyrite (ore)	100–5,000
Marble	−9.4	Pyrrhotite	10^3–10^5
Dolomite (pure)	−12.5 to +44	Hematite (ore)	420–10,000
Granite (without magnetite)	10–65	Ilmenite (ore)	3×10^5 to 4×10^6
Granite (with magnetite)	25–50,000	Magnetite (ore)	7×10^4 to 1×10^7
Basalt	1,500–25,000	Magnetite (pure)	1.5×10^7
Pegmatite	3,000–75,000		

motion of elementary particles they cannot align along the external magnetic field. For instance, in the case of magnetite the vector of magnetization becomes very small if the temperature is higher than 580°C. A decrease in the temperature below the Curie point results in a restoration of ferromagnetic properties of a material. In this light it is worth noticing that due to an increase of temperature with depth, at distances exceeding 20 km from the earth's surface a medium becomes practically nonmagnetic.

The residual magnetization $\boldsymbol{P}_r$ can be less or greater than the induced magnetization and in a ferromagnetic may reach 10^6 A/m or greater, while in rocks it can usually vary from 10 to 100 A/m. As was mentioned earlier, in general, the induced and residual magnetizations may have different directions. As was pointed out earlier the ordered orientation of magnetization currents, that is magnetization, is the result of an action of the magnetic field $\boldsymbol{B}$ and therefore it would be more natural, instead of Equation (2.26), to consider the equation

$$\boldsymbol{P} = k\boldsymbol{B} + \boldsymbol{P}_r$$

However, paying tribute to an old tradition, we will use Equation (2.26).

2.6.3. *Magnetic permeability*

Now we are ready to establish the relationship between the vectors $\boldsymbol{B}$ and $\boldsymbol{H}$. Substituting Equation (2.26) into (2.23), we have

$$\boldsymbol{B} = \mu_0(\boldsymbol{H} + \boldsymbol{P}) = \mu_0(\boldsymbol{H} + \chi\boldsymbol{H} + \boldsymbol{P}_r), \quad \text{or} \quad \boldsymbol{B} = \mu\boldsymbol{H} + \mu_0\boldsymbol{P}_r \tag{2.29}$$

In particular, if residual magnetization is absent

$$\boldsymbol{B} = \mu\boldsymbol{H} \tag{2.30}$$

where

$$\mu = \mu_r\mu_0 \quad \text{and} \quad \mu_r = 1 + \chi \tag{2.31}$$

The parameter μ is called the magnetic permeability of a medium, and its value approximately changes within the range:

$$4\pi \times 10^{-7}\ \text{H/m} < \mu < 1.0\ \text{H/m}$$

At the same time μ_r is the relative magnetic permeability, and it is obvious that for diamagnetic and paramagnetic materials μ_r is close to unity, while in ferromagnetic media it can be very large. In the practical system of units the parameter μ is measured in henries per meter. In solving the forward problem we will assume that the magnetic permeability is given and therefore, substituting Equation (2.29) into (2.22), obtain the system of equations either with respect to the magnetic field $\boldsymbol{B}$ or a fictitious field $\boldsymbol{H}$. Thus, we have solved the problem caused by the fact that at the beginning the right-hand side of the system of equations with respect to the field $\boldsymbol{B}$ contains unknown magnetization currents.

Using the system to determine these fields we imply that the magnetic permeability μ, residual magnetization $\boldsymbol{P}_{\mathrm{r}}$, and the density of conduction currents, $\boldsymbol{j}_{\mathrm{c}}$, $\boldsymbol{i}_{\mathrm{c}}$ are known. For instance, a distribution of these currents is defined by the electric field but is independent of a constant magnetic field. Taking into account the different nature of the fields $\boldsymbol{B}$ and $\boldsymbol{H}$, and the relatively complicated relationship between them, it seems that it is more appropriate to consider systems of equations for these fields separately. Of course, it is obvious that solving a system with respect the field $\boldsymbol{H}$ and making use Equation (2.29), we can determine the field $\boldsymbol{B}$ too. Also there are cases when it is convenient to deal with the system of equations which contains two vectors, $\boldsymbol{H}$ and $\boldsymbol{B}$, as well as a relation between them.

2.7. SYSTEM OF EQUATIONS FOR THE MAGNETIC FIELD *B*

Substitution of Equation (2.29) into the set (2.22) gives

$$\operatorname{curl}\frac{\boldsymbol{B}}{\mu} = \boldsymbol{j}_{\mathrm{c}} + \mu_0 \operatorname{curl}\frac{\boldsymbol{P}_{\mathrm{r}}}{\mu} \qquad \operatorname{div}\boldsymbol{B} = 0$$

and

$$\operatorname{Curl}\frac{\boldsymbol{B}}{\mu} = \boldsymbol{i}_{\mathrm{c}} + \mu_0 \operatorname{Curl}\frac{\boldsymbol{P}_{\mathrm{r}}}{\mu} \qquad \operatorname{Div}\boldsymbol{B} = 0 \tag{2.32}$$

These form the system of equations of the magnetic field $\boldsymbol{B}$ in the presence of a magnetic medium. Equations (2.32) are based on the Biot–Savart law, the principle of charge conservation:

$$\operatorname{div}\boldsymbol{j}_{\mathrm{c}} = 0$$

and the relation (2.29), and they clearly show that sources (charges) of the magnetic field are absent and that the conduction and magnetization currents are the sole generators (vortices) of the field $\boldsymbol{B}$. Consider one special but very important case of a piece-wise uniform medium, where in each homogeneous portion $\boldsymbol{P}_{\mathrm{r}} = \text{constant}$ and $\mu = \text{constant}$. Also, conduction currents are absent. Then, the system (2.32) is simplified, and we have

$$\operatorname{curl}\boldsymbol{B} = 0 \qquad \operatorname{div}\boldsymbol{B} = 0$$

and

$$\operatorname{Curl}\left(\frac{\boldsymbol{B} - \mu_0\boldsymbol{P}_{\mathrm{r}}}{\mu}\right) = 0 \qquad \operatorname{Div}\boldsymbol{B} = 0 \tag{2.33}$$

One more simplification takes place, if $\boldsymbol{P}_r = 0$. This gives

$$\text{curl}\,\boldsymbol{B} = 0 \quad \text{div}\,\boldsymbol{B} = 0$$

and

$$\text{Curl}\frac{\boldsymbol{B}}{\mu} = 0 \quad \text{Div}\,\boldsymbol{B} = 0 \tag{2.34}$$

2.8. DISTRIBUTION OF MAGNETIZATION CURRENTS

Now we will study the distribution of magnetization currents, which in accordance with Equation (2.32), depend on μ and $\boldsymbol{P}_r$ as well as the field $\boldsymbol{B}$. In other words, our goal is to find such places in a magnetic medium, where the density of magnetization currents differs from zero. Earlier we pointed out that as in the case of the conduction currents the density of magnetization currents is a macroscopic concept, and in many cases in order to understand the field behavior it is very convenient to know a distribution of these currents.

2.8.1. Volume density

First, consider their behavior at regular points of a medium. Making use of the equality again

$$\text{curl}\,\varphi\boldsymbol{a} = \varphi\text{curl}\,\boldsymbol{a} + (\text{grad}\,\varphi \times \boldsymbol{a})$$

we have

$$\text{curl}\frac{\boldsymbol{B}}{\mu} = \frac{1}{\mu}\text{curl}\,\boldsymbol{B} + \left(\text{grad}\frac{1}{\mu} \times \boldsymbol{B}\right)$$

and

$$\text{curl}\frac{\boldsymbol{P}_r}{\mu} = \frac{1}{\mu}\text{curl}\,\boldsymbol{P}_r + \left(\text{grad}\frac{1}{\mu} \times \boldsymbol{P}_r\right)$$

Then, the first equation of the field can be written as

$$\text{curl}\,\boldsymbol{B} = \mu\boldsymbol{j}_c - \mu\left(\text{grad}\frac{1}{\mu} \times \boldsymbol{B}\right) + \mu_0\text{curl}\,\boldsymbol{P}_r + \mu_0\mu\left(\text{grad}\frac{1}{\mu} \times \boldsymbol{P}_r\right)$$

Since

$$\text{grad}\frac{1}{\mu} = -\frac{1}{\mu^2}\text{grad}\,\mu$$

we have

$$\operatorname{curl} \boldsymbol{B} = \mu \boldsymbol{j}_c + \frac{1}{\mu}(\operatorname{grad} \mu \times \boldsymbol{B}) + \mu_0 \operatorname{curl} \boldsymbol{P}_r - \frac{\mu_0}{\mu}(\operatorname{grad} \mu \times \boldsymbol{P}_r) \tag{2.35}$$

At the same time, in terms of the conduction and magnetization currents the first equation of the field is

$$\operatorname{curl} \boldsymbol{B} = \mu_0(\boldsymbol{j}_c + \boldsymbol{j}_m)$$

Comparing the last two equations we conclude that the volume density of magnetization currents is

$$\boldsymbol{j}_m = \frac{\mu - \mu_0}{\mu_0}\boldsymbol{j}_c + \frac{1}{\mu\mu_0}(\operatorname{grad} \mu \times \boldsymbol{B}) + \operatorname{curl} \boldsymbol{P}_r - \frac{1}{\mu}(\operatorname{grad} \mu \times \boldsymbol{P}_r) \tag{2.36}$$

Thus, in general there are four types of magnetization currents.

The first type

$$\boldsymbol{j}_{1m} = \frac{\mu - \mu_0}{\mu_0}\boldsymbol{j}_c \tag{2.37}$$

arises in the vicinity of points where the density of conduction currents is not equal to zero, and both vectors $\boldsymbol{j}_c$ and $\boldsymbol{j}_m$ have the same direction, if $\mu > \mu_0$.

The second type

$$\boldsymbol{j}_{2m} = \frac{1}{\mu\mu_0}(\operatorname{grad} \mu \times \boldsymbol{B}) \tag{2.38}$$

appears in parts of a medium where the component of the field perpendicular to the direction of the maximal change of magnetic permeability is not equal to zero. The direction of these currents depends on the mutual position of the field $\boldsymbol{B}$ and grad μ. This type of current may appear in an inhomogeneous medium, if the vectors $\nabla\mu$ and $\boldsymbol{B}$ are not parallel to each other. It is proper to notice that only a change of μ is not sufficient to generate a magnetization current.

The third type

$$\boldsymbol{j}_{3m} = \operatorname{curl} \boldsymbol{P}_r \tag{2.39}$$

is entirely defined by the behavior of the residual magnetization and it arises in the vicinity of points where curl $\boldsymbol{P}_r \neq 0$. For instance, if we assume that the residual magnetization is absent or it is constant, the density of these currents is equal to zero.

The fourth type of the current density is

$$\mathbf{j}_{4\mathrm{m}} = -\frac{1}{\mu}(\mathrm{grad}\,\mu \times \mathbf{P}_\mathrm{r}) \tag{2.40}$$

and they appear in places where the residual magnetization and grad μ are not parallel to each other.

2.8.2. Surface density

These generators of the magnetic field are the most useful for understanding a field behavior since we usually deal with a piece-wise uniform medium, where both the magnetic permeability and residual magnetization are constants inside of each homogeneous portion of a medium, and at the same time they can be discontinuous functions at interfaces. By definition, from Equation (2.32), we have

$$\mathbf{n} \times \left(\frac{\mathbf{B}_2}{\mu_2} - \frac{\mathbf{B}_1}{\mu_1}\right) = \mathbf{i}_\mathrm{c} + \mu_0 \mathbf{n} \times \left(\frac{\mathbf{P}_{2\mathrm{r}}}{\mu_2} - \frac{\mathbf{P}_{1\mathrm{r}}}{\mu_1}\right) \tag{2.41}$$

Here, $\mathbf{B}_2, \mathbf{P}_{2\mathrm{r}}$ and $\mathbf{B}_1, \mathbf{P}_{1\mathrm{r}}$ are the magnetic field and residual magnetization at the front and back sides of the interface, respectively. The normal $\mathbf{n}$ is directed from the back to front side. Making use of the equality

$$\left(\frac{\mathbf{a}_2}{\beta_2} - \frac{\mathbf{a}_1}{\beta_1}\right) = \frac{1}{2}\left[\left(\frac{1}{\beta_2} - \frac{1}{\beta_1}\right)(\mathbf{a}_2 + \mathbf{a}_1) + \left(\frac{1}{\beta_2} + \frac{1}{\beta_1}\right)(\mathbf{a}_2 - \mathbf{a}_1)\right]$$

and letting $\mathbf{a}_1 = \mathbf{B}_1$, $\mathbf{a}_2 = \mathbf{B}_2$, $\beta_1 = \mu_1$, $\beta_2 = \mu_2$, we represent Equation (2.41) as

$$\mathrm{Curl}\,\mathbf{B} = \frac{1}{\beta^{\mathrm{av}}}\mathbf{i}_\mathrm{c} - \frac{\Delta\beta}{\beta^{\mathrm{av}}}\mathbf{n} \times \mathbf{B}^{\mathrm{av}} + \mu_0 \mathrm{Curl}\,\mathbf{P}_\mathrm{r} + \mu_0 \frac{\Delta\beta}{\beta^{\mathrm{av}}}\mathbf{n} \times \mathbf{P}_\mathrm{r}^{\mathrm{av}} \tag{2.42}$$

where

$$\beta^{\mathrm{av}} = \frac{1}{2}\left(\frac{1}{\mu_2} + \frac{1}{\mu_1}\right), \quad \Delta\beta = \frac{1}{\mu_2} - \frac{1}{\mu_1}$$

$$\mathbf{B}^{\mathrm{av}} = \frac{\mathbf{B}_1 + \mathbf{B}_2}{2}, \quad \mathbf{P}_\mathrm{r}^{\mathrm{av}} = \frac{\mathbf{P}_{1\mathrm{r}} + \mathbf{P}_{2\mathrm{r}}}{2}$$

As we know

$$\mathrm{Curl}\,\mathbf{B} = \mu_0(\mathbf{i}_\mathrm{c} + \mathbf{i}_\mathrm{m})$$

Therefore, the surface density of magnetization currents is

$$\boldsymbol{i}_{\rm m} = \left(\frac{1}{\mu_0 \beta^{\rm av}} - 1\right)\boldsymbol{i}_{\rm c} + 2\frac{K_{12}}{\mu_0}\boldsymbol{n} \times \boldsymbol{B}^{\rm av} + {\rm Curl}\, \boldsymbol{P}_{\rm r} - 2K_{12}\boldsymbol{n} \times \boldsymbol{P}_{\rm r}^{\rm av} \tag{2.43}$$

Here

$$K_{12} = \frac{\mu_2 - \mu_1}{\mu_2 + \mu_1} \tag{2.44}$$

As in the case of the volume density, we distinguish four types of surface currents.

The first type

$$\boldsymbol{i}_{1\rm m} = \frac{1}{\mu_0}\left(\frac{2\mu_1\mu_2}{\mu_1 + \mu_2} - \mu_0\right)\boldsymbol{i}_{\rm c} \tag{2.45}$$

occurs in the vicinity of conduction currents at interfaces of media with different magnetic permeability. The current density of *the second type* is

$$\boldsymbol{i}_{2\rm m} = 2\frac{K_{12}}{\mu_0}\boldsymbol{n} \times \boldsymbol{B}^{\rm av} \tag{2.46}$$

and it is directly proportional to the contrast coefficient K_{12} and the average value of the tangential component of the field at some point q, caused by all currents except those in the vicinity of this point. *The third type* of the surface current is

$$\boldsymbol{i}_{3\rm m} = {\rm Curl}\, \boldsymbol{P}_{\rm r}$$

and it is defined by the difference between tangential components of the residual magnetization. Finally, *the fourth type* of currents arises in places on the interface where the average tangential component of the vector $\boldsymbol{P}_{\rm r}$ differs from zero:

$$\boldsymbol{i}_{4\rm m} = -2K_{12}\boldsymbol{n} \times \boldsymbol{P}_{\rm r}^{\rm av}$$

It may be natural to consider together the last two types of currents:

$$\boldsymbol{i}_{3\rm m} + \boldsymbol{i}_{4\rm m} = {\rm Curl}\, \boldsymbol{P}_{\rm r} - 2K_{12}\boldsymbol{n} \times \boldsymbol{P}_{\rm r}^{\rm av} \tag{2.47}$$

2.9. SYSTEM OF EQUATIONS FOR THE FICTITIOUS FIELD *H* AND DISTRIBUTION OF ITS GENERATORS

Earlier we introduced a fictitious field $\boldsymbol{H}$ as a combination of the magnetic field and the vector of magnetization but now consider its main features. First of all let us derive the system of equations of this auxiliary field. Substituting Equation (2.29) into set (2.22) we obtain

$$\begin{aligned}\text{Curl}\,\boldsymbol{H} &= \boldsymbol{j}_c \quad \text{div}(\mu\boldsymbol{H}) = -\mu_0 \text{div}\,\boldsymbol{P}_r \\ \text{Curl}\,\boldsymbol{H} &= \boldsymbol{i}_c \quad \text{Div}(\mu\boldsymbol{H}) = -\mu_0 \text{Div}\,\boldsymbol{P}_r\end{aligned} \tag{2.48}$$

Consequently, the generators of the field $\boldsymbol{H}$ consist of conduction currents and fictitious sources (magnetic charges). To describe the latter we will proceed from the fact that the divergence of any field may characterize the density of its sources, regardless of whether this field is real or a pure mathematical concept. Therefore, we will introduce the density of magnetic charges as

$$\delta_m = \text{div}\,\boldsymbol{H} \quad \sigma_m = \text{Div}\,\boldsymbol{H} \tag{2.49}$$

2.9.1. *Volume density*

First, consider their distribution at regular points of the medium. Inasmuch as

$$\text{div}\,\varphi\boldsymbol{a} = \varphi\text{div}\,\boldsymbol{a} + \boldsymbol{a}\cdot\text{grad}\,\varphi$$

and

$$\text{div}\,\boldsymbol{B} = \text{div}(\mu\boldsymbol{H} + \mu_0\boldsymbol{P}_r) = 0$$

we have

$$\text{div}\mu\boldsymbol{H} = \mu\text{div}\,\boldsymbol{H} + \boldsymbol{H}\cdot\text{grad}\,\mu = -\mu_0\text{div}\,\boldsymbol{P}_r$$

Making use of Equation (2.49), the latter gives

$$\delta_m = -\frac{\boldsymbol{H}\cdot\text{grad}\,\mu}{\mu} - \frac{\mu_0\text{div}\,\boldsymbol{P}_r}{\mu} \tag{2.50}$$

Thus, we distinguish two types of sources of the field $\boldsymbol{H}$. One of them

$$\delta_{1m} = -\frac{\boldsymbol{H}\cdot\text{grad}\,\mu}{\mu} \tag{2.51}$$

"arises" in the vicinity of points where there is a component of the field $\boldsymbol{H}$ along grad μ. Correspondingly, this type of charge vanishes if the field $\boldsymbol{H}$ is perpendicular to the direction of the maximal change of the magnetic permeability. Also, $\delta_{1\mathrm{m}}$ equals zero at places where a medium is uniform.

The second type of source is

$$\delta_{2\mathrm{m}} = -\frac{\mu_0}{\mu}\mathrm{div}\,\boldsymbol{P}_\mathrm{r} \tag{2.52}$$

and it is related to the behavior of the residual magnetization only.

2.9.2. *Surface density*

Now consider a distribution of surface magnetic charges introduced by Equation (2.49):

$$\sigma_\mathrm{m} = H_{2\boldsymbol{n}} - H_{1\boldsymbol{n}}$$

where the normal $\boldsymbol{n}$ is directed toward the medium with the index "2." Since

$$\begin{aligned}\mathrm{Div}\,\mu\boldsymbol{H} &= \mu_2\boldsymbol{H}_2 - \mu_1\boldsymbol{H}_1 = \frac{1}{2}\left[(\mu_2-\mu_1)(H_{2\boldsymbol{n}}+H_{1\boldsymbol{n}}) + (\mu_2+\mu_1)(H_{2\boldsymbol{n}}-H_{1\boldsymbol{n}})\right]\\ &= -\mu_0\mathrm{Div}\,\boldsymbol{P}_\mathrm{r}\end{aligned}$$

we have

$$\sigma_\mathrm{m} = -2K_{12}H_{\boldsymbol{n}}^{\mathrm{av}} - \frac{\mu_0}{\mu^{\mathrm{av}}}\mathrm{Div}\,\boldsymbol{P}_\mathrm{r} \tag{2.53}$$

Consequently, there are two types of surface fictitious magnetic charges, namely,

$$\sigma_{1\mathrm{m}} = -2K_{12}H_{\boldsymbol{n}}^{\mathrm{av}} \quad \text{and} \quad \sigma_{2\mathrm{m}} = -\frac{\mu_0}{\mu^{\mathrm{av}}}\mathrm{Div}\,\boldsymbol{P}_\mathrm{r} \tag{2.54}$$

One of them is "located" in the vicinity of points where the average normal component of the field $\boldsymbol{H}$ differs from zero. The other is defined by the behavior of the normal component of the residual magnetization.

2.10. DIFFERENCE BETWEEN THE FIELDS *B* AND *H*

Bearing in mind that the behavior of a field is defined by its generators let us describe the main features of fields $\boldsymbol{B}$ and $\boldsymbol{H}$ and they are

1. The magnetic field $\boldsymbol{B}$ is caused by vortices only, and these include both conduction and magnetization currents.

2. In general, the field $\boldsymbol{H}$ has two different types of generators, the conduction currents, but not magnetization ones, and fictitious sources.
3. The magnetic field $\boldsymbol{B}$ obeys the Biot–Savart law, but this law does not describe the behavior of the field $\boldsymbol{H}$.
4. The force acting on a moving electric charge is defined by the magnetic field $\boldsymbol{B}$, but not the field $\boldsymbol{H}$.
5. In essence, the field $\boldsymbol{H}$ is an auxiliary field, which was introduced to modify the system of equations of the magnetic field $\boldsymbol{B}$ in order to facilitate a solution of the boundary-value problems.

Now we will consider several examples illustrating the difference in a behavior between fields $\boldsymbol{B}$ and $\boldsymbol{H}$.

2.10.1. Example 1: Current loop in a homogeneous medium

Assume that a current loop with a density $\boldsymbol{j}_c$ is placed in a uniform medium where the magnetic permeability is equal to μ and residual magnetization $\boldsymbol{P}_r$ is absent. Then, magnetization currents appear in the vicinity of the conduction current and in accordance with Equation (2.37) their density is

$$\boldsymbol{j}_m = \frac{\mu - \mu_0}{\mu_0}\boldsymbol{j}_c$$

Thus, the total current density becomes equal to

$$\boldsymbol{j} = \boldsymbol{j}_c + \boldsymbol{j}_m = \frac{\mu}{\mu_0}\boldsymbol{j}_c \tag{2.55}$$

and its magnitude is μ_r times larger than in a nonmagnetic medium where only conduction currents are present, $(\mu > \mu_0)$. Correspondingly, we observe a great increase of the current density, if $\mu_r \gg 1$. At the same time, at other places in spite of the inductive magnetization of every elementary volume of a medium, $\boldsymbol{P}_{in} \neq 0$, the density of magnetization currents is zero. This shows that the presence of a homogeneous medium results in a change of the magnetic field by μ_r times, while the field geometry remains the same. Taking into account the fact that fictitious charges are absent, such medium does not affect the field $\boldsymbol{H}$, and it is the same as in a free space.

2.10.2. Example 2: Uniform fields B and H in a medium with one plane interface

Suppose that there is a planar interface between two media having magnetic permeability μ_1 and μ_2, respectively (Fig. 2.2(a)). It is also assumed that the uniform magnetic field $\boldsymbol{B}$ is perpendicular to this boundary and the residual magnetization and conduction currents are absent in a volume where the field is considered:

$$\boldsymbol{P}_r = 0, \quad \boldsymbol{j}_c = \boldsymbol{i}_c = 0$$

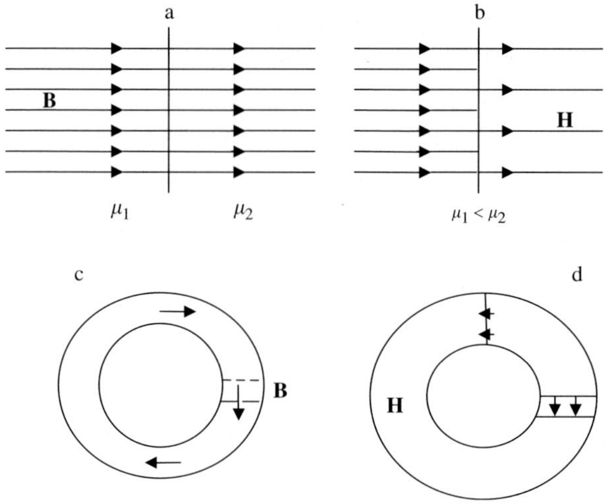

Fig. 2.2. (a) Field ***B*** in the presence of the plane interface. (b) Field ***H*** in the presence of plane interface. (c) Field ***B*** inside a magnetic toroid with small gap. (d) Field ***H*** inside the magnetic toroid with small gap.

First of all, it is clear that in such a volume vortices of the magnetic field are absent. In fact, as follows from examples (2.36) and (2.43) the volume and surface densities of magnetization currents vanish. At the same time, the magnetic field does not have sources

$$\text{div}\,\boldsymbol{B} = 0 \quad \text{and} \quad \text{Div}\,\boldsymbol{B} = 0$$

and therefore its vector lines are always closed. In particular, they do not break at the interface. Since the field ***B*** is perpendicular to the boundary, we have to conclude that the density of its vector lines remains the same in both media; that is, this interface does not influence ***B***. However, the field ***H*** behaves in a different way. In fact, ***H*** is related to ***B*** by

$$\boldsymbol{H} = \frac{\boldsymbol{B}}{\mu}$$

and therefore it is uniform at each part of a medium but has different values at both sides of the interface. For instance, in a medium with greater magnetic permeability, the field ***H*** is smaller. Consequently, the vector lines of this field break off at the interface (Fig. 2.2(b)) and fictitious magnetic charges "arise." In accordance with Equation (2.54) their density is

$$\sigma_{1\text{m}} = -2K_{12}H_{\boldsymbol{n}}^{\text{av}} \quad \text{or} \quad \sigma_{1\text{m}} = \left(\frac{1}{\mu_2} - \frac{1}{\mu_1}\right)B_{\boldsymbol{n}} \tag{2.56}$$

where the normal component is positive if it is directed from the medium with magnetic permeability μ_1 to that with permeability μ_2 and vice versa. It is obvious that

$$H_{\boldsymbol{n}}^{\text{av}} = \frac{1}{2}\left(\frac{B_{2n}}{\mu_2} + \frac{B_{1n}}{\mu_1}\right) = \left(\frac{1}{\mu_2} + \frac{1}{\mu_1}\right)\frac{B_n}{2}$$

As follows from Equation (2.56) it follows that if $B_n > 0$ and $\mu_2 > \mu_1$, negative surface charges with constant density "appear" at the interface, while their volume density is zero.

2.10.3. Example 3: Fields B and H inside the toroid with a small gap

Now we assume that a uniform magnetic medium has a shape of a toroid with a very small gap, and it is surrounded by a nonmagnetic medium (Fig. 2.2(c)). Also suppose that the medium was earlier subjected to a magnetic field so that now it possesses the residual magnetization $\boldsymbol{P}_{\text{r}}$. It has everywhere the same magnitude and is directed along the toroid axis. Inasmuch as conduction currents are absent and every elementary volume of the uniform medium has the same magnetization, the volume density of magnetization currents equals zero (Equation (2.36)). At the lateral surface of the toroid the density of magnetization currents does not vanish. In fact, taking into the account the fact that $\boldsymbol{P}_{2\text{r}} = 0$ and making use of Equation (2.43), we obtain for the current density related to the residual magnetization

$$\boldsymbol{i}_{3\text{m}} = -\boldsymbol{n} \times \boldsymbol{P}_{\text{r}} \quad \text{and} \quad \boldsymbol{i}_{4\text{m}} = -\boldsymbol{n} \times K_{12}\boldsymbol{P}_{\text{r}}$$

and

$$\boldsymbol{i}_{3\text{m}} + \boldsymbol{i}_{4\text{m}} = -\boldsymbol{n} \times (1 + K_{12})\boldsymbol{P}_{\text{r}} = -\frac{2\mu_0}{\mu_0 + \mu}\boldsymbol{n} \times \boldsymbol{P}_{\text{r}} \tag{2.57}$$

As is seen from Fig. 2.2(c), the vectors of the surface current density, given by Equation (2.57), and the residual magnetization $\boldsymbol{P}_{\text{r}}$ are oriented in agreement with the right-hand rule. These currents form a system of current loops with the same density uniformly distributed on the lateral surface of the toroid. It is obvious that such currents create a practically uniform magnetic field inside the magnet and the magnetization vector is directed along the toroid axis. Until now we have considered magnetization currents related to the residual magnetization. Besides, there are surface currents associated with the field $\boldsymbol{B}$, and in accordance with Equation (2.46)

$$\boldsymbol{i}_{2\text{m}} = -\frac{\mu - \mu_0}{\mu + \mu_0}\boldsymbol{n} \times \frac{\boldsymbol{B}}{\mu_0}$$

which also describes a system of current loops with the same direction as the current. Therefore, we can say that the magnetic field of the permanent magnet, having only surface magnetization currents, is equivalent to that of a solenoid with the same distribution of conduction currents. It is easy to predict that if the gap width is small with respect to the toroid diameter, then the vector lines of $\boldsymbol{B}$ are almost parallel to each other. This means that inside the toroid and within the gap the field remains the same. Now we will consider the behavior of the field $\boldsymbol{H}$. By definition, we have

$$\boldsymbol{B} = \mu\boldsymbol{H} + \mu_0\boldsymbol{P}_{\mathrm{r}}$$

That is, $\boldsymbol{H}$ is uniform and directed along the toroid axis. Therefore, fictitious magnetic charges are absent at the lateral surface of the magnet (Equation (2.53)). Also, the conduction current density equals zero. However, sources of the field $\boldsymbol{H}$ arise at the two boundaries between the toroid and its gap (Fig. 2.2(d)). Assuming that the normal directed from the magnet to the gap space, we have

$$\sigma_{1\mathrm{m}} = 2\frac{\mu - \mu_0}{\mu + \mu_0} H_{\boldsymbol{n}}^{\mathrm{av}}, \quad \text{and} \quad \sigma_{2\mathrm{m}} = \frac{2\mu_0}{\mu + \mu_0} P_{\boldsymbol{n}}$$

By definition, the vector lines of the field $\boldsymbol{H}$ start from positive sources and finish at negative ones. Inasmuch as the fields $\boldsymbol{B}$ and $\boldsymbol{H}$ have the same direction within the gap, the positive charges "appear" at the face where the magnetic field is directed from the magnet to the gap and vice versa. The boundary with positive charges is usually associated with the North Pole, while the opposite side of the gap is related to the South Pole. Therefore, inside the toroid the magnetic field $\boldsymbol{B}$ and fictitious field $\boldsymbol{H}$ have opposite directions, but within the gap they have the same direction and differ from each other by the constant μ_0:

$$\boldsymbol{B} = \mu_0\boldsymbol{H}$$

Now suppose that the toroid does not have a gap, and still the magnetic field and magnetization vector are tangential to the lateral surface. Since conduction currents are absent, as well as magnetic charges, we have to conclude that the field $\boldsymbol{H}$ within the toroid and outside equals zero:

$$\boldsymbol{H} = 0$$

Consequently, inside this permanent magnet Equation (2.29) is simplified and it becomes

$$\boldsymbol{B} = \mu_0\boldsymbol{P}_{\mathrm{r}}$$

but outside the toroid

$$\boldsymbol{B} = \boldsymbol{P}_{\mathrm{r}} = 0$$

Certainly, this is a vivid example which emphasizes the difference between the fields $\boldsymbol{B}$ and $\boldsymbol{H}$.

2.10.4. Example 4: Fields B and H inside the solenoid

Next we will consider a solenoid that has the same dimensions and shape as the toroid with the small gap (shown in Fig. 2.2(c)). Inside and outside the solenoid a magnetic medium is absent, and therefore the field $\mathbf{B}$ is caused by the conduction currents in the coil only. Since the gap width is small compared to the solenoid diameter, the magnetic field is practically uniform inside the solenoid and in the gap. The field $\mathbf{H}$ is also "generated" by the coil current only and

$$\mathbf{B} = \mu_0 \mathbf{H}$$

Unlike the previous case these fields differ by a constant, in particular, they have everywhere the same direction. Suppose that the current density in the coil has a magnitude and direction such that the magnetic fields coincide in both gaps of the solenoid and the toroid. Then, due to uniformity of these fields we can state that inside the solenoid and toroid they are also equal to each other. This happens in spite of the fact that in one case the field is caused by conduction currents, while in the other magnetization currents are sole generators of the magnetic field. Now compare the field $\mathbf{H}$ in both models. It is clear that within the gaps of the permanent magnet and the solenoid they are equal to each other, since magnetic fields coincide. However, inside the toroid and the coil the behavior of the magnetic field $\mathbf{H}$ does not have common features. In fact, inside the solenoid we have

$$\mathbf{H} = \frac{1}{\mu_0}\mathbf{B} \tag{2.58}$$

but inside the toroid $\mathbf{H}$ is caused by fictitious sources in the vicinity of the poles and is directed opposite to the magnetic field $\mathbf{B}$. Besides, with a decrease of the gap width of the toroid $\mathbf{H}$ tends to zero.

2.10.5. Example 5: Fields B and H inside the magnetic solenoid

Suppose that a toroid with a very small gap is wound by a current coil and that both the conduction and surface magnetization currents have the same direction. Consequently, the magnetic field becomes stronger. If the current in the coil is sufficiently large, then the field $\mathbf{H}$ is mainly caused by this current, and therefore both fields $\mathbf{B}$ and $\mathbf{H}$ have the same direction inside the toroid and in a gap.

2.10.6. Example 6: Influence of a thin magnetic shell

Let us assume that a magnetic field $\mathbf{B}$ was caused by either conduction currents or magnets or both of them and surrounding medium is not magnetic. Now we surround these currents by a thin and closed shell of an arbitrary shape and dimensions with a magnetic permeability, μ (Fig. 2.3(a)). For instance, we can

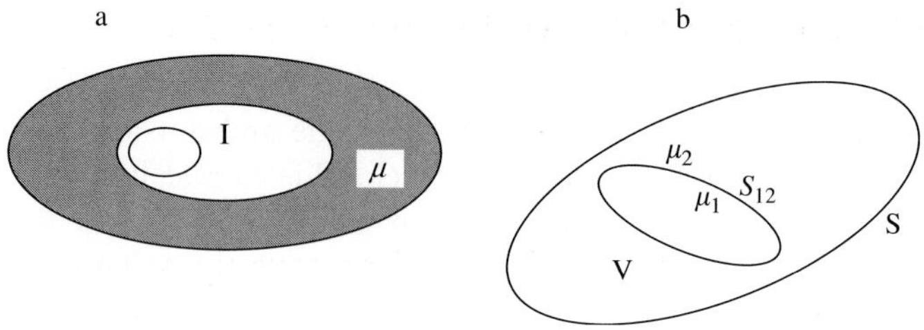

Fig. 2.3. (a) Influence of a thin shell on the magnetic field outside. (b) The field in piece-wise magnetic medium.

imagine that a coil with constant current is placed in a borehole with casing which may have a large magnetic permeability. It is useful to raise the following question. What happens to the field $\boldsymbol{B}$ outside the shell? At the first glance, it seems that the magnetic flux will be concentrated inside the thin shell and the field outside vanishes; in other words, the magnetic shell plays a role of a screen. However, this assumption is incorrect. Indeed, in reality magnetization currents arise at the internal and external surfaces of the shell. In the vicinity of the same point they have opposite directions and almost the same magnitude, because of practically equal distances to the primary currents creating the field. For this reason, at relatively large distances from the shell the field of its currents is negligible. Therefore, thin shell, regardless of the value of μ, does not produce any screening, and the field $\boldsymbol{B}$ remains almost the same as before. It is proper to notice that if constant electric charges are surrounded by a thin conducting shell, then the electric field outside remains also the same provided that the distance to an observation point is sufficiently large in comparison with the shell thickness.

2.11. THE SYSTEM OF EQUATIONS FOR THE FIELDS *B* AND *H* IN SPECIAL CASES

Now we will return to the system of field Equation (2.32) and consider several models of a medium where this system is greatly simplified.

2.11.1. Case 1: A nonmagnetic medium

In this simplest model conduction currents are the sole generators of the field $\boldsymbol{B}$, and Equation (2.32) gives

$$\begin{aligned} \text{curl}\,\boldsymbol{B} &= \mu_0 \boldsymbol{j}_c \quad \text{div}\,\boldsymbol{B} = 0 \\ \text{Curl}\,\boldsymbol{B} &= \mu_0\, \boldsymbol{i}_c \quad \text{Div}\,\boldsymbol{B} = 0 \end{aligned} \tag{2.59}$$

At the same time the system of equations for the field $\boldsymbol{H}$ is

$$\text{curl } \boldsymbol{H} = \boldsymbol{j}_c \quad \text{div } \boldsymbol{H} = 0$$

and

$$\text{Curl } \boldsymbol{H} = \boldsymbol{i}_c \quad \text{Div } \boldsymbol{H} = 0 \tag{2.60}$$

that is, these fields differ from each other by the constant μ_0.

2.11.2. Case 2: Conduction currents are absent

This means that in a volume where we study the field, the density of conduction currents is zero but outside of it the conduction currents may exist, and also we assume that, in general, both the inductive and residual magnetizations are present. Then, the set (2.32) gives

$$\text{curl}\frac{\boldsymbol{B}}{\mu} = \mu_0 \text{curl}\frac{\boldsymbol{P}_r}{\mu} \quad \text{div } \boldsymbol{B} = 0$$

and

$$\text{Curl}\frac{\boldsymbol{B}}{\mu} = \mu_0 \text{Curl}\frac{\boldsymbol{P}_r}{\mu} \quad \text{Div } \boldsymbol{B} = 0 \tag{2.61}$$

As follows from Equations (2.36) and (2.43) the density of currents generating this field is

$$\boldsymbol{j}_m = \frac{1}{\mu\mu_0}(\text{grad } \mu \times \boldsymbol{B}) + \text{curl } \boldsymbol{P}_r - \frac{1}{\mu}(\text{grad } \mu \times \boldsymbol{P}_r) \tag{2.62}$$

and

$$\boldsymbol{i}_m = 2\frac{K_{12}}{\mu_0}\boldsymbol{n} \times \boldsymbol{B}^{av} + \text{Curl } \boldsymbol{P}_r - 2K_{12}\boldsymbol{n} \times \boldsymbol{P}_r^{av}$$

It is obvious that the first term of the volume and surface density of currents cannot be determined if the field $\boldsymbol{B}$ is unknown. As was pointed out earlier, we faced with the problem of "closed circle": in order to determine the field we have to know its generators, but the latter can be specified if the field is given. Therefore, the Biot–Savart law cannot be used to calculate the magnetic field and instead of it we have to formulate a boundary-value problem. In this connection it is useful to consider the system of equations for the field $\boldsymbol{H}$:

$$\text{curl } \boldsymbol{H} = 0 \quad \text{div } \mu\boldsymbol{H} = -\mu_0 \text{div } \boldsymbol{P}_r$$

and

$$\text{Curl } \boldsymbol{H} = 0 \quad \text{Div } \mu\boldsymbol{H} = -\mu_0 \text{Div } \boldsymbol{P}_r \tag{2.63}$$

In the absence of conduction currents the field $\boldsymbol{H}$ has a source origin only. In accordance with Equations (2.50) and (2.53) one type of fictitious magnetic charges depends on the field $\boldsymbol{H}$ and this fact also requires the formulation of a boundary-value problem.

Thus, a determination of the magnetic field can be, in principle, accomplished in two ways. One of them is based on the solution of the system (2.61), while the other allows us to find the field $\boldsymbol{H}$, provided that $\boldsymbol{P}_r$ is given, and then, making use of Equation (2.29), to find $\boldsymbol{B}$. Taking into account the fact that $\boldsymbol{H}$ is the source field, ($\boldsymbol{j}_c = \boldsymbol{i}_c = 0$), the second approach is sometimes preferable since it permits us to introduce a scalar potential U, which essentially simplifies the determination of the field.

2.11.3. Case 3: Residual magnetization and conduction currents are absent

Suppose that a magnetic medium is placed in an external magnetic field $\boldsymbol{B}_0$ which is given. Then magnetization currents arise and they generate a secondary magnetic field $\boldsymbol{B}_s$. Therefore, the total magnetic field $\boldsymbol{B}$ consists of two parts:

$$\boldsymbol{B} = \boldsymbol{B}_0 + \boldsymbol{B}_s$$

Since conduction currents and residual magnetization are absent, the system of field equations is markedly simplified and we have

$$\text{curl}\,\frac{\boldsymbol{B}}{\mu} = 0 \quad \text{div}\,\boldsymbol{B} = 0$$

and

$$\boldsymbol{n}x\left(\frac{\boldsymbol{B}_2}{\mu_2} - \frac{\boldsymbol{B}_1}{\mu_1}\right) = 0 \quad \boldsymbol{n}\cdot(\boldsymbol{B}_2 - \boldsymbol{B}_1) = 0 \tag{2.64}$$

Consequently, the volume and surface densities of currents are

$$\boldsymbol{j} = \frac{1}{\mu\mu_0}(\text{grad}\,\mu \times \boldsymbol{B}), \quad \text{and} \quad \boldsymbol{i} = 2\frac{K_{12}}{\mu_0}\boldsymbol{n}\times\boldsymbol{B}^{\text{av}} \tag{2.65}$$

At the same time, the system of equations for the field $\boldsymbol{H}$ is

$$\text{curl}\,\boldsymbol{H} = 0 \quad \text{div}\,\mu\boldsymbol{H} = 0$$

and

$$\boldsymbol{n}x(\boldsymbol{H}_2 - \boldsymbol{H}_1) = 0 \quad \boldsymbol{n}\cdot(\mu_2\boldsymbol{H}_2 - \mu_1\boldsymbol{H}_1) = 0 \tag{2.66}$$

and the density of its sources is

$$\delta_m = -\frac{\boldsymbol{H}\cdot\text{grad}\,\mu}{\mu} \quad \text{and} \quad \sigma_m = -2K_{12}H_{\boldsymbol{n}}^{\text{av}}$$

2.11.4. Case 4: Uniform piece-wise medium where conduction current and residual magnetization are absent

In this very important case the systems of equations are greatly simplified, since the volume density of magnetization currents, as well as the volume density of magnetic charges, vanishes and we obtain

$$\text{curl}\,\boldsymbol{B} = 0 \quad \text{and} \quad \text{div}\,\boldsymbol{B} = 0$$

and

$$\boldsymbol{n}x\left(\frac{\boldsymbol{B}_2}{\mu_2} - \frac{\boldsymbol{B}_1}{\mu_1}\right) = 0 \quad \boldsymbol{n}\cdot(\boldsymbol{B}_2 - \boldsymbol{B}_1) = 0 \tag{2.67}$$

For the field $\boldsymbol{H}$ we have

$$\text{curl}\,\boldsymbol{H} = 0 \quad \text{div}\,\mu\boldsymbol{H} = 0$$

and

$$\boldsymbol{n}x(\boldsymbol{H}_2 - \boldsymbol{H}_1) = 0 \quad \boldsymbol{n}\cdot(\mu_2\boldsymbol{H}_2 - \mu_1\boldsymbol{H}_1) = 0 \tag{2.68}$$

Both systems are sufficiently simple and they show that fields $\boldsymbol{B}$ and $\boldsymbol{H}$ are caused by surface currents and surface charges, respectively, and for this reason the approaches based on a calculation of either field $\boldsymbol{B}$ or $\boldsymbol{H}$ are equivalent to each other. As will be shown later these fields can be expressed in terms of a scalar potential that greatly facilitates a solution of the boundary-value problems.

We have considered several cases in which the system of field equations can be simplified. In general, the magnetic field $\boldsymbol{B}$ can be represented as the sum of three fields:

$$\boldsymbol{B} = \boldsymbol{B}^{(1)} + \boldsymbol{B}^{(2)} + \boldsymbol{B}^{(3)}$$

and each of them satisfies one of the following equations:

$$\text{curl}\frac{\boldsymbol{B}^{(1)}}{\mu} = \boldsymbol{j}_{\text{c}} \quad \text{div}\,\boldsymbol{B}^{(1)} = 0$$

$$\text{Curl}\frac{\boldsymbol{B}^{(1)}}{\mu} = \boldsymbol{i}_{\text{c}} \quad \text{Div}\,\boldsymbol{B}^{(1)} = 0$$

$$\text{curl}\frac{\boldsymbol{B}^{(2)}}{\mu} = \mu_0\text{curl}\frac{\boldsymbol{P}_{\text{r}}}{\mu} \quad \text{div}\,\boldsymbol{B}^{(2)} = 0$$

$$\text{Curl}\,\frac{\boldsymbol{B}^{(2)}}{\mu} = \mu_0 \text{Curl}\,\frac{\boldsymbol{P}_{\mathrm{r}}}{\mu} \quad \text{Div}\,\boldsymbol{B}^{(2)} = 0$$

and finally

$$\text{curl}\,\frac{\boldsymbol{B}^{(3)}}{\mu} = 0 \quad \text{div}\,\boldsymbol{B}^{(3)} = 0$$

$$\text{Curl}\,\frac{\boldsymbol{B}^{(3)}}{\mu} = 0 \quad \text{Div}\,\boldsymbol{B}^{(3)} = 0$$

Chapter 3

Magnetic Field in the Presence of Magnetic Medium

3.1. SOLUTION OF THE FORWARD PROBLEM IN A PIECE-WISE UNIFORM MEDIUM WHEN CONDUCTION CURRENTS AND RESIDUAL (REMANENT) MAGNETIZATION ARE ABSENT

Now we start to study the influence of a piece-wise uniform magnetic medium placed in a given field $\boldsymbol{B}_0$. This field can be, for instance, the field of the earth. An example of such a medium is shown in Fig. 2.3(b). Due to the field $\boldsymbol{B}_0$, magnetization currents arise in the medium and they generate a secondary magnetic field $\boldsymbol{B}_s$. Consequently, the total field at each point is

$$\boldsymbol{B} = \boldsymbol{B}_0 + \boldsymbol{B}_s$$

Of course, the magnitude and direction of magnetization currents are defined at every point by the resultant field $\boldsymbol{B}$. Inasmuch as the medium is piece-wise uniform, the volume density of these currents vanishes, and the field $\boldsymbol{B}_s$ is generated by surface magnetization currents only. This fact greatly simplifies the solution of the forward problem, that is, the determination of the secondary field. As we know, the density of these currents is usually unknown prior to the calculation of the field, since their distribution depends on the total magnetic field $\boldsymbol{B}$. In other words, the interaction between currents can be significant and often it cannot be ignored. Therefore, in order to determine the secondary field, $\boldsymbol{B}_s$, in general, it is necessary to solve a boundary-value problem, and with this purpose in mind we will first describe equations which characterize the behavior of the field at all points of space.

3.1.1. Equations for the scalar potential

Earlier, we demonstrated that at regular points of each homogeneous medium, the field obeys equations

$$\operatorname{curl} \boldsymbol{B} = 0 \quad \text{and} \quad \operatorname{div} \boldsymbol{B} = 0 \tag{3.1}$$

At the interfaces of media with different magnetic permeability, we also have

$$\frac{B_{2t}}{\mu_2} - \frac{B_{1t}}{\mu_1} = 0 \quad \text{and} \quad B_{2n} - B_{1n} = 0 \tag{3.2}$$

since residual magnetization and conduction currents are absent. Thus, four equations with respect to the vector field $\boldsymbol{B}$ characterize the field behavior everywhere. This is a rather complicated system, but fortunately there is much simpler way to describe the field. With this purpose in mind, proceeding from the first equation of the set (3.1), we will introduce the scalar potential U as

$$\boldsymbol{B} = -\text{grad}\, U \tag{3.3}$$

In fact, from vector analysis, it follows that

$$\text{curl grad}\, U = 0$$

Let us note that the choice of sign in the equality (3.3) is not important, and we selected the negative only by analogy with the electric field. Then, substitution of Equation (3.3) into the second equation of the system:

$$\text{div}\, \boldsymbol{B} = 0$$

gives Laplace's equation

$$\text{div grad}\, U = 0 \quad \text{or} \quad \Delta U = 0 \tag{3.4}$$

By definition of the gradient, we have for the tangential and normal components of the field:

$$B_t = -\frac{\partial U}{\partial t} \quad \text{and} \quad B_n = -\frac{\partial U}{\partial n} \tag{3.5}$$

and consequently, in terms of the potential, the conditions at the interface of media with different values of the magnetic permeability are

$$\frac{1}{\mu_2}\frac{\partial U_2}{\partial t} - \frac{1}{\mu_1}\frac{\partial U_1}{\partial t} = 0, \quad \frac{\partial U_2}{\partial n} - \frac{\partial U_1}{\partial n} = 0$$

or

$$\frac{U_2}{\mu_2} = \frac{U_1}{\mu_1} \quad \text{and} \quad \frac{\partial U_2}{\partial n} = \frac{\partial U_1}{\partial n} \tag{3.6}$$

since from continuity of function U/μ, the continuity of the derivative $(1/\mu)(\partial U/\partial t)$ in the tangential direction to the interface follows. The same approach is not applied to the normal derivatives, because their calculation requires values of the potential above and beneath the interface and these can be different. Thus, the system of equations for the scalar potential, describing its behavior in a piece-wise uniform medium, is

$$\nabla^2 U = 0$$

and

$$\frac{U_2}{\mu_2} - \frac{U_1}{\mu_1} = 0, \quad \frac{\partial U_2}{\partial n} = \frac{\partial U_1}{\partial n} \tag{3.7}$$

It is obvious that if there are several interfaces, we have to ensure continuity of functions U/μ and $\partial U/\partial n$ at all of these surfaces. Note that introduction of the scalar potential is possible in more general case when conduction currents are absent and the vector of the remanent magnetization is such that curl $\boldsymbol{P}_r = 0$.

3.2. THEOREM OF UNIQUENESS AND BOUNDARY-VALUE PROBLEMS

Suppose that we have found the function U which obeys Equation (3.7). Inasmuch as this system includes Laplace's equation, which is a partial differential equation of the second order, it is natural to expect that this system has an infinite number of solutions. Certainly, it is not surprising because even an ordinary linear differential equation of the first order with a constant coefficient has unlimited number of solutions. In other words, the system (3.7) alone does not allow us to determine the potential U and we need an additional information about its behavior to determine the field $\boldsymbol{B}$ uniquely. In order to find these conditions, we prove the theorem of uniqueness and start from the simple case, when all points of the volume V surrounded by the surface S have the same magnetic permeability. In other words, these points are regular ones. Then, we can apply Gauss formula

$$\int_V \nabla \boldsymbol{N} \, dV = \oint_S \boldsymbol{N} \cdot d\boldsymbol{S} \tag{3.8}$$

where $\boldsymbol{N}$ is any vector function which has derivatives inside the volume V and $\nabla \boldsymbol{N}$ its divergence. Suppose that

$$\boldsymbol{N} = U \nabla U \tag{3.9}$$

and U is the potential of the magnetic field. Taking into account the fact that

$$\nabla(U \nabla U) = (\nabla U)^2 + U \nabla^2 U$$

and that the potential obeys Laplace's equation, the latter gives

$$\nabla(U\nabla U) = (\nabla U)^2$$

Then, in place of Equation (3.8) we have

$$\int_V (\nabla U)^2 dV = \oint_S U \frac{\partial U}{\partial n} dS \tag{3.10}$$

because

$$\nabla U \cdot d\boldsymbol{S} = \text{grad } U \cdot d\boldsymbol{S} = \frac{\partial U}{\partial n} dS \tag{3.11}$$

It is essential that the integrand at each point of the volume, $(\nabla U)^2$ in Equation (3.10), cannot be negative number. Now assume that the surface integral on the right-hand side of this equality is equal to zero. This means that the volume integral also vanishes:

$$\int_V (\nabla U)^2 dV = 0$$

but this is possible only if at each point of the volume

$$\nabla U = 0$$

This is a very important conclusion and it will be used in formulating two boundary conditions.

3.2.1. The first boundary-value problem

Let us assume that there are two different functions U_1 and U_2 which obey Laplace's equation inside the volume V, but at the surface S they coincide, that is

$$\Delta U_1 = \Delta U_2 = 0,$$

and at points of S:

$$U_1 = U_2 = U = \varphi(S) \tag{3.12}$$

Now we demonstrate that these functions also coincide at each point inside the volume V. With this purpose in mind, introduce the difference

$$U_3 = U_2 - U_1 \tag{3.13}$$

Inasmuch as Laplace's equation is a linear equation, the function U_3 is a solution too:

$$\Delta U_3 = 0$$

and, therefore, we can use Equation (3.10)

$$\int_V (\nabla U_3)^2 dV = \oint_S U_3 \frac{\partial U_3}{\partial n} dS \tag{3.14}$$

By definition (Equation (3.12)):

$$U_3 = 0, \quad \text{on } S$$

and we conclude that at each point inside the volume V:

$$\text{grad } U_3 = 0$$

This means that this function is constant: $U_3 = C$ and, moreover, we know its value. In fact, at the surface S it is equal to zero and therefore $U_3 = 0$ everywhere inside the volume, that is

$$U_1 = U_2$$

Thus, we have proved that the boundary condition:

$$U = \varphi(q), \quad \text{on } S \tag{3.15}$$

uniquely defines a solution of Laplace's equation in the volume V. We can say that the theorem of uniqueness allows us to select among an infinite number of solutions of Laplace's equation only one, which obeys Equation (3.15) and, correspondingly, to formulate the first boundary-value problem:

$$\Delta U = 0, \quad \text{in } V$$

and

$$U(q) = \varphi(q), \quad \text{on } S \tag{3.16}$$

where $\varphi(q)$ is a given function. In accordance with the theorem of uniqueness, the set (3.16) uniquely defines the potential of the magnetic field. Now let us make two comments:

(a) The volume V can be surrounded by several surfaces and at each of them the functions $\varphi(q)$ can be different.
(b) The first boundary-value problem is also called Dirichlet's problem.

Since the potential on the surface S is given, we can calculate the tangential component of the magnetic field

$$B_{\mathrm{t}} = -\frac{\partial U}{\partial t}$$

Respectively, this boundary-value problem can be formulated in terms of the magnetic field as

$$\operatorname{curl} \boldsymbol{B} = 0, \quad \operatorname{div} \boldsymbol{B} = 0, \quad \text{in } V$$

and

$$B_{\mathrm{t}}(q) = \phi(q), \quad \text{on } S \tag{3.17}$$

3.2.2. The second boundary-value problem

Next we will describe a different condition on the surface S, which ensures uniqueness of the solution of the forward problem. Suppose that we know the derivatives of the potential on the surface S:

$$\frac{\partial U}{\partial n} = \xi(S) \tag{3.18}$$

and consider two solutions of the Laplace's equation: U_1 and U_2 inside the volume V, which have equal derivatives along the normal at points of the surface S:

$$\frac{\partial U_1}{\partial n} = \frac{\partial U_2}{\partial n} = \xi(S)$$

Correspondingly, their difference

$$\frac{\partial U_3}{\partial n} = \frac{\partial (U_2 - U_1)}{\partial n} = 0, \quad \text{on } S \tag{3.19}$$

and in accordance with Equation (3.10) inside the volume V, we have

$$\nabla U_3 = 0 \quad \text{or} \quad U_3 = C \tag{3.20}$$

where C is unknown constant. We have proved that the potential of the magnetic field is defined inside the volume V with an accuracy of a constant, if it satisfies Laplace's equation and the condition (3.18) at points of the surface S. Thus, the second boundary-value problem is written as

$$\Delta U = 0 \quad \text{in } V,$$

and

$$\frac{\partial U}{\partial n} = \xi(S) \tag{3.21}$$

It is essential that the set (3.21) uniquely defines the magnetic field since a gradient from the constant is zero, and in terms of the field $\boldsymbol{B}$ the system is written as

$$\operatorname{curl} \boldsymbol{B} = 0 \quad \text{and} \quad \operatorname{div} \boldsymbol{B} = 0, \quad \text{in } V$$

and

$$B_{\rm n} = -\xi(S) \tag{3.22}$$

Thus, the magnetic field is defined uniquely in V if we know its either tangential or normal component on the surface S, surrounding this volume. Note that the last boundary-value problem is called the Neumann problem.

3.2.3. Boundary-value problem in the presence of an interface of media with different μ

Until now the theorem of uniqueness was formulated for a homogeneous medium; next suppose that there is an interface between media with different but constant values of magnetic permeability, μ_1 and μ_2 (Fig. 2.3(b)). Of course, inside the volume surrounded by the surface S, the potential U obeys the system of equations

$$\Delta U = 0$$

and

$$\frac{U_1}{\mu_1} = \frac{U_2}{\mu_2} \quad \text{and} \quad \frac{\partial U_1}{\partial n} = \frac{\partial U_2}{\partial n}, \quad \text{on } S_{12}$$

In order to determine conditions, which uniquely define the field, suppose that there are two different solutions of this system: $U^{(1)}$ and $U^{(2)}$. This means that

$$\Delta U^{(1)} = 0, \quad \Delta U^{(2)} = 0$$

and

$$\frac{U_1^{(1)}}{\mu_1} = \frac{U_2^{(1)}}{\mu_2}, \quad \frac{\partial U_1^{(1)}}{\partial n} = \frac{\partial U_2^{(1)}}{\partial n} \quad \text{and} \quad \frac{U_1^{(2)}}{\mu_1} = \frac{U_2^{(2)}}{\mu_2}, \quad \frac{\partial U_1^{(2)}}{\partial n} = \frac{\partial U_2^{(2)}}{\partial n}$$

where μ_1, U_1 and μ_2, U_2 are the magnetic permeability and potentials at back and front sides of the surface S_{12} correspondingly. Consider a difference of two solutions

$$U^{(3)} = U^{(2)} - U^{(1)}$$

It is obvious that this function satisfies the following conditions:

$$\Delta U^{(3)} = 0, \quad \frac{U_1^{(3)}}{\mu_1} = \frac{U_2^{(3)}}{\mu_2}, \quad \frac{\partial U_1^{(3)}}{\partial n} = \frac{\partial U_2^{(3)}}{\partial n} \tag{3.23}$$

Next we introduce the vector $\boldsymbol{N}$:

$$\boldsymbol{N} = U^{(3)} \frac{1}{\mu} \nabla U^{(3)} \tag{3.24}$$

and again make use of Gauss theorem

$$\int_{V^*} \operatorname{div} \boldsymbol{N}\, dV = \oint_S \boldsymbol{N}\, d\boldsymbol{S} + \oint_{S^*} \boldsymbol{N}\, d\boldsymbol{S}$$

Here S^* is a surface of "safety" which surrounds the interface S_{12} where the vector $\boldsymbol{N}$ is a discontinuous function, since values of magnetic permeability μ_1 and μ_2 are different. Correspondingly, Gauss theorem is applied to the volume surrounded by surfaces S and S^*. As follows from last two equations

$$\int_{V^*} U^{(3)} \nabla \left(\frac{1}{\mu} \nabla U^{(3)} \right) dV + \int_{V^*} \frac{1}{\mu} (\nabla U^{(3)})^2 dV = \oint_S \frac{U^{(3)}}{\mu} \frac{\partial U^{(3)}}{\partial n} dS + \oint_{S^*} \frac{U^{(3)}}{\mu} \frac{\partial U^{(3)}}{\partial n} dS \tag{3.25}$$

In approaching S^* to S_{12}, integration over S^* is reduced to integration at both sides of the interface S_{12}. Taking into account that at the back and front sides of this surface

$$\boldsymbol{n}_1 = \boldsymbol{n}^* \quad \text{and} \quad \boldsymbol{n}_2 = -\boldsymbol{n}^*$$

we have

$$\oint_{S^*} \frac{U^{(3)}}{\mu} \frac{\partial U^{(3)}}{\partial n} dS = \int_{S_{12}} \left[\frac{U_1^{(3)}}{\mu_1} \frac{\partial U_1^{(3)}}{\partial n} - \frac{U_2^{(3)}}{\mu_2} \frac{\partial U_2^{(3)}}{\partial n} \right] dS \tag{3.26}$$

and $V^* \to V$.

Bearing in mind that at regular points the magnetic permeability is constant and taking into account Equations (3.23) in place of Equation (3.25), we obtain

$$\int_V \frac{1}{\mu} (\nabla U^{(3)})^2 dV = \oint_S \frac{1}{\mu} U^{(3)} \frac{\partial U^{(3)}}{\partial n} dS$$

Suppose that surface integral vanishes, then we have

$$\int_V \frac{1}{\mu} (\nabla U^{(3)})^2 dV = 0$$

This means that at each point of a volume, $\nabla U^{(3)} = 0$ since $\mu > 0$. In other words, the function $U^{(3)}$ is constant inside a volume and correspondingly two solutions for potential may differ by a constant only. Earlier considering simpler models of a medium, we found conditions when surface integral in the last equation becomes equal to zero. This allowed us to formulate two boundary-value problems. Now, considering a piece-wise uniform medium, we slightly generalize these problems. In the first problem, potential has to satisfy the following conditions:

1. $\Delta U = 0$ at regular points.
2. $(U_1/\mu_1) = (U_2/\mu_2)$, $(\partial U_1/\partial n) = (\partial U_2/\partial n)$ at interface S_{12}.
3. $U(q) = \phi(q)$ on S.

In the second boundary-value problem, it is assumed that in place of potential we know its normal derivative at all points of the surface S.

Thus, we found out conditions which uniquely define the magnetic field $\boldsymbol{B}$ and this shows a fundamental meaning of the theorem of uniqueness. Later we will often use the theorem of uniqueness to prove that our assumptions about a field behavior are correct. Now let us start to study the magnetic field and with this purpose in mind consider several examples.

3.3. A CYLINDER IN A UNIFORM MAGNETIC FIELD

3.3.1. Solution of the boundary problem

Suppose that a cylinder with radius a and magnetic permeability μ_i is placed in a uniform magnetic field $\boldsymbol{B}_0$, which is perpendicular to the cylinder axis (Fig. 3.1(a)). Remanent magnetization is absent, and the magnetic permeability of the surrounding medium is μ_e. Because of the external magnetic field, magnetization currents become orderly oriented and inductive magnetization arises. Taking into account the fact that the medium is uniform inside and outside of the cylinder, magnetization does not produce volume magnetization currents. At the same time, the surface density of these currents is not zero, and as was shown in Chapter 2:

$$\boldsymbol{i} = \frac{2K_{12}}{\mu_0} \boldsymbol{n} \times \boldsymbol{B}^{\text{av}} \tag{3.27}$$

where

$$K_{12} = \frac{\mu_e - \mu_i}{\mu_e + \mu_1} \quad \text{and} \quad \boldsymbol{B}^{\text{av}} = \frac{\boldsymbol{B}^i + \boldsymbol{B}^e}{2}$$

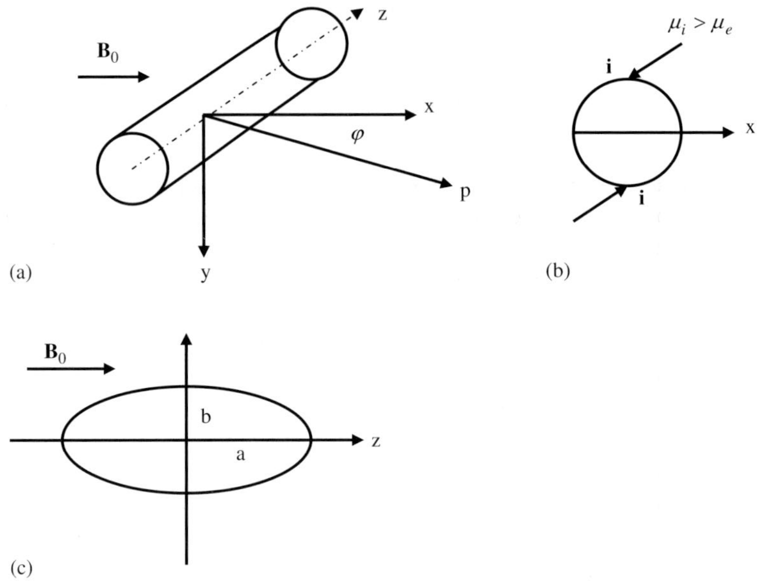

Fig. 3.1. (a) Cylinder in a magnetic field. (b) Distribution of currents. (c) Elongated ellipsoid in a uniform magnetic field.

and $\boldsymbol{n}$ is the unit vector, normal to the cylinder surface and directed outward. $\boldsymbol{B}^{av}$ is the average value of the field on the cylinder surface. Inasmuch as the primary field $\boldsymbol{B}_0$ is uniform and has only the component B_{0x}, we may expect that the vector of magnetization current density is oriented along the cylinder axis, and it does not change in this direction. In other words, the secondary field $\boldsymbol{B}_s$ is caused by linear current filaments located on the cylinder surface. To determine this field, we will first find a solution of Laplace's equation. Let us choose a cylindrical system of coordinates: r, φ, and z, so that the z-axis coincides with the cylindrical axis. As is well known, in this system of coordinates we have

$$\Delta U = \frac{\partial^2 U}{\partial r^2} + \frac{1}{r}\frac{\partial U}{\partial r} + \frac{1}{r^2}\frac{\partial^2 U}{\partial \varphi^2} = 0 \tag{3.28}$$

since the field and its potential are independent of coordinate z. To solve this partial differential equation of the second order, we will apply the method of separation of variables and represent the potential as a product of two functions

$$U(r, \varphi) = T(r)F(\varphi) \tag{3.29}$$

Substitution of Equation (3.29) into Equation (3.28) and multiplication of all terms by

$$\frac{r^2}{TF}$$

yields

$$\frac{r^2}{T(r)}\frac{\partial^2 T}{\partial r^2}+\frac{r}{T(r)}\frac{\partial T(r)}{\partial r}+\frac{1}{F(\varphi)}\frac{\partial^2 F(\varphi)}{\partial \varphi^2}=0 \tag{3.30}$$

It seems that the first two terms depend on r but the last is a function of φ. However, this is impossible, since their sum is always equal to zero regardless of the values of their arguments. Therefore, these two groups of terms can be only constants which differ by sign, and we obtain

$$\frac{1}{F}\frac{d^2 F}{d\varphi^2}=\pm n^2 \quad \text{and} \quad \frac{r^2}{T}\frac{d^2 T}{dr^2}+\frac{r}{T}\frac{dT}{dr}=\mp n^2 \tag{3.31}$$

This means that instead of the partial differential equation we have obtained two ordinary differential equations, solutions of which are well known, and this is the main purpose of the method of separation of variables. To choose the proper sign on the right-hand side of these equations, we will make use of the fact that the field $\boldsymbol{B}$ is a periodic function of the argument φ with period equal 2π. Otherwise, it would become a many-valued function. For this reason, we will select the negative sign on the right-hand side of the first equation of the set (3.31) and assume that n is integer. This gives

$$\frac{d^2 F}{d\varphi^2}+n^2 F=0$$

We have the equation of the harmonic oscillator and its solution is

$$F(n,\varphi)=A_n^* \sin n\varphi + B_n^* \cos n\varphi \tag{3.32}$$

where $F(n, \varphi)$ is a partial solution for a given integer value of n, and A_n^* and B_n^* the constants. It is appropriate to notice that if we choose the positive sign, then the function F would not be periodic since in this case

$$F(n,\varphi)=A_n^* \exp(n\varphi)+B_n^* \exp(-n\varphi)$$

This analysis also shows that on the right-hand side of the second equation of the set (3.31), we have to select positive sign, and consequently

$$\frac{d^2 T}{dr^2}+\frac{1}{r}\frac{dT}{dr}-\frac{n^2}{r^2}T=0$$

The latter is an ordinary differential equation that has also been studied in detail, and its solution is

$$T(n,r)=C_n^* r^n + D_n^* r^{-n} \tag{3.33}$$

Therefore, the general solution of Laplace's equation, represented as a sum of partial solutions, is

$$U(r,\varphi) = \sum_{n=0}^{\infty} (C_n^* r^n + D_n^* r^{-n})(A_n^* \sin n\varphi + B_n^* \cos n\varphi) \tag{3.34}$$

It is obvious that after a summation, we obtain a function $U(r, \varphi)$ that is independent on a separation of variable n. To satisfy the other conditions of the boundary-value problem, let us find an expression for the potential of the primary magnetic field. By definition, we have

$$\boldsymbol{B}_0 = -\text{grad } U_0 \quad \text{or} \quad B_0 = -\frac{\partial U_0}{\partial x}$$

since the field $\boldsymbol{B}_0$ has only an x-component. Performing integration, we obtain

$$U_0 = -B_0 x + C$$

Bearing in mind that a constant does not affect the field, we let $C = 0$ and this gives

$$U_0 = -B_0 x = -B_0 r \cos \varphi \tag{3.35}$$

It is convenient to represent the potential of the magnetic field inside and outside of the cylinder as

$$U(r,\varphi) = \begin{cases} U_i \\ U_e = U_0 + U_s \end{cases} \tag{3.36}$$

where U_i and U_e are potentials inside and outside of the cylinder, respectively. Taking into account the fact that the secondary magnetic field has everywhere a finite value and decreases with an increase of the distance from the cylinder, the functions U_i and U_e are

$$\begin{aligned} U_i(r,\varphi) &= \sum_{n=0}^{\infty} (A_n^i \sin n\varphi + B_n^i \cos n\varphi) r^n, \quad \text{if } r < a \\ U_e(r,\varphi) &= -B_0 r \cos \varphi + \sum_{n=0}^{\infty} (A_n^e \sin n\varphi + B_n^e \cos n\varphi) r^{-n}, \quad \text{if } r > a \end{aligned} \tag{3.37}$$

It is essential to note that U_i and U_e, given by Equation (3.37), satisfy Laplace's equation and the boundary condition at infinity, since

$$U_e(r,\varphi) \to U_0(r,\varphi), \quad \text{as } r \to \infty$$

3.3.2. Determination of unknown coefficients and field expressions

Next, we will determine the unknown coefficients A_n and B_n, and with this purpose in mind it is natural to apply conditions at the cylinder's surface

$$\frac{U_e}{\mu_e} = \frac{U_i}{\mu_i} \quad \text{and} \quad \frac{\partial U_e}{\partial r} = \frac{\partial U_i}{\partial r}, \quad \text{if } r = a$$

From Equation (3.37), it follows that

$$-\frac{B_0 a \cos\varphi}{\mu_e} + \frac{1}{\mu_e}\sum_{n=0}^{\infty}(A_n^e \sin n\varphi + B_n^e \cos n\varphi)a^{-n} = \frac{1}{\mu_i}\sum_{n=0}^{\infty}(A_n^i \sin n\varphi + B_n^i \cos n\varphi)a^n$$

and

$$-B_0 \cos\varphi - \sum_{n=0}^{\infty} n(A_n^i \sin n\varphi + B_n^i \cos n\varphi)a^{-n-1} = \sum_{n=0}^{\infty} n(A_n^e \sin n\varphi + B_n^e \cos n\varphi)a^{n-1} \tag{3.38}$$

As is well known, one of the most remarkable features of the trigonometric functions $\sin n\varphi$ and $\cos n\varphi$ is the fact that they are orthogonal to each other, and therefore the equalities

$$\int_0^{2\pi} \sin m\varphi \sin n\varphi \, d\varphi = \begin{cases} 0, & m \neq n \\ \pi, & m = n \end{cases}$$
$$\int_0^{2\pi} \cos m\varphi \cos n\varphi \, d\varphi = \begin{cases} 0, & m \neq n \\ \pi, & m = n \end{cases} \tag{3.39}$$

and

$$\int_0^{2\pi} \cos m\varphi \sin n\varphi \, d\varphi = 0$$

hold. Here m and n are arbitrary integers.

Multiplying both Equations (3.38) by $\sin n\varphi$ and integrating with respect to φ from zero to 2π, we obtain

$$\frac{A_n^e a^{-n}}{\mu_e} = \frac{A_n^i a^n}{\mu_i}$$

and

$$-nA_n^e a^{-n-1} = nA_n^i a^{n-1} \tag{3.40}$$

where n is any positive integer including zero. It is clear that the system (3.40) has only the zero solution; that is

$$A_n^e = A_n^i = 0$$

In a similar manner, we can prove that

$$B_n^e = B_n^i = 0, \quad \text{if } n \neq 1$$

Certainly, this is an important result, since we have demonstrated that the secondary field inside and outside of the cylinder is described, as well as the primary field, by the first cylindrical harmonic, $n = 1$, only. This is an amazingly great simplification. In accordance with Equation (3.37), the coefficients B_1^e and B_1^i are determined from the system:

$$\frac{1}{\mu_e}(-B_0 a + B_1^e a^{-1}) = \frac{1}{\mu_i} B_1^i a$$

and

$$-B_0 - B_1^e a^{-2} = B_1^i \tag{3.41}$$

Its solution is

$$B_1^e = K_{21} a^2 B_0 = \frac{\mu_i - \mu_e}{\mu_i + \mu_e} a^2 B_0$$

and

$$B_1^i = -\frac{2\mu_i}{\mu_i + \mu_e} B_0$$

Thus, we have derived the following expressions for the potential:

$$U_e(r, \varphi) = -B_0 r \cos\varphi + \frac{\mu_i - \mu_e}{\mu_i + \mu_e} \frac{a^2}{r} B_0 \cos\varphi, \quad \text{if } r > a$$

and

$$U_i(r, \varphi) = -\frac{2\mu_i}{\mu_i + \mu_e} B_0 r \cos\varphi, \quad \text{if } r < a \tag{3.42}$$

which satisfy all the conditions of the boundary-value problem and therefore uniquely describe the magnetic field for this model.

Since

$$B_r = -\frac{\partial U}{\partial r} \quad \text{and} \quad B_\varphi = -\frac{1}{r}\frac{\partial U}{\partial \varphi}$$

the secondary field outside of the cylinder is

$$B^e_{rs}(r, \varphi) = \frac{\mu_i - \mu_e}{\mu_i + \mu_e} B_0 \frac{a^2}{r^2} \cos \varphi$$

and

$$B^e_{\varphi s}(r, \varphi) = \frac{\mu_i - \mu_e}{\mu_i + \mu_e} B_0 \frac{a^2}{r^2} \sin \varphi \qquad (3.43)$$

Comparison of the potential U_i with that of the primary field shows that the magnetic field inside the cylinder remains uniform and has only an x-component. As follows from the second equation of the set (3.42), this field is

$$B^i_x = \frac{2\mu_i}{\mu_i + \mu_e} B_0, \quad \text{if } r < a \qquad (3.44)$$

3.3.3. *Distribution of magnetization currents*

In accordance with Equation (3.27), we have

$$i_z(a, \varphi) = \frac{K_{12}}{\mu_0}(B^e_\varphi + B^i_\varphi) \qquad (3.45)$$

since

$$\boldsymbol{n} = \boldsymbol{i}_r, \quad \boldsymbol{i}_r x \boldsymbol{i}_r = 0 \quad \text{and} \quad \boldsymbol{i}_r x \boldsymbol{i}_\varphi = \boldsymbol{i}_z$$

Here $\boldsymbol{i}_r$, $\boldsymbol{i}_\varphi$, and $\boldsymbol{i}_z$ are unit vectors along coordinate lines. As follows from Equations (3.42)–(3.44), at the cylinder surface

$$B^e_\varphi = -B_0 \sin \varphi + K_{12} B_0 \sin \varphi$$

and

$$B^i_\varphi = -\frac{2\mu_i}{\mu_i + \mu_e} B_0 \sin \varphi \qquad (3.46)$$

and consequently the surface current density is

$$i_z(a, \varphi) = -2\frac{K_{12}}{\mu_0} B_0 \sin \varphi$$

or

$$i_z(a, \varphi) = \frac{2}{\mu_0} \frac{\mu_i - \mu_e}{\mu_i + \mu_e} B_0 \sin \varphi \tag{3.47}$$

Therefore, the currents generating the secondary magnetic field are distributed in such a way that in one half of the surface ($0 < \varphi < \pi$) they are directed along the z-axis, $\mu_i > \mu_e$, while in the other part ($\pi < \varphi < 2\pi$) the currents have opposite direction (Fig. 3.1(b)). In particular, the current density reaches a maximal value along two lines of the plane $x = 0$, and it vanishes at the plane $y = 0$. It is natural that the current density is directly proportional to the primary magnetic field $\boldsymbol{B}_0$. At the same time, its dependence on the magnetic permeability of the medium is defined by the contrast coefficient

$$K_{12} = \frac{\mu_e - \mu_i}{\mu_e + \mu_i}$$

which varies from -1 to $+1$.

3.3.4. Behavior of the magnetic field inside the cylinder

Consider the behavior of the magnetic field caused by these currents. In accordance with Equation (3.44), the field inside the cylinder is uniform and has the same direction as the primary field. In other words, the secondary field cannot exceed the primary one. With an increase in magnetic permeability μ_i ($\mu_i > \mu_e$), B_x^i also increases and for sufficiently large values of the ratio μ_i/μ_e we have

$$B_x^i \approx 2B_0, \quad \text{if } \frac{\mu_i}{\mu_e} \gg 1$$

That is, the field of the surface currents coincides with $\boldsymbol{B}_0$. In the opposite case, when the surrounding medium has a greater magnetic permeability, $\mu_e > \mu_i$, the surface currents have such a direction that the primary and secondary fields are opposite to each other inside the cylinder. Consequently, the total field B_x^i is smaller than the primary one, and in particular when $\mu_i/\mu_e \ll 1$, it is almost zero.

3.3.5. Induced magnetization vector

It is also useful to determine the induced magnetization vector. By definition, this is

$$\boldsymbol{P} = \chi_i \boldsymbol{H} = \frac{\chi_i}{\mu_i} \boldsymbol{B}^i = \frac{1}{\mu_0}\left(1 - \frac{\mu_0}{\mu_i}\right) \boldsymbol{B}^i$$

since

$$\mu_i = \mu_0(1 + \chi_i) \quad \text{and} \quad \boldsymbol{P}_r = 0$$

Taking into account Equation (3.44), we have

$$\boldsymbol{P} = \frac{2\chi_i}{\mu_i + \mu_e}\boldsymbol{B}_0 \tag{3.48}$$

Thus, the density of dipole moments is defined by the primary and secondary fields and they are uniformly distributed within the cylinder. Due to this fact, the volume density of magnetization currents is equal to zero. It is appropriate to notice that the induced magnetization $\boldsymbol{P}$ has the same direction as the field $\boldsymbol{B}_0$ ($\mu_i > \mu_0$), and the presence of the coefficient of proportionality between vectors $\boldsymbol{P}$ and $\boldsymbol{B}_0$ indicates that the dipole moments are directly proportional to the total field $\boldsymbol{B}$.

3.3.6. Medium of small susceptibility

Now suppose that susceptibility of the medium is much less than unity

$$\chi_e \ll 1 \quad \text{and} \quad \chi_i \ll 1$$

As will be shown later, this case is of a great practical interest in magnetic prospecting. Substitution of χ into Equation (3.48) gives

$$\boldsymbol{P} = \frac{\chi_i \boldsymbol{B}_0}{\mu_0[1 + ((\chi_i + \chi_e)/2)]} \tag{3.49}$$

Taking into account the fact that

$$\frac{\chi_i + \chi_e}{2} \ll 1$$

and expanding the right-hand side of Equation (3.49) in a series, we obtain

$$\boldsymbol{P} = \frac{\chi_i}{\mu_0}\boldsymbol{B}_0 - \frac{\chi_i(\chi_i + \chi_e)}{2\mu_0}\boldsymbol{B}_0 + \cdots$$

It is clear that the second term, as well as the following ones, is very small and therefore it can be neglected. Then

$$\boldsymbol{P} = \frac{\chi_i}{\mu_0}\boldsymbol{B}_0, \quad \text{if } \chi \ll 1 \tag{3.50}$$

This means that in this approximation, the density of dipole moments is defined by the primary field only. In other words, we assume that the interaction between magnetization currents is negligible, and this happens when the secondary field is much smaller than the primary one. Correspondingly, the density of surface magnetization currents is approximately equal to

$$i_z(a, \varphi) = (\chi_i - \chi_e)\frac{B_0}{\mu_0} \sin \varphi, \quad \text{if } \chi \ll 1 \tag{3.51}$$

In this approximation, Equation (3.27) becomes

$$\boldsymbol{i} = \frac{\chi_e - \chi_i}{\mu_0} \boldsymbol{n} x \boldsymbol{B}_0 \tag{3.52}$$

and this allows us, applying Biot–Savart law, to calculate the secondary magnetic field for practically any shape of a body.

3.3.7. Secondary field outside the cylinder

Let us determine the Cartesian components of the field outside the cylinder. We have

$$B_y = B_r \sin \varphi + B_\varphi \cos \varphi, \quad B_x = B_r \cos \varphi - B_\varphi \sin \varphi$$

where

$$\sin \varphi = \frac{y}{r}, \quad \cos \varphi = \frac{x}{r}, \quad r = (x^2 + y^2)^{1/2}$$

Therefore

$$B_y^e = 2\frac{\mu_i - \mu_e}{\mu_i + \mu_e} a^2 \frac{xy}{(x^2 + y^2)^2}$$

and

$$B_x^e = \frac{\mu_i - \mu_e}{\mu_i + \mu_e} a^2 \frac{x^2 - y^2}{(x^2 + y^2)^2} \tag{3.53}$$

As an example, consider the behavior of these components along the profile $y = y_0$ and $z = 0$. Features of the field curves characterize the position and parameters of the cylinder. For instance, the observation point, where $B_y^e = 0$ and B_x^e has a maximal magnitude, is located above the cylinder axis. At the same time, the x-coordinate of the point where the horizontal component B_x^e changes sign equals the distance y_0 between the profile and the cylinder axis. It is a simple matter to show that this field is equivalent to that of a linear dipole (two lines with opposite direction of currents located at the cylinder axis). For illustration, assume that

$$\chi_i - \chi_e = 10^{-3}, \quad \frac{r}{a} = 3 \quad \text{and} \quad B_0 = 50,000\gamma$$

then we find that the field magnitude is around 2.0γ, that constitutes a very small portion of the primary magnetic field. It is proper to notice that in practice of magnetic methods, often even much smaller fields are measured.

3.3.8. The primary field is directed along the cylinder axis

Until now we have considered only the case when the primary field $\boldsymbol{B}_0$ is perpendicular to the cylinder axis. Next, let us suppose that this field is oriented along the cylinder axis. To determine the influence of such a cylinder on the magnetic field, we will use the following approach. The normal field $\boldsymbol{B}_0$ is accompanied by the field $\boldsymbol{H}_0$, which is

$$\boldsymbol{H}_0 = \frac{\boldsymbol{B}_0}{\mu_0}$$

The presence of the cylinder does not change this field. In fact, earlier we demonstrated that the field $\boldsymbol{H}$ can be caused only by the conduction currents, as well as fictitious charges. Then, taking into account the fact that the cylinder is uniform and $\boldsymbol{H}_0$ is everywhere tangential to its lateral surface, we conclude that the volume and surface magnetic charges are absent. Therefore, inside and outside the cylinder, we have

$$\boldsymbol{H}^e = \boldsymbol{H}^i = \boldsymbol{H}_0 = \frac{\boldsymbol{B}_0}{\mu_e}$$

Consequently, we arrive at the conclusion that outside the cylinder, the magnetic field does not change and it is equal to the primary field

$$\boldsymbol{B}^e = \mu_e \boldsymbol{H}^e = \boldsymbol{B}_0$$

However, inside the cylinder, the field is different, and we have

$$\boldsymbol{B}^i = \mu_i \boldsymbol{H}^i = \frac{\mu_i}{\mu_e} \boldsymbol{B}_0$$

Thus, the secondary magnetic field $\boldsymbol{B}_s$ can be written as

$$\boldsymbol{B}_s = \left(\frac{\mu_i}{\mu_e} - 1\right) \boldsymbol{B}_0, \quad \text{if } r < a, \quad \text{and} \quad \boldsymbol{B}_s = 0, \quad \text{if } r > a \tag{3.54}$$

It is obvious that this result is easily generalized to a cylinder with an arbitrary and constant cross-section. As follows from Equation (3.27), the density of surface currents generating this field is

$$\boldsymbol{i}_\varphi = \frac{1}{\mu_0}\frac{\mu_e - \mu_i}{\mu_e + \mu_i}(B^i + B_0)(\boldsymbol{i}_r x \boldsymbol{i}_z)$$

or

$$i_\varphi = \frac{1}{\mu_0}\left(\frac{\mu_i - \mu_e}{\mu_e}\right)B_0 \tag{3.55}$$

since

$$\boldsymbol{i}_r x \boldsymbol{i}_z = -\boldsymbol{i}_\varphi \quad \text{and} \quad B^i + B_0 = \left(\frac{\mu_i}{\mu_e} + 1\right)B_0$$

Therefore, the surface magnetization currents form a system of circular current loops, located in planes perpendicular to the z-axis. Their density is everywhere the same. It is essential to note that such a distribution of currents is able to create a very strong magnetic field inside of the cylinder if $\mu_i \gg \mu_e$. The analogy with the solenoid is obvious. In contrast, if the magnetic permeability of the surrounding medium is much greater than that of the cylinder, the field of these currents almost cancels the normal field and, consequently, the total field $\boldsymbol{B}^i$ tends to zero ($\mu_e \gg \mu_i$).

3.4. AN ELONGATED SPHEROID IN A UNIFORM MAGNETIC FIELD $\boldsymbol{B}_0$

Suppose that an elongated spheroid with semi-axes a and b ($a>b$) and magnetic permeability μ_i is placed in a uniform magnetic field $\boldsymbol{B}_0$ directed along the major axis (Fig. 3.1(c)). The magnetic permeability of the surrounding medium is μ_e. As in the previous example, magnetization currents arise on the spheroid surface and they create a secondary magnetic field. To find this field, we will use the fact that volume currents are absent (curl $\boldsymbol{B} = 0$) and again introduce the potential U:

$$\boldsymbol{B} = -\text{grad}\, U$$

Our goal is to solve the boundary-value problem and we start from Laplace's equation. The shape of the body suggests making use of a spheroidal system of coordinates.

3.4.1. Laplace's equation and its solution in spheroidal system of coordinates

Taking into account the relatively simple shape of the body, it is natural to apply the method of separation of variables and find an expression for the potential of the magnetic field. For this purpose, let us introduce a prolate spheroidal system of coordinates ξ, η, and φ related to cylindrical coordinates by

$$r = c[(1-\xi^2)(\eta^2-1)]^{1/2} \quad \text{and} \quad z = c\xi\eta \tag{3.56}$$

where

$$c = (a^2 - b^2)^{1/2}$$

and

$$-1 \le \xi \le +1, \quad 1 \le \eta < \infty$$

In particular, the surface of the spheroid with semi-axes a and b is the coordinate surface $\eta_0 =$ constant, and

$$a = c\eta_0, \quad b = c(\eta_0^2 - 1)^{1/2} \tag{3.57}$$

The metric coefficients of this system are

$$h_1 = c\left(\frac{\eta^2-\xi^2}{1-\xi^2}\right)^{1/2}, \quad h_2 = c\left(\frac{\eta^2-\xi^2}{\eta^2-1}\right)^{1/2}, \quad h_3 = r \tag{3.58}$$

Then, bearing in mind the fact that the field possesses an axial symmetry with respect to the z-axis, Laplace's equation for the potential U is

$$\frac{\partial}{\partial\xi}\left[(1-\xi^2)\frac{\partial U}{\partial\xi}\right] + \frac{\partial}{\partial\eta}\left[(\eta^2-1)\frac{\partial U}{\partial\eta}\right] = 0 \tag{3.59}$$

As in the previous case, it is convenient to represent the potential inside and outside the spheroid as

$$U_i(\xi,\eta), \quad \text{if } \eta < \eta_0$$

and

$$U_e(\xi,\eta) = U_0 + U_e^s, \quad \text{if } \eta > \eta_0 \tag{3.60}$$

Here U_0 and U_e^s are the potentials of the primary and secondary fields, respectively. As the distance from the spheroid increases, the field of the magnetization currents decreases, and therefore the boundary condition at infinity is

$$U(\xi,\eta) \to U_0, \quad \text{if } \eta \to \infty \tag{3.61}$$

Thus, the potential should satisfy the following conditions:

1. At regular points

$$\Delta U = 0$$

2. At the spheroid surface

$$\frac{U_e}{\mu_e} = \frac{U_i}{\mu_i} \quad \text{and} \quad \frac{\partial U_e}{\partial \eta} = \frac{\partial U_i}{\partial \eta}, \quad \text{at } \eta = \eta_0$$

3. At infinity

$$U_e \to U_0, \quad \text{if } \eta \to \infty$$

As we know, these conditions uniquely define the magnetic field. First, applying the method of separation of variables, we will find a solution of Laplace's equation. Representing the potential as

$$U(\xi, \eta) = T(\eta)F(\xi)$$

and substituting it into Equation (3.59), we obtain two ordinary differential equations of second order:

$$\frac{d}{d\xi}\left[(1-\xi^2)\frac{dF}{d\xi}\right] + n(n+1)F = 0$$

and

$$\frac{d}{d\eta}\left[(\eta^2-1)\frac{dT}{d\eta}\right] - n(n+1)T = 0 \tag{3.62}$$

Here n is integer. These equations are well known and they are called Legendre equations (Chapter 4). Their solutions are Legendre functions of the first and second kinds, P_n and Q_n. Correspondingly, we have

$$T_n(\eta) = A_n P_n(\eta) + B_n Q_n(\eta), \quad F(\xi) = C_n P_n(\xi) + D_n Q_n(\xi) \tag{3.63}$$

Legendre functions are another example of orthogonal functions and they are widely used in mathematics and applied physics. As an illustration, expressions for the functions $P_n(x)$ and $Q_n(x)$ for the first three values of n are given as follows:

$$\begin{gathered} P_0(x) = 1, \quad P_1(x) = x, \quad P_2(x) = \frac{1}{2}(3x^2 - 1) \\ Q_0(x) = \frac{1}{2}\ln\frac{x+1}{x-1}, \quad Q_1(x) = \frac{1}{2}x\ \ln\frac{x+1}{x-1} \end{gathered} \tag{3.64}$$

$$Q_2(x) = \frac{1}{4}(3x^2 - 1)\ln\frac{x+1}{x-1} - \frac{3x}{2}$$

Thus, the general solution of Laplace's equation is

$$U(\xi, \eta) = \sum_{n=0}^{\infty}[A_n P_n(\eta) + B_n Q_n(\eta)][C_n P_n(\xi) + D_n Q_n(\xi)] \tag{3.65}$$

Before we continue our search for a solution of the boundary-value problem, let us express the potential of the primary field in terms of Legendre functions. Since $\boldsymbol{B}_0$ is uniform and directed along the z-axis, we have

$$B_0 = -\frac{\partial U_0}{\partial z} \quad \text{or} \quad U_0 = -B_0 z$$

Then, making use of Equations (3.56) and (3.64), we obtain

$$U_0(\xi, \eta) = -B_0 c P_1(\eta) P_1(\xi) \tag{3.66}$$

That is, the potential of the primary field is expressed with the help of Legendre functions of the first kind and first order. Let us notice that function $P_1(\xi)$ describes the change of the potential U_0 at the coordinate surface where $\eta =$ constant, and in particular at the spheroid surface $\eta = \eta_0$. Therefore, it is natural to assume that the potential of the secondary field depends on the coordinate ξ in the same manner. Bearing in mind the fact that the function $Q_1(\eta)$ (Equations (3.64)) decreases with an increase of the distance, we will represent the potential outside the spheroid as

$$U_e(\xi, \eta) = -B_0 c[P_1(\eta) + AQ_1(\eta)]P_1(\xi), \quad \text{if } \eta > \eta_0 \tag{3.67}$$

Also, we suppose that the field inside the spheroid remains uniform and directed along the z-axis:

$$U_i(\xi, \eta) = -B_0 c D P_1(\eta) P_1(\xi), \quad \text{if } \eta < \eta_0 \tag{3.68}$$

where A and D are unknown coefficients. It is clear that U_i and U_e satisfy Laplace's equation and U_e tends to U_0 as the distance from the spheroid increases. To determine these coefficients, we will make use of conditions at the interface $\eta = \eta_0$ and obtain

$$\frac{1}{\mu_i}[P_1(\eta_0) + AQ_1(\eta_0)] = \frac{1}{\mu_e} D P_1(\eta_0)$$

and

$$P_1'(\eta_0) + AQ_1'(\eta_0) = DP_1'(\eta_0) \tag{3.69}$$

Here $P_1'(\eta_0)$ and $Q_1'(\eta_0)$ are first derivatives of Legendre functions with respect to η

$$P_1'(\eta_0) = 1 \quad \text{and} \quad Q_1'(\eta_0) = \frac{1}{2}\ln\frac{\eta_0 + 1}{\eta_0 - 1} - \frac{\eta_0}{\eta_0^2 - 1}$$

Solving this system, we obtain

$$A = -\frac{((\mu_i/\mu_e) - 1)ab^2}{c^3[1 + ((\mu_i/\mu_e) - 1)L]} \quad \text{and} \quad D = \frac{\mu_i/\mu_e}{1 + ((\mu_i/\mu_e) - 1)L} \tag{3.70}$$

where

$$L = \frac{1 - e^2}{2e^3}\left(\ln\frac{1 + e}{1 - e} - 2e\right) \tag{3.71}$$

and

$$e = \frac{c}{a}$$

The function L characterizes the influence of finite dimensions. For instance, for a markedly elongated spheroid ($e \to 1$)

$$L \approx \frac{b^2}{a^2}\left(\ln\frac{2a}{b} - 1\right) \ll 1 \tag{3.72}$$

On the contrary, in the case of a sphere ($e \to 0$)

$$L = \frac{1}{3} \tag{3.73}$$

In accordance with Equation (3.70), the uniform magnetic field inside spheroid ($a > b$) is

$$B_z = DE_0 = \frac{\mu_i}{\mu_e}\frac{B_0}{1 + ((\mu_i/\mu_e) - 1)L} \tag{3.74}$$

Inasmuch as with an increase of the ratio a/b, the function L tends to zero rapidly (Equation (3.72)) in the limit when the spheroid coincides with an infinitely long cylinder, the field $\boldsymbol{B}_i$ again becomes equal to

$$\boldsymbol{B}_i = \frac{\mu_i}{\mu_e}\boldsymbol{B}_0$$

At the same time, as follows from Equation (3.70), the secondary field vanishes outside an infinitely long cylinder. The fact that the magnetic field inside of a

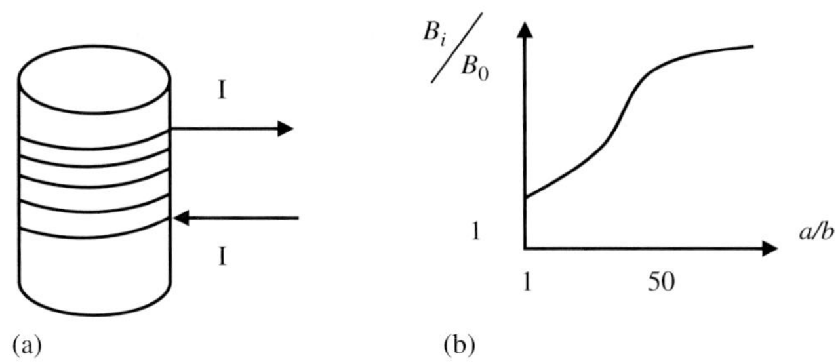

Fig. 3.2. (a) Coil with a magnetic core. (b) Behavior of the magnetic field inside the spheroid.

spheroid, elongated along the field, can be much stronger than the primary one plays a fundamental role in measurements since it essentially allows us to increase the moment of receiver coils. Consider a coil with a magnetic core and having the shape of the cylinder, as is shown in Fig. 3.2(a). As is well known, such coils are often used for measuring alternating magnetic fields, because the electromotive force induced in the coil is directly proportional to the rate of change of the field with time. Therefore, the increase of the field $\boldsymbol{B}$ inside the coil due to the presence of the core strongly increases the sensitivity of the receiver.

The behavior of the field B_i as a function of the ratio of semi-axes of the spheroid is shown in Fig. 3.2(b). It is clear that the right asymptote of the curve corresponds to the case of an infinitely long cylinder, when the maximal increase of the field is observed. Cores are usually made from ferrites with relative magnetic permeability reaching several thousands. For instance, if we assume that $\mu_i/\mu_e = 5000$, then, as is seen from Fig. 3.2(b), a maximal increase of the field B_i almost takes place, provided that $a/b > 400$. To satisfy this inequality, it is usually necessary to use very long cores, which are inconvenient for geophysical applications. Correspondingly, shorter cores are used that still provide a strong increase of the field $\boldsymbol{B}_i$. For example, if $\mu_i/\mu_e = 5000$ and $a/b = 20$, we have

$$\frac{B_i}{B_0} = 100$$

Note that the results of the field calculation inside a markedly elongated spheroid can be applied for the central part of a relatively long cylinder. Now let us consider the behavior of the field $\boldsymbol{B}_i$ when the spheroid is transformed into a sphere with radius a. As follows from Equation (3.73)

$$B_i = \frac{3\mu_i}{2\mu_e + \mu_i} B_0 \tag{3.75}$$

That is, even in the case of the sphere, the field B_i can be almost three times greater than the primary field.

Earlier we demonstrated that the potential U is independent of the coordinate φ, and correspondingly the component of the field $\boldsymbol{B}$ in this direction equals zero. This means that surface currents have only the component i_φ (Chapter 2):

$$\boldsymbol{i}_\varphi = \frac{K_{12}}{\mu_0}\boldsymbol{i}_\eta \times \boldsymbol{i}_\xi (B_\xi^e + B_\xi^i)$$

Since

$$K_{12} = \frac{\mu_e - \mu_i}{\mu_e + \mu_i}, \qquad \boldsymbol{i}_\varphi = \boldsymbol{i}_\eta x \boldsymbol{i}_\xi$$

and

$$B_\xi = -\frac{1}{h_1}\frac{\partial U}{\partial \xi}, \qquad h_1 = c\left(\frac{\eta^2 - \xi^2}{1 - \xi^2}\right)^{1/2}$$

the magnitude of the current density is

$$i_\varphi = \frac{(\mu_i - \mu_e)(1 - \xi^2)^{1/2}}{c\mu_0(\mu_i + \mu_e)(\eta_0^2 - \xi^2)^{1/2}}\frac{\partial}{\partial \xi}(U_e + U_i)$$

Substituting the expressions for the potential, we obtain

$$i_\varphi = \frac{1}{\mu_0}\left(\frac{\mu_i}{\mu_e} - 1\right)\frac{B_0\eta_0(1 - \xi^2)^{1/2}}{[1 + ((\mu_i/\mu_e) - 1)L](\eta_0^2 - \xi^2)^{1/2}}$$

Thus, the current density reaches its maximal magnitude in the plane $z = 0$ and then gradually decreases in both directions when ξ approaches either $+1$ or -1.

3.5. FIELD OF A MAGNETIC DIPOLE LOCATED AT THE CYLINDER AXIS

Next suppose that the center of a small horizontal loop (magnetic dipole) has a current I and it is located at the cylinder axis. The magnetic permeability of the cylinder and the surrounding medium are μ_i and μ_e, respectively. The cylinder radius is a. This problem is of some practical interest in borehole geophysics. The influence of the medium on the magnetic field can be described in the following way. Due to the primary field of the current loop, magnetization currents arise in the medium, in particular, in its vicinity as well as at the cylinder surface. Consequently, at every point, the magnetic field consists of the primary and secondary fields, and the latter is caused by magnetization currents. It is essential to remember that the density of these currents is defined by the total magnetic field. This is the reason why we have to formulate a boundary-value problem to determine the field $\boldsymbol{B}$. With this purpose in mind, let us introduce a cylindrical system of coordinates r, φ, and z, so that the magnetic dipole is located at its origin, and represent the potential U ($\boldsymbol{B} = -\nabla U$) as

$$U_1 = U_0 + U_s, \quad r<a$$

and

$$U_e, \quad r>a \tag{3.76}$$

Here U_0 is the potential of the primary field, caused by the current loop in a uniform medium with magnetic permeability μ_i. Thus, the potential U has to satisfy following conditions:

1. At regular points of the medium

$$\Delta U = 0$$

2. Near the current loop

$$U_i \to U_0, \quad \text{if } R \to 0$$

3. At the interface

$$\frac{U_i}{\mu_i} = \frac{U_e}{\mu_e} \quad \text{and} \quad \frac{\partial U_i}{\partial r} = \frac{\partial U_e}{\partial r}, \quad \text{if } r = a$$

4. At infinity

$$U \to 0, \quad \text{if } R \to \infty$$

To facilitate the derivations, we will take into account the axial symmetry of the magnetic field and potential. In other words, U, as well as the field $\boldsymbol{B}$, is independent of the coordinate φ, and therefore

$$\frac{\partial U}{\partial \varphi} = 0$$

3.5.1. Solution of Laplace's equation in the cylindrical coordinates

Inasmuch as the potential U is a function of the coordinates: r and z, we have for Laplace's equation:

$$\frac{\partial^2 U}{\partial r^2} + \frac{1}{r}\frac{\partial U}{\partial r} + \frac{\partial^2 U}{\partial z^2} = 0 \tag{3.77}$$

This is a differential equation of second order with partial derivatives, and in order to solve it we will suppose that its solution can be represented as the product of two functions, so that each function depends on one argument only. Consequently, we have

$$U(r,z) = T(r)F(z) \tag{3.78}$$

Substitution of Equation (3.78) into Equation (3.77) gives

$$F\frac{d^2T}{dr^2} + \frac{F}{r}\frac{dT}{dr} + T\frac{d^2F}{dz^2} = 0$$

Dividing both sides of the equation by the product TF, we have

$$\frac{1}{T}\frac{d^2T}{dr^2} + \frac{1}{rT}\frac{dT}{dr} + \frac{1}{F}\frac{d^2F}{dz^2} = 0 \tag{3.79}$$

On the left-hand side of Equation (3.79), it is natural to distinguish two terms

$$\text{Term1} = \frac{1}{T}\frac{d^2T}{dr^2} + \frac{1}{rT}\frac{dT}{dr}$$

and

$$\text{Term2} = \frac{1}{F}\frac{d^2F}{dz^2}$$

As before, at the first glance it seems that they depend on the arguments r and z, respectively, and Equation (3.79) can be represented as

$$\text{Term1}(r) + \text{Term2}(z) = 0$$

However, such equality is impossible, since changing one of the arguments, for example, r, the first term varies while the second one remains the same, and correspondingly the sum of these terms cannot be equal to zero for arbitrary values of r and z. Therefore, we have to conclude that each term does not depend on the coordinates and is constant. As was pointed out in the previous sections, this fact constitutes the key point of the method of separation of variables, allowing us to describe the potential as a product of two functions. For convenience let us represent this constant in the form $\pm m^2$, where m is called a constant of separation. Thus, instead of Laplace's equation we have

$$\frac{1}{T}\frac{d^2T}{dr^2} + \frac{1}{rT}\frac{dT}{dr} = \pm m^2 \quad \text{and} \quad \frac{1}{F}\frac{d^2F}{dz^2} = \mp m^2 \tag{3.80}$$

It may be proper to emphasize again that the replacement of the differential equation with partial derivatives by two ordinary differential equations is the main purpose of the method of separation of variables, since the solution of the latter is known. To choose the proper sign on the right-hand side of Equation (3.80), we will take into account the fact that the potential U inside and outside the cylinder is an even function with respect to the coordinate z

$$U(r,z) = U(r,-z)$$

For this reason, we choose the minus sign on the right-hand sign of the equation for F, and correspondingly it gives

$$F''(z) + m^2 F(z) = 0 \tag{3.81}$$

Here

$$F''(z) = \frac{d^2F}{dz^2}$$

As is well known, the latter has two independent solutions, $\sin mz$ and $\cos mz$; thus, the function $F(z)$ can be written as

$$F(z) = C_{1m} \sin mz + C_{2m} \cos mz \tag{3.82}$$

where C_{1m} and C_{2m} are arbitrary constants independent of z. As follows from Equations (3.80), on the right-hand side of the equation for function $T(r)$, we have to take the sign "+" and therefore

$$T''(r) + \frac{1}{r}T'(r) - m^2 T(r) = 0 \tag{3.83}$$

where

$$T' = \frac{dT}{dr} \quad \text{and} \quad T'' = \frac{d^2T}{dr^2}$$

Introducing the variable $x = mr$, we have

$$\frac{dT}{dr} = \frac{dT}{dx}\frac{dx}{dr} = m\frac{dT}{dx} \quad \text{and} \quad \frac{d^2T}{dr^2} = m^2\frac{d^2T}{dx^2}$$

Substitution of these equalities into Equation (3.83) gives

$$\frac{d^2T}{dx^2} + \frac{1}{x}\frac{dT}{dx} - T = 0 \tag{3.84}$$

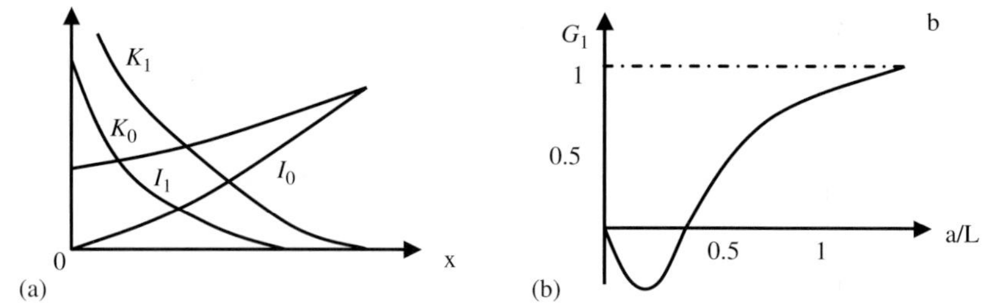

Fig. 3.3. (a) Modified Bessel functions. (b) Geometric factor of the cylinder.

This equation is also well known and is often used in various boundary-value problems with cylindrical interfaces. Its solutions are modified Bessel functions of the first and second type but zero order, $I_0(x)$ and $K_0(x)$, respectively. Their behavior is shown in Fig. 3.3(a) and they have been studied in detail along with other Bessel functions. Also we will use modified Bessel functions of the first order $I_1(x)$ and $K_1(x)$, which describe derivatives of functions $I_0(x)$ and $K_0(x)$. Relations between them are

$$\frac{dI_0(x)}{dx} = I_1(x) \quad \text{and} \quad \frac{dK_0}{dx} = -K_1(x) \tag{3.85}$$

These functions are also shown in Fig. 3.3(a). It is useful to demonstrate the asymptotic behavior of these functions

$$I_0(x) \to 1, \quad K_0(x) \to -\ln x$$
$$I_1(x) \to \frac{x}{2}, \quad K_1(x) \to \frac{1}{x}, \quad \text{as } x \to 0$$

and

$$\begin{aligned} I_0(x) &\to \frac{\exp(x)}{(2\pi x)^{1/2}}, \quad K_0(x) \to \left(\frac{\pi}{2x}\right)^{1/2} \exp(-x) \\ I_1(x) &\to \frac{\exp(x)}{(2\pi x)^{1/2}}, \quad K_1(x) \to \left(\frac{\pi}{2x}\right)^{1/2} \exp(-x), \quad \text{as } x \to \infty \end{aligned} \tag{3.86}$$

Let us notice that modified Bessel functions are described in numerous monographs; there are many tables of their values, different representations of these functions, relations between them, polynomial approximations and so on. Certainly, application of these functions is as convenient as that of elementary functions. Thus, a solution of Equation (3.84) can be represented as

$$T(x) = D_1 I_0(x) + D_2 K_0(x) \quad \text{or} \quad T(mr) = D_{1m} I_0(mr) + D_{2m} K_0(mr) \tag{3.87}$$

where D_{1m} and D_{2m} are arbitrary constants that are independent of r.

3.5.2. Expressions for the potential of the magnetic field

Now, making use of Equations (3.78), (3.82), and (3.87) for each value of m, we have

$$U(r,z,m) = [A_m I_0(mr) + B_m K_0(mr)][C_{1m} \sin mz + C_{2m} \cos mz] \tag{3.88}$$

It is clear that the function $U(r, z, m)$ satisfies Laplace's equation and we might think that the first step of solving the boundary-value problem is accomplished. However, this assumption is incorrect, since the function $U(r, z, m)$ depends on m, which appears as a result of the transformation of Laplace's equation into two ordinary differential equations, while the potential U describing the magnetic field in a medium is independent of m. Inasmuch as the function $U(r, z, m)$ given by Equation (3.88) satisfies Laplace's equation for any m, we will represent U as definite integral

$$U(r,z) = \int_0^\infty [A_m I_0(mr) + B_m K_0(mr)][C_{1m} \sin mz + C_{2m} \cos mz]dm \tag{3.89}$$

that is independent of m. Thus, we have arrived at the general solution of Laplace's equation, which includes an infinite number of solutions corresponding to different coefficients A_m and B_m, as well as C_{1m} and C_{2m}. Now we are ready to perform the second step in solving the boundary-value problem: to choose among the functions $U(r, z)$ solutions, which obey the boundary conditions near the magnetic dipole and at infinity. With this purpose in mind, we will take into account the asymptotic behavior of functions $I_0(mr)$ and $K_0(mr)$. As was shown earlier, $K_0(mr)$ tends to infinity as its argument approaches zero, and therefore this function cannot describe the potential of the secondary field inside the cylinder (borehole). At the same time, the function $I_0(mr)$ increases unlimitedly with an increase of r and, correspondingly, an expression for the field should not contain this function outside the borehole. Thus, instead of Equation (3.89), we write

$$U_1(r,z) = U_0(r,z) + \int_0^\infty A_m I_0(mr)[C_{1m} \sin mz + C_{2m} \cos mz]dm, \quad \text{if } r<a$$

and

$$U_2(r,z) = \int_0^\infty B_m K_0(mr)[C_{1m} \sin mz + C_{2m} \cos mz]dm, \quad \text{if } r>a \tag{3.90}$$

Here U_0 is the potential of the magnetic dipole in a uniform medium with magnetic permeability μ_i. It is clear that these functions satisfy both Laplace's equation and the boundary conditions near the dipole and at infinity. In fact, in approaching the small current loop the function U_1 tends to the potential of the dipole, while with an increase of r the function U_2, due to the presence of $K_0(mr)$, decreases. Also both integrals in Equations (3.90) contain the oscillating factor $\cos mz$ and therefore the functions U_1 and U_2 tend to zero as the distance along the z-axis increases. Before we proceed, let us represent the potential U_0 in the same form as the function $U(r, z)$. As was demonstrated in Chapter 2, the scalar potential of the magnetic dipole is

$$U_0(r,z) = \frac{\mu_i}{4\pi}\frac{M\ \cos\theta}{R^2} \tag{3.91}$$

where $M = ISn$, $\cos\theta = z/R$, $R = (r^2 + z^2)^{1/2}$, and I, S, and n are the current, area of the loop, and the number of turns, respectively. First, we represent U_0 as

$$U_0 = -\frac{\mu_i M}{4\pi}\frac{\partial}{\partial z}\frac{1}{R}$$

Then, making use of Sommerfeld integral

$$\frac{1}{R} = \frac{2}{\pi}\int_0^\infty K_0(mr)\cos mz\, dm$$

and performing a differentiation, we obtain

$$U_0(r,z) = \frac{\mu_i M}{2\pi^2}\int_0^\infty mK_0(mr)\sin mz\, dm \tag{3.92}$$

Correspondingly, we will write expressions for the resultant potential inside and outside of the cylinder in a similar manner

$$U_1 = \frac{\mu_0 M}{2\pi^2}\int_0^\infty m\left[\frac{\mu_i}{\mu_0}K_0(mr) + A_m I_0(mr)\right]\sin mz\, dm$$

and

$$U_2 = \frac{\mu_0 M}{2\pi^2}\int_0^\infty mB_m K_0(mr)\sin mz\, dm \tag{3.93}$$

3.5.3. *Coefficients A_m and B_m*

It is clear that these functions (Equation (3.93)) obey Laplace's equation and boundary conditions. Now we will demonstrate that for certain values of coefficients A_m and B_m, they also satisfy two conditions of the boundary-value problem at the interface of the cylinder (borehole). With this purpose in mind, we will use one of the remarkable features of the Fourier integral, namely, from the equality

$$\int_0^\infty \varphi_1(m)\sin mz\, dm = \int_0^\infty \varphi_2(m)\sin mz\, dm$$

it follows that

$$\varphi_1(m) = \varphi_2(m)$$

that is, an equality of functions leads to an equality of their spectra. Then applying the conditions for the potential at the cylinder surface, we obtain a system of equations with respect to A_m and B_m:

$$\frac{1}{\mu_0} K_0(ma) + \frac{1}{\mu_i} A_m I_0(ma) = \frac{B_m K_0(ma)}{\mu_e}$$

$$-\frac{\mu_i}{\mu_0} K_1(ma) + A_m I_1(ma) = -B_m K_1(ma)$$

Its solution is

$$A_m = \frac{\mu_i}{\mu_0} \frac{(\mu_i - \mu_e) K_0 K_1}{\mu_i I_1 K_0 + \mu_e I_0 K_1}$$

and

$$B_m = \frac{\mu_i \mu_e}{\mu_0} \frac{I_0 K_1 + I_1 K_0}{\mu_i I_1 K_0 + \mu_e I_0 K_1}$$

Inasmuch as

$$I_0(ma) K_1(ma) + I_1(ma) K_0(ma) = \frac{1}{ma}$$

we finally obtain

$$A_m = \frac{\mu_i}{\mu_0} \left(\frac{\mu_i}{\mu_e} - 1 \right) \frac{ma K_0(ma) K_1(ma)}{1 + ((\mu_i/\mu_e) - 1) ma K_0(ma) I_1(ma)}$$

and

$$B_m = \frac{\mu_i}{\mu_0} \frac{1}{1 + ((\mu_i/\mu_e) - 1) ma K_0(ma) I_1(ma)} \tag{3.94}$$

Thus, we have solved the boundary-value problem for the potential, and the components of the magnetic field are defined as

$$B_r = -\frac{\partial U}{\partial r}, \quad B_z = -\frac{\partial U}{\partial z}, \quad B_\varphi = 0$$

3.5.4. The current density

From Chapter 2, it follows that the surface current has only a φ-component and its density is

$$i_\varphi = \frac{1}{\mu_0}\frac{\mu_i - \mu_e}{\mu_i + \mu_e}(B_z^e + B_z^i)$$

since

$$\boldsymbol{n} = \boldsymbol{i}_r, \quad \boldsymbol{i}_r x \boldsymbol{i}_z = -\boldsymbol{i}_\varphi$$

Then, substituting expressions for the vertical component of the field, we obtain

$$i_\varphi = -\frac{M}{2\pi^2}\frac{\mu_i}{\mu_0}\left(\frac{\mu_i}{\mu_e} - 1\right)\int_0^\infty m^2 \frac{K_0(ma)\cos mz\, dm}{1 + ((\mu_i/\mu_e) - 1)maK_0(ma)I_1(ma)}$$

It is clear that the surface currents form a system of circular current loops located symmetrically with respect to the plane $z = 0$ and their density depends on the coordinate z. Of course, they create the same field as that caused by magnetic dipoles at each elementary volume of the medium and their moment is equal to $\boldsymbol{P}(q)dV$.

3.5.5. Asymptotic behavior of the field on the cylinder (borehole) axis

Next consider the behavior of the field at points of the cylinder axis. By definition, we have:

$$B_r = -\frac{\partial U}{\partial r}$$

and since $I_1(0) = 0$, the field has only a vertical component at the borehole axis. Of course, this also follows from the fact that the observation point is located on the axis of the current loops. In accordance with Equation (3.91), the primary magnetic field caused by the dipole in a uniform medium is

$$B_z^0 = \frac{\mu_i M}{2\pi z^3}, \quad \text{if } r = 0$$

Appling Equations (3.93) and (3.94), the total magnetic field on the borehole axis, $r = 0$, can be represented as

$$B_z(0, z) = \frac{\mu_i M}{2\pi}\left[\frac{1}{L^3} - \left(\frac{\mu_i}{\mu_e} - 1\right)\frac{1}{\pi a^3}\int_0^\infty \frac{x^3 K_0(x)K_1(x)\cos \alpha x\, dx}{1 + ((\mu_i/\mu_e) - 1)I_1(x)K_0(x)}\right] \tag{3.95}$$

Here $x = ma$, $\alpha = L/a$, and $L = z$ is the distance between the coil and the observation point, usually called the probe length in the borehole geophysics. It is convenient to normalize the total field by the primary one, B_z^0. As a result, we obtain

$$b_z = \frac{B_z}{B_z^0} = 1 - \left(\frac{\mu_i}{\mu_e} - 1\right)\frac{\alpha^3}{\pi}\int_0^\infty \frac{x^3 K_0(x)K_1(x)\cos\alpha x\, dx}{1 + x((\mu_i/\mu_e) - 1)I_1(x)K_0(x)} \tag{3.96}$$

The function b_z depends only two parameters, μ_i/μ_e and α. Let us briefly study the asymptotic behavior of the field as a function of α. As the parameter α decreases, the integral on the right-hand side of Equation (3.96) tends to some constant, and therefore

$$b_z \to 1, \quad \text{if } \alpha \to 0$$

In other words, in the vicinity of the dipole (near zone), the field B_z coincides with the field in a uniform medium with the magnetic permeability of the borehole, μ_i. This field is practically caused by the conduction currents in the loop and magnetization currents in its vicinity, while the influence of magnetization currents on the interface $r = a$ is negligible. To find an asymptotic expression for the field in the opposite case, when the probe length is much greater than the borehole radius, we will use the following approach. Let us focus our attention on the integral at the right-hand side of Equation (3.96). Its integrand has the form

$$A(x)\cos\alpha x$$

One of these functions $A(x)$ changes gradually and for sufficiently large values of its argument it decreases. At the same time, $\cos\alpha x$ is an oscillating function. The interval Δx within which it does not change sign is defined by the condition

$$\Delta x = \frac{\pi}{\alpha}$$

With an increase of the parameter α, this interval decreases and, correspondingly, $A(x)$ becomes practically constant within every interval ($x \gg \Delta x$). Taking into account the fact that $A(x)$ is a continuous function of x, we can say that with a decrease of Δx the integrals over neighboring intervals are almost equal in magnitude, but have opposite sign. In other words, they cancel each other, and with an increase of α this behavior manifests itself for smaller x. This means that in the limit as α tends to infinity, the integral in Equation (3.96) is defined by very small values of x. Taking this fact into account, we replace functions I_1, K_0, and K_1 by their asymptotic expressions

$$I_1(x) \to \frac{x}{2}, \quad K_0(x) \to -\ln x, \quad K_1(x) \to \frac{1}{x}, \quad \text{if } x \to 0$$

except for $K_0(x)$ in the numerator. This gives

$$\int_0^\infty \frac{x^3 K_0(x) K_1(x) \cos \alpha x \, dx}{1 + ((\mu_i/\mu_e) - 1) x I_1(x) K_0(x)} \to \int_0^\infty x^2 K_0(x) \cos \alpha x \, dx$$
$$= -\frac{\partial^2}{\partial \alpha^2} \int_0^\infty K_0(x) \cos \alpha x \, dx$$

From Sommerfeld integral

$$\int_0^\infty K_0(x) \cos \alpha x \, dx = \frac{\pi}{2} \frac{1}{(1+\alpha^2)^{1/2}} \approx \frac{\pi}{2\alpha}, \quad \text{if } \alpha \gg 1$$

it follows that

$$\frac{\partial^2}{\partial \alpha^2} \int_0^\infty x^2 K_0(x) \cos \alpha x \, dx = \frac{\pi}{\alpha^3}, \quad \text{if } \alpha \gg 1 \tag{3.97}$$

Substituting this result into Equation (3.96), we have

$$b_z \approx 1 + \left(\frac{\mu_i}{\mu_e} - 1 \right) = \frac{\mu_i}{\mu_e}$$

Thus, in the far zone, $\alpha \gg 1$, the magnetic field is inversely proportional to μ_e.

3.5.6. Concept of geometric factor

Suppose that both susceptibilities χ_i and χ_e are very small. Then, by neglecting the second term in the denominator of the integrand (Equation (3.96)), we obtain

$$b_z = 1 - (\chi_i - \chi_e) \frac{\alpha^3}{\pi} \int_0^\infty x^3 K_0(x) K_1(x) \cos \alpha x \, dx$$

Taking into account the fact that $\mu_i = \mu_0(1+\chi_i)$ and neglecting products of χ, we have

$$B_z = B_0 + B_s = \frac{\mu_0 M}{2\pi L^3} [1 + \chi_i (1 - G_2) + \chi_e G_2] \tag{3.98}$$

where

$$B_0 = \frac{\mu_0 M}{2\pi L^3}$$

is the magnetic field of the dipole in a nonmagnetic medium when $r = 0$, while B_s is the secondary field, caused by surface currents:

$$B_s(0,\alpha) = \frac{\mu_0 M}{2\pi L^3}\{\chi_i[1 - G_2(\alpha)] + \chi_e G_2(\alpha)\} \tag{3.99}$$

and

$$G_2(\alpha) = \frac{\alpha^3}{\pi}\int_0^\infty x^3 K_0(x)K_1(x)\cos\alpha x\,dx \tag{3.100}$$

Let us write Equation (3.99) as

$$B_s(0,L) = \frac{\mu_0 M}{2\pi L^3}[\chi_i G_1(\alpha) + \chi_e G_2(\alpha)] \tag{3.101}$$

where

$$G_1(\alpha) + G_2(\alpha) = 1 \tag{3.102}$$

The functions G_1 and G_2 are usually called the geometric factors of the cylinder and the surrounding medium, respectively. In accordance with Equation (3.101), the secondary field is a sum of two terms, provided that the induced magnetization is defined by the primary field $\boldsymbol{B}_0$ only. In other words, the interaction between magnetization currents is neglected, and for this reason each term in parentheses in Equation (3.101) is the product of the susceptibility and the corresponding geometric factor. The terms containing a product of susceptibilities, χ_i and χ_e, are absent. As follows from Equations (3.100) and (3.102), the geometric factor of the borehole G_1 is a function of the parameter of α only, and its behavior is shown in Fig. 3.3(b). It is obvious that

$$G_1(\alpha) \to 1, \quad \text{if } \alpha \to 0$$

and

$$G_1(\alpha) \to 0, \quad \text{if } \alpha \to \infty \tag{3.103}$$

Hence, with an increase of the probe length L the influence of the borehole decreases, and the field approaches that corresponding to a uniform medium with susceptibility χ_e. In conclusion, it is proper to make two comments:

1. Applying the principle of superposition and neglecting the interaction of molecular currents, Equation (3.101) can easily be generalized to a model with several coaxial cylindrical interfaces. Then, we have for the vertical component of the magnetic field on the borehole axis

$$B_z = \frac{\mu_0 M}{2\pi L^3}\sum_{n=1}^{N}\chi_n G_n \tag{3.104}$$

where χ_n and G_n are the susceptibility and geometric factor of n-cylindrical layer, respectively. The function G_n is expressed in terms of the geometric factor of the borehole.

2. When the system with two coils (one is a transmitter, the other is a receiver) is placed into a borehole, the magnetic field is usually generated by an alternating current. However, if the frequency is chosen in such a way that an influence of electromagnetic induction is very small, we can still use the theory of the constant magnetic field.

3.6. ELLIPSOID IN A UNIFORM MAGNETIC FIELD

Next, we will assume that an ellipsoid with semi-axes a, b, and c is placed in a uniform magnetic field $\boldsymbol{B}_0$. At the beginning we will assume that it is directed along the x-axis (Fig. 3.4(a)) and then consider the general case. The magnetic permeabilities of the ellipsoid and surrounding medium are μ_i are μ_e, respectively, and our goal is to find the magnetic field, caused by magnetization currents arising on the ellipsoid surface. As before, first we will solve Laplace's equation for the potential. Then we will choose among its solutions those which satisfy the boundary conditions and finally select functions that obey the conditions at the interface.

3.6.1. System of ellipsoidal coordinates

By definition, the equation

$$\frac{x^2}{a^2}+\frac{y^2}{b^2}+\frac{z^2}{c^2}=1 \quad (a>b>c) \tag{3.105}$$

is the equation of an ellipsoid with semi-axes a, b, and c.

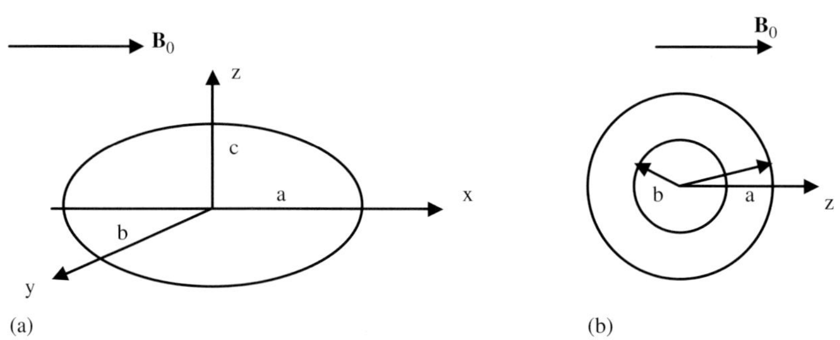

Fig. 3.4. (a) Ellipsoid in a uniform magnetic field. (b) Spherical layer.

Then

$$\begin{aligned}
&\frac{x^2}{a^2+\xi}+\frac{y^2}{b^2+\xi}+\frac{z^2}{c^2+\xi}=1 \quad (\xi>-c^2)\\
&\frac{x^2}{a^2+\eta}+\frac{y^2}{b^2+\eta}+\frac{z^2}{c^2+\eta}=1 \quad (-c^2>\eta>-b^2)\\
&\frac{x^2}{a^2+\zeta}+\frac{y^2}{b^2+\zeta}+\frac{z^2}{c^2+\zeta}=1 \quad (-b^2>\zeta>-a^2)
\end{aligned} \tag{3.106}$$

are the equations of an ellipsoid, and hyperboloids of one and two sheets, respectively, and they are confocal with the ellipsoid. Each surface of every type passes through any point of a space and its position is characterized by three values: ξ, η, and ζ. The variables

$$u_1=\xi, \quad u_2=\eta, \quad u_3=\zeta \tag{3.107}$$

are called ellipsoidal coordinates, and the surface $\xi=$ constant is an ellipsoid, the surface $\eta=$ constant is a hyperboloid of one sheet and $\zeta=$ constant is a hyperboloid of two sheets.

Solution of Equations (3.105) and (3.106) with respect to Cartesian coordinates gives

$$\begin{aligned}
x&=\pm\left[\frac{(\xi+a^2)(\eta+a^2)(\zeta+a^2)}{(b^2-a^2)(c^2-a^2)}\right]^{1/2}\\
y&=\pm\left[\frac{(\xi+b^2)(\eta+b^2)(\zeta+b^2)}{(c^2-b^2)(a^2-b^2)}\right]^{1/2}\\
z&=\pm\left[\frac{(\xi+c^2)(\eta+c^2)(\zeta+c^2)}{(a^2-c^2)(b^2-c^2)}\right]^{1/2}
\end{aligned} \tag{3.108}$$

The latter allows us to find metric coefficients of the system of coordinates, and they are

$$\begin{aligned}
h_1&=\frac{1}{2}\left[\frac{(\xi-\eta)(\xi-\zeta)}{(\xi+a^2)(\xi+b^2)(\xi+c^2)}\right]^{1/2}\\
h_2&=\frac{1}{2}\left[\frac{(\eta-\zeta)(\eta-\xi)}{(\eta+a^2)(\eta+b^2)(\eta+c^2)}\right]^{1/2}\\
h_3&=\frac{1}{2}\left[\frac{(\zeta-\xi)(\zeta-\eta)}{(\zeta+a^2)(\zeta+b^2)(\zeta+c^2)}\right]^{1/2}
\end{aligned} \tag{3.109}$$

As is known, Laplace's equation can be written in any orthogonal system of coordinates as

$$\Delta U = \frac{1}{h_1 h_2 h_3}\left[\frac{\partial}{\partial u_1}\left(\frac{h_2 h_3}{h_1}\frac{\partial U}{\partial u_1}\right) + \frac{\partial}{\partial u_2}\left(\frac{h_1 h_3}{h_2}\frac{\partial U}{\partial u_2}\right) + \frac{\partial}{\partial u_3}\left(\frac{h_1 h_2}{h_3}\frac{\partial U}{\partial u_3}\right)\right] = 0$$

Introducing the notation

$$R_s = [(s + a^2)(s + b^2)(s + c^2)]^{1/2} \tag{3.110}$$

where $s = \xi, \eta, \zeta$, the Laplacian in an ellipsoidal system of coordinates is

$$\begin{aligned}\Delta U = &\frac{4}{(\xi-\eta)(\xi-\zeta)(\eta-\zeta)}\left[(\eta-\zeta)R_\xi\frac{\partial}{\partial\xi}\left(R_\xi\frac{\partial U}{\partial\xi}\right)\right.\\ &\left.+(\zeta-\xi)R_\eta\frac{\partial}{\partial\eta}\left(R_\eta\frac{\partial U}{\partial\eta}\right) + (\xi-\eta)R_\zeta\frac{\partial}{\partial\zeta}\left(R_\zeta\frac{\partial U}{\partial\zeta}\right)\right]\end{aligned} \tag{3.111}$$

3.6.2. Expressions for the potential of the primary field

If the primary field is directed along the x-axis, then by definition

$$B_0 = -\frac{\partial U_0}{\partial x}$$

and taking into account Equations (3.108), we have

$$U_0 = -B_0 x = -B_0\left[\frac{(\xi + a^2)(\eta + a^2)(\zeta + a^2)}{(b^2 - a^2)(c^2 - a^2)}\right]^{1/2} \tag{3.112}$$

The potential is represented as a product of three functions

$$U_0 = CF_1(\xi)F_2(\eta)F_3(\zeta) \tag{3.113}$$

where

$$C = -\frac{B_0}{[(b^2 - a^2)(c^2 - a^2)]^{1/2}} \tag{3.114}$$

$$F_1(\xi) = (\xi + a^2)^{1/2}, \quad F_2 = (\eta + a^2)^{1/2}, \quad F_3(\zeta) = (\zeta + a^2)^{1/2} \tag{3.115}$$

3.6.3. Solutions of Laplace's equation

It is a simple matter to show that function, given by Equation (3.113), is a solution of Laplace's equation, and by definition it has a singularity at infinity ($x \to \infty$). This is the first type of solution. In order to find the secondary field, we have to make use of the solution of Laplace's equation which decreases with an increase of the distance from the ellipsoid. Let us assume that the second type of solution has the form

$$U_s(\xi, \eta, \zeta) = AG_1(\xi)F_2(\eta)F_3(\zeta) \tag{3.116}$$

This presentation is very natural because the primary and secondary potentials have to change on the ellipsoid surface $\xi = \xi_0$ in the same manner. Otherwise, it is impossible to satisfy the boundary conditions. Equation (3.116) contains the unknown function $G_1(\xi)$ and our goal is to find its expression. Inasmuch as the potential $U_s(\xi, \eta, \zeta)$ has to satisfy the Laplace equation, we substitute Equation (3.116) into Equation (3.111) and it gives

$$R_\xi \frac{d}{d\xi}\left(R_\xi \frac{dG_1}{d\xi}\right) - \left(\frac{b^2 + c^2}{4} + \frac{\xi}{2}\right)G_1 = 0 \tag{3.117}$$

Thus, applying the method of separation of variables we have arrived at an ordinary linear differential equation of the second order. From the theory of these equations, it follows that if one solution is known, then the second independent solution can be obtained by integration. For instance, if $y_1(x)$ is a solution of the equation

$$\frac{d^2y}{dx^2} + p(x)\frac{dy}{dx} + q(x)y = 0 \tag{3.118}$$

then the second solution y_2 is

$$y_2 = y_1 \int \frac{\exp(-\int p\, dx)}{y_1^2}\, dx \tag{3.119}$$

In our case

$$p(x) = \frac{1}{R_\xi}\frac{dR_\xi}{d\xi} = \frac{d}{d\xi}\ln R_\xi$$

Since

$$\exp\left(-\int p\, dx\right) = \frac{1}{R_\xi}$$

we have

$$G_1(\xi) = F_1 \int \frac{d\xi}{F_1^2 R_\xi} \tag{3.120}$$

In principle, limits of integration are arbitrary, but in order to satisfy a condition at infinity, we represent the function $G_1(\xi)$ as

$$G_1(\xi) = F_1 \int_\xi^\infty \frac{d\xi}{F_1^2 R_\xi} \tag{3.121}$$

Here ξ is the coordinate of an observation point. Suppose that $\xi \to \infty$. Then, in accordance with Equations (3.110) and (3.115), we have

$$G_1(\xi) \to \xi^{1/2} \int_\xi^\infty \frac{d\xi}{\xi \xi^{3/2}} \to \frac{2}{3}\frac{1}{\xi} \tag{3.122}$$

On the other hand, the equation of ellipsoid can be written in the form

$$\frac{x^2}{1+(a^2/\xi)} + \frac{y^2}{1+(b^2/\xi)} + \frac{z^2}{1+(c^2/\xi)} = \xi$$

If the distance from the observation point to the ellipsoid increases, $\xi \gg 1$, the latter gives

$$r^2 = x^2 + y^2 + z^2 \to \xi \tag{3.123}$$

and, correspondingly the function $G_1(\xi)$ decreases inversely proportional to the square of the distance r. The same behavior holds for the potential of the system of magnetization currents when $r \to \infty$. This is the reason why this function (Equation (3.121)) will be used to describe the secondary field outside the ellipsoid. Let us notice that the integral

$$E_1 = \int_\xi^\infty \frac{d\xi}{(\xi + a^2) R_\xi} \tag{3.124}$$

is called the elliptic integral of second order, and this function has been investigated in detail.

3.6.4. Potential inside and outside an ellipsoid

As follows from a comparison of Equations (3.105) and (3.106), the surface of the ellipsoid with given semi-axes a, b, and c is characterized by the coordinate

$\xi = 0$. Assuming that the field inside the ellipsoid is uniform, we will represent the potential inside and outside as

$$U_1 = AF_1(\xi)F_2(\eta)F_3(\zeta), \quad \text{if } \xi < 0$$

and

$$U_2 = [CF_1(\xi) + DG_1(\xi)]F_2(\eta)F_3(\zeta), \quad \text{if } \xi > 0 \tag{3.125}$$

These functions satisfy Laplace's equation and conditions at infinity. Next, we will find coefficients A and D, which allow us to satisfy also conditions at the interface

$$\frac{U_1}{\mu_i} = \frac{U_2}{\mu_e} \quad \text{and} \quad \frac{1}{h_1}\frac{\partial U_1}{\partial \xi} = \frac{1}{h_1}\frac{\partial U_2}{\partial \xi}, \quad \text{if } \xi = 0 \tag{3.126}$$

Substitution of Equations (3.125) into the first equation of this set yields

$$\frac{1}{\mu_i} AF_1(\xi_0) = \frac{1}{\mu_e}[CF_1(\xi_0) + DG_1(\xi_0)]$$

Taking into account Equation (3.120), the latter becomes

$$A = \frac{\mu_i}{\mu_e}\left[C + D\int_0^\infty \frac{ds}{(s+a^2)R_s}\right] \tag{3.127}$$

The equality of normal derivatives gives

$$AF_1'(0) = [C + DE_1(0)]F_1'(0) - D\frac{1}{a^2abc}F_1(0) \tag{3.128}$$

since

$$F_1(0) = a \quad \text{and} \quad F_1'(0) = \frac{1}{2a}$$

Then Equation (3.128) gives

$$A = C + D\left[E_1(0) - \frac{2}{abc}\right] \tag{3.129}$$

From Equations (3.127) and (3.129), we have

$$\frac{\mu_i}{\mu_e}[C + DE_1(0)] = C + D\left[E_1(0) - \frac{2}{abc}\right]$$

hence

$$D = \frac{(\mu_i/\mu_e) - 1}{-(2/abc) + (1 - (\mu_i/\mu_e))E_1(0)} C = -\frac{abc}{2} \frac{(\mu_i/\mu_e) - 1}{1 + (abc/2)((\mu_i/\mu_e) - 1)E_1(0)} C \tag{3.130}$$

From Equation (3.127), we obtain

$$\begin{aligned} A &= \frac{\mu_1}{\mu_2}\left[1 - \frac{abc}{2}\left(\frac{\mu_i}{\mu_e} - 1\right)\frac{E_1(0)}{1 + (abc/2)((\mu_i/\mu_e) - 1)E_1(0)}\right] C \\ &= \frac{\mu_i}{\mu_e} \frac{1}{1 + (abc/2)((\mu_i/\mu_e) - 1)E_1(0)} C \end{aligned} \tag{3.131}$$

Thus, the field inside of the ellipsoid is uniform and directed along the primary field; that is, along the x-axis:

$$B_{1x} = \frac{\mu_i}{\mu_e} \frac{B_{0x}}{1 + (abc/2)((\mu_i/\mu_e) - 1)E_1(0)} \tag{3.132}$$

In the same manner, we find the field inside an ellipsoid when the primary field is directed along the two other axes

$$B_{1y} = \frac{\mu_i}{\mu_e} \frac{B_{0y}}{1 + (abc/2)((\mu_i/\mu_e) - 1)E_2(0)}$$

and

$$B_{1z} = \frac{\mu_i}{\mu_e} \frac{B_{0z}}{1 + (abc/2)((\mu_i/\mu_e) - 1)E_3(0)} \tag{3.133}$$

Here

$$E_2(0) = \int_0^\infty \frac{ds}{(s + b^2)R_s} \quad \text{and} \quad E_3(0) = \int_0^\infty \frac{ds}{(s + c^2)R_s}$$

We see that as long as the primary field is uniform, the total field inside is also uniform. Now assume that the primary field is uniform but arbitrarily oriented with respect to the ellipsoid. In this case, it can be represented as

$$\boldsymbol{B}_0 = \boldsymbol{B}_{0x}\boldsymbol{i} + B_{0y}\boldsymbol{j} + B_{0z}\boldsymbol{k}$$

Each component generates a uniform field inside the ellipsoid directed along a corresponding axis, and therefore the total field $\boldsymbol{B}_1$ remains also uniform. However, its direction may not coincide with the primary field $\boldsymbol{B}_0$, and this follows from the fact that the functions $E_1(0)$, $E_2(0)$, and $E_3(0)$ are different in

Equations (3.132) and (3.133). Suppose that in the past an ellipsoid was placed in the uniform magnetic field $\boldsymbol{B}_0$. Correspondingly, magnetization takes place and for a ferromagnetic material it remains even if the primary field is removed. Inasmuch as the field $\boldsymbol{B}_1$ has in general a different direction than $\boldsymbol{B}_0$, the vector of magnetization $\boldsymbol{P}$ ($\boldsymbol{P} = (\chi/\mu_i)\boldsymbol{B}_1$) is not oriented along the primary field. However, if in place of the ellipsoid we deal with a sphere, fields $\boldsymbol{B}_0$ and $\boldsymbol{B}_1$, as well as $\boldsymbol{P}$, have the same direction. This subject has some relation to paleomagnetism, which will be briefly discussed later. It may be proper to make one comment. We have demonstrated that the field inside of an ellipsoid is uniform, provided that the primary field is also constant. If $\boldsymbol{B}_0$ is an arbitrary function of a point, then the field inside becomes also nonuniform. At the same time, if a body differs from an ellipsoid, then it is natural to expect that the field inside and the magnetization vector may vary from point to point even when the primary field is uniform. In other words, we may not observe a uniform magnetization. It is obvious that Equations (3.125) and (3.130) allow us to find the magnetic field caused by magnetization currents outside of the ellipsoid.

3.7. SPHERICAL LAYER IN A UNIFORM MAGNETIC FIELD

During some time in the past there were attempts to explain the origin of the magnetic field of the earth by the magnetization of its upper part. To evaluate this effect, we will solve the following problem. Suppose that a spherical layer with magnetic permeability μ is surrounded by a nonmagnetic medium and its external and internal radii are equal to a and b, respectively (Fig. 3.4(b)). As before, the primary magnetic field $\boldsymbol{B}_0$ is uniform and in the spherical system of coordinates it is directed along the z-axis. Due to magnetization, currents arise at both interfaces, and the potential U of the resultant field is described by three functions:

$$\begin{aligned} &U_1, \quad \text{if } R > a \\ &U_2, \quad \text{if } b < R < a \\ &U_3, \quad \text{if } R < B \end{aligned}$$

They have to obey the following conditions of the boundary-value problem:

1. At regular points

$$\Delta U = 0$$

2. At interfaces

$$\frac{U_1}{\mu_0} = \frac{U_2}{\mu} \quad \text{and} \quad \frac{\partial U_1}{\partial R} = \frac{\partial U_2}{\partial R}, \quad \text{if } R = a$$

and

$$\frac{U_2}{\mu} = \frac{U_3}{\mu_0} \quad \text{and} \quad \frac{\partial U_2}{\partial R} = \frac{\partial U_3}{\partial R}, \quad \text{if } R = b$$

3. At infinity when $R \to \infty$, $U_1 \to U_0$.
   Here U_0 is the potential of the primary field and since

$$B_0 = -\frac{\partial U_0}{\partial z}$$

we have in terms of the spherical coordinates

$$U_0 = -B_0 z = B_0 R \cos\theta \tag{3.134}$$

As we know, if the function U satisfies all the conditions of the boundary-value problem, then it uniquely represents the magnetic field in this case. First, we will find a solution of the Laplace equation. Taking into account the fact that the field is independent of the coordinate φ, this equation has the form

$$\frac{\partial}{\partial R}\left(R^2 \frac{\partial U}{\partial R}\right) + \frac{1}{\sin\theta}\frac{\partial}{\partial\theta}\left(\sin\theta \frac{\partial U}{\partial\theta}\right) = 0 \tag{3.135}$$

Let us assume that the potential is a product of two functions

$$U(R,\theta) = T(R)S(\theta)$$

Its substitution into Equation (3.135) gives two ordinary differential equations of the second order

$$\frac{1}{R}\frac{d}{dR}\left(R^2 \frac{dT}{dR}\right) = n(n+1) \quad \text{and} \quad \frac{1}{S\sin\theta}\frac{d}{d\theta}\left(\sin\theta \frac{dS}{d\theta}\right) = -n(n+1)$$

Here n is integer number. The solution of the first equation is

$$T(R) = A_n R^n + B_n R^{-n-1}$$

As is known, the second equation is called the Legendre equation and, correspondingly, its solution is Legendre functions. From Equation (3.134), it follows that the potential of the primary field is represented as a product of solutions of both equations when $n = 1$. In order to satisfy the boundary conditions, we have to expect that both the primary and secondary potentials depend in the same manner from the azimuth θ. For this reason, we describe the potentials in each medium as

$$
\begin{aligned}
U_1(R,\theta) &= -B_0R\cos\theta + AR^{-2}B_0\cos\theta, \quad \text{if } R>a \\
U_2(R,\theta) &= CRB_0\cos\theta + DR^{-2}B_0\cos\theta, \quad \text{if } b<R<a \\
U_3(R,\theta) &= FB_0R\cos\theta, \quad \text{if } R<b
\end{aligned}
\tag{3.136}
$$

Here A, C, D, and F are unknown constants which we will determine from conditions at interfaces. Before this, let us make two comments: (a) While selecting solutions of Laplace equation, we took into account the fact that potential has to have finite values. (b) We wrote expressions for potential rather arbitrarily but the theorem of uniqueness allows us to prove that this choice is correct. The conditions at the interfaces give four equations with four unknowns and they are

$$\frac{1}{\mu_0}(-1 + Aa^{-3}) = \frac{1}{\mu}(C + Da^{-3})$$

$$-1 - 2Aa^{-3} = C - 2Da^{-3}$$

$$\frac{1}{\mu}(C + Db^{-3}) = \frac{1}{\mu_0}F, \quad C - 2Db^{-3} = F$$

Solving this system, we have

$$A = \frac{(\mu - \mu_0) - (\mu_0 + 2\mu)K(b^3/a^3)}{(\mu + 2\mu_0) + 2(\mu_0 - \mu)K(b^3/a^3)}a^3 \tag{3.137}$$

where

$$K = \frac{\mu - \mu_0}{2\mu + \mu_0}$$

Also we obtain the following relations, allowing us to derive everywhere expressions for the field:

$$C = -\frac{1 + 2Aa^{-3}}{1 - 2K(b^3/a^3)} \quad \text{and} \quad D = KCb^3, \quad F = C - 2Db^{-3}$$

Thus, we have found functions for the potential at each part of medium, which obey all conditions of the boundary-value problem, and in accordance with the theorem of uniqueness we have correctly determined the magnetic field. Consider several special cases and, first, suppose that $b = a$, that is, the spherical layer disappears. Then

$$A = \frac{(\mu - \mu_0) - (\mu_0 + 2\mu)((\mu - \mu_0)/(2\mu + \mu_0))}{(\mu + 2\mu_0) + 2(\mu_0 - \mu)K} a^3 = 0$$

Hence

$$C = \frac{1}{2K - 1} \quad \text{and} \quad F = -1$$

It is useful to study two more cases.

3.7.1. Spherical magnetic body in a uniform field

In this case $b = 0$ and as follows from Equation (3.137)

$$A = \frac{\mu - \mu_0}{\mu + 2\mu_0} a^3 \tag{3.138}$$

and, correspondingly

$$C = -\left(1 + 2\frac{\mu - \mu_0}{\mu + 2\mu_0}\right) = \frac{3\mu}{\mu + 2\mu_0} \quad \text{and} \quad D = 0 \tag{3.139}$$

Thus, expressions of the potential outside and inside the magnetic sphere are

$$U_1(R, \theta) = -B_0 R \cos\theta + \frac{\mu - \mu_0}{\mu + 2\mu_0} a^3 B_0 \frac{\cos\theta}{R^2}, \quad \text{if } R > a$$

and

$$U_2(R, \theta) = -\frac{3\mu}{2\mu + 2\mu_0} B_0 R \cos\theta, \quad \text{if } R < a \tag{3.140}$$

Taking into account the fact that

$$B_R = -\frac{\partial U}{\partial R} \quad \text{and} \quad B_\theta = -\frac{1}{R}\frac{\partial U}{\partial \theta}$$

we see that inside the sphere the field is uniform and directed, as is the primary one, along the z-axis:

$$B_i = \frac{3\mu}{\mu + 2\mu_0} B_0 \tag{3.141}$$

while outside the sphere the secondary magnetic field is equivalent to that of a magnetic dipole and in a spherical system of coordinates

$$B_R^s(R,\theta) = 2\frac{\mu-\mu_0}{\mu+2\mu_0}B_0\frac{a^3}{R^3}\cos\theta, \quad B_\theta^s(R,\theta) = \frac{\mu-\mu_0}{\mu+2\mu_0}\frac{a^3}{R^3}\sin\theta \tag{3.142}$$

The moment of this dipole is directed as the primary field along the z-axis. It is obvious that this field is caused by magnetization currents on the surface of the sphere which have only a φ-component and they reach a maximum at the plane $z=0$. At the same time, their density is equal to zero when $\theta=0$. Orientation of these currents and that of the magnetic moment

$$\boldsymbol{M} = \frac{4\pi}{\mu_0}\frac{\mu-\mu_0}{\mu+2\mu_0}a^3\boldsymbol{B}_0 \tag{3.143}$$

obeys the right-hand rule.

3.7.2. *Thin spherical shell in a uniform field*

Next we assume that a spherical layer is very thin, that is, its thickness h is much smaller than b ($b<a$). Then we have

$$\frac{b^3}{a^3} = \frac{(a-h)^3}{a^3} = \left(1-\frac{h}{a}\right)^3 \approx 1-3\frac{h}{a}$$

Substitution of the latter into Equation (3.137) gives

$$A \approx \frac{\mu-\mu_0}{(\mu+2\mu_0)+2(\mu_0-\mu)K}\frac{3h}{a}a^3 \tag{3.144}$$

We still observe a uniform field inside the shell when $R<b$, and the field of the magnetic dipole if $R>a$. For instance, making use of Equation (3.136), we have for the radial component of the magnetic field on the external surface of the shell

$$B_R^s(a,\theta) = 2AB_0\cos\theta \tag{3.145}$$

Let us consider one numerical example which illustrates the fact that magnetization of the upper part of the earth's crust is not able to create the magnetic field of the earth. Assume that

$$B_0 = 50{,}000\ \text{nT}, \quad \mu = 10\mu_0, \quad h = 20\ \text{km}, \quad a = 6300\ \text{km}$$

Then for the secondary field on the shell surface, we have

$$B_R^s \approx \frac{6 \cdot 20 \cdot 2}{6300} 50,000 \text{ nT} \approx 2000 \text{ nT}$$

Note that we consider a shell which has magnetic permeability in many thousands times exceeding that of paramagnetic, but the secondary field constitutes a very small portion of the earth's field. This result can be interpreted in a different manner. Suppose that in the past the earth was placed in an external field $\boldsymbol{B}_0$ which produced the magnetization of the upper part of the earth. Inasmuch as the radial component of the field is a continuous function at the interface, we can say that the vector $\boldsymbol{P}$ at points of the z-axis is equal to

$$\boldsymbol{P} = \chi H_R = \frac{\chi \boldsymbol{B}_R}{\mu_0} \tag{3.146}$$

Now we may treat this vector as a remanent magnetization and, as our calculations show, the intensity of the dipole moments creates only a very small portion of the earth field, even if we assumed an extremely high susceptibility for the upper crust. Returning back to the general case of the spherical layer, let us make two comments:

1. As follows from Equation (3.136), the field inside the layer changes and therefore the vector of magnetization is not constant. In other words, we are dealing with inhomogeneous magnetization.
2. The secondary magnetic field is caused by two systems of magnetization currents on the external and internal surfaces of the layer which have opposite directions. Inasmuch as these currents are located on surfaces with different radii, their magnetic field has a dipole character as in the case of a uniform magnetic sphere.

3.8. THE MAGNETIC FIELD DUE TO PERMANENT MAGNET

Our next subject is a study of the field caused by a magnetic material (magnet) which has a given magnetization $\boldsymbol{P}_r(q)$. With this purpose in mind, let us recall some physical aspects of the problem described in Chapter 2. Suppose that a ferromagnetic material is placed in the external magnetic field, which produces a magnetization and the latter remains even when this field is removed. Correspondingly, the magnetic dipoles and magnetization currents are distributed in an orderly fashion and they generate a magnetic field inside and outside the magnetic material. In order to determine the field, we will use either the vector or scalar potentials and, of course, both approaches give us the same result.

3.8.1. The vector potential

As follows from Biot–Savart law, the vector potential of this field can be represented in the form

$$A(p) = \frac{\mu_0}{4\pi}\int_V \frac{\boldsymbol{j}_{\rm m}(q)}{L_{qp}}\,dV + \frac{\mu_0}{4\pi}\int_S \frac{\boldsymbol{i}_{\rm m}(q)}{L_{qp}}\,dS \tag{3.147}$$

Here $\boldsymbol{j}_{\rm m}(q)$ and $\boldsymbol{i}_{\rm m}(q)$ are the vectors of the volume and surface current density in the elementary volume and elementary surface, which correctly represent a field of the elementary magnetic dipoles. It is clear that within such a volume we can imagine a small loop of a current, and its magnetic field is equivalent to that of a magnetic dipole. As was shown in Chapter 1, the vector potential of this elementary dipole is

$$d\boldsymbol{A} = \frac{\mu_0}{4\pi}\frac{\boldsymbol{P}_r(q)\times \boldsymbol{L}_{qp}}{L_{qp}^3}\,dV \tag{3.148}$$

Thus, the vector potential caused by all dipoles is equal to

$$\boldsymbol{A}(p) = \frac{\mu_0}{4\pi}\int_V \frac{\boldsymbol{P}_r(q)\times \boldsymbol{L}_{qp}}{L_{0P}^3}\,dV$$

Making use of a known relation of the vector analysis (Chapter 2), the last equation can be represented as

$$\boldsymbol{A}(p) = \frac{\mu_0}{4\pi}\int_V \frac{\text{curl}\,\boldsymbol{P}_r}{L_{qp}}\,dV + \frac{\mu_0}{4\pi}\oint_S \frac{\text{Curl}\,\boldsymbol{P}_r}{L_{qp}}\,dS \tag{3.149}$$

Here S is the boundary of the magnetic body with a nonmagnetic medium. Comparison of Equations (3.147) and (3.149) permits us to find the relations between the density of magnetization currents and the vector of magnetization, that is, the density of the dipole moments, and they are

$$\boldsymbol{j} = \text{curl}\,\boldsymbol{P}_r \quad \text{and} \quad \boldsymbol{i} = \text{Curl}\,\boldsymbol{P}_r = -\boldsymbol{n}\times\boldsymbol{P}_r \tag{3.150}$$

since magnetization is absent in the surrounding medium and the unit vector $\boldsymbol{n}$ is perpendicular to the boundary and directed outward. The importance of Equations (3.150) is difficult to overestimate. In fact, they define the density of magnetization currents and therefore we can determine the vector potential at each point inside and outside of the permanent magnet (Equation (3.147)). Then, from the definition of the function $\boldsymbol{A}(p)$:

$$\boldsymbol{B}(p) = \text{curl}\,\boldsymbol{A}(p) \tag{3.151}$$

the magnetic field can be calculated. Now we focus on the special case when the vector of magnetization does not have vortices inside the magnet:

$$\text{curl}\, \boldsymbol{P}_r = 0 \tag{3.152}$$

Of course, this condition implies that the vector function $\boldsymbol{P}_r$ may vary from point to point, that is, we may deal with some cases of a nonuniform magnetization. In accordance with Equation (3.152), the volume density of magnetization currents is zero and the magnetic field is caused by the surface currents, and Equation (3.147) gives

$$\boldsymbol{A}(p) = \frac{\mu_0}{4\pi} \oint_S \frac{\boldsymbol{i}}{L_{qp}} dS \tag{3.153}$$

or

$$\boldsymbol{A}(p) = -\frac{\mu_0}{4\pi} \oint_S \frac{\boldsymbol{n} \times \boldsymbol{P}_r}{L_{qp}} dS \tag{3.154}$$

or

$$\boldsymbol{A}(p) = \frac{\mu_0}{4\pi} \oint_S \frac{\boldsymbol{P}_r \times d\boldsymbol{S}(q)}{L_{qp}} \tag{3.155}$$

Here $d\boldsymbol{S} = \boldsymbol{n}\, dS$ and its orientation depends on a point q, and dS is an elementary surface of the magnetic medium. Thus, Equations (3.153)–(3.155) allow us to find the vector potential caused by surface currents at any point inside and outside of the permanent magnet. Applying Equation (3.151), we find the magnetic field. It may be proper to emphasize again that the field is caused by elementary magnetic dipoles, which are continuously distributed inside the magnet. At the same time, surface currents permit us to replace volume integration by the surface one. For illustration, consider one example.

3.8.2. The field outside of a thin cylinder

Suppose that the radius of the cylinder is much smaller than its extension, and the vector of magnetization $\boldsymbol{P}_r$ is constant and directed along the cylinder's axis. Taking into account the fact that an ellipsoid is not transformed into such a cylinder, we may think that only at the central part of this body there is a uniform magnetization but in the vicinity of the ends this assumption is hardly correct. Our goal is to find the magnetic field outside the magnet in a nonmagnetic medium. Let us choose a cylindrical system of coordinates with its z-axis directed along the body (Fig. 3.5(a)). In this case

$$\boldsymbol{P}_r = P_r \boldsymbol{i}_z \quad \text{and} \quad \boldsymbol{i}_r = \boldsymbol{i}_\varphi x \boldsymbol{i}_z \tag{3.156}$$

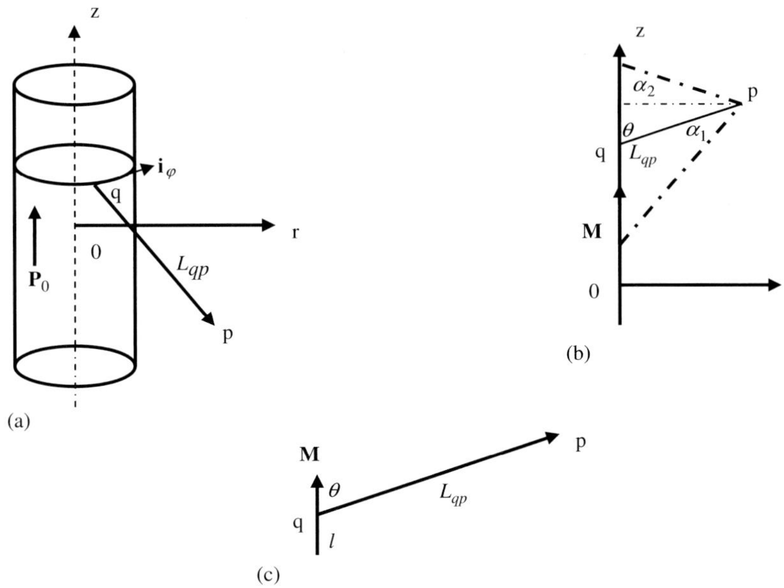

Fig. 3.5. (a) Field of cylindrical magnet. (b) A system of magnetic dipoles. (c) Illustration of Equation (3.159).

Here $\boldsymbol{i}_r$, $\boldsymbol{i}_\varphi$, and $\boldsymbol{i}_z$ are unit vectors of the cylindrical system of coordinates. As follows from Equation (3.150)

$$\boldsymbol{i} = -P_r \boldsymbol{i}_r \times \boldsymbol{i}_z = P_r \boldsymbol{i}_\varphi$$

Thus, surface magnetization currents have a φ-component only, and they form a system of circular currents with constant density located at the horizontal planes of the lateral surface of the body. At the same time they are absent at the top and bottom of the cylinder, since at such points the cross-product in the second equation of the set (3.150) is zero. It is obvious that that all currents generate only an azimuth component of the vector potential and its vector lines are closed circles with their centers at the z-axis. If a point of observation is placed close to the magnet, then the distance to each current element of the horizontal circuit varies and, as was shown in Chapter 1, the vector potential A_φ is expressed in terms of elliptical integrals. With an increase of the distance L_{qp}, these variations become smaller and we can treat each element of the magnet's volume as a magnetic dipole and mentally replace the distribution of currents by a continuous system of dipoles located at the z-axis (Fig. 3.5(b)). In accordance with Equation (3.148), the vector potential caused by such a dipole is

$$d\boldsymbol{A}(p) = \frac{\mu_0}{4\pi} \frac{\boldsymbol{M}_r \times \boldsymbol{L}_{qp}}{L_{qp}^3} dz \qquad (3.157)$$

Here $\boldsymbol{M}_r$ is the dipole moment per unit length of the magnet and it is equal to the product of the vector of magnetization and the area of the cross-section:

$$\boldsymbol{M}_r = \boldsymbol{P}_r S$$

In the cylindrical system of coordinates (Fig. 3.5(b)), we have

$$dA_\varphi(r,z) = \frac{\mu_0}{4\pi}\frac{M_r\,dz}{(z-z_q)^2+r^2}\sin\theta = \frac{\mu_0}{4\pi}\frac{M_r r\,dz}{[(z-z_q)^2+r^2]^{3/2}}$$

Here z_q and z are coordinates of the elementary dipole and an observation point, respectively. Performing integration along the dipole line, we obtain

$$A_\varphi(r,0) = \frac{\mu_0}{4\pi} r M_r \int_{z_1}^{z_2} \frac{dz_q}{[(z-z_q)^2+r^2]^{3/2}} \tag{3.158}$$

After a replacement of variable: $z - z_q = r\tan\alpha$ and taking into account the fact that

$$dz_q = -r\sec^2\alpha\,d\alpha$$

we have

$$A_\varphi(r,z) = -\frac{\mu_0 M_r}{4\pi}\frac{1}{r}\int_{\alpha_1}^{\alpha_2} \cos\alpha\,d\alpha$$

Thus

$$A_\varphi(r,z) = -\frac{\mu_0 M_r}{4\pi r}(\sin\alpha_2 - \sin\alpha_1)$$

or

$$A_\varphi = \frac{\mu_0 M_r}{4\pi}\frac{1}{r}\left[\frac{z-z_1}{[(z-z_1)^2+r^2]^{1/2}} - \frac{z-z_2}{[(z-z_2)^2+r^2]^{1/2}}\right] \tag{3.159}$$

and the limits of integration are shown in Fig. 3.5(b). Suppose that the observation point is located far away from the magnet and correspondingly its distance from any point q remains practically the same:

$$L_{qp} \approx (z^2+r^2)^{1/2}$$

Assuming that the point 0 coincides with the magnet middle, Equation (3.158) gives

$$A_\varphi(r,z) = \frac{\mu_0 M_r (z_2 - z_1) r}{4\pi L_{qp}^3}$$

Thus, we obtain an approximate expression for the vector potential

$$\boldsymbol{A}_\varphi(r,z) = \frac{\mu_0 M_r l \sin\theta}{4\pi}\frac{}{L_{qp}^2}\boldsymbol{i}_\varphi = \frac{\mu_0 \boldsymbol{M} x \boldsymbol{L}_{qp}}{4\pi L_{qp}^3}$$

which coincides with Equation (3.148), where $\boldsymbol{M} = \boldsymbol{M}_r l$ is the dipole moment of the magnet. As is well known (Chapter 1), the vector potential is equal to zero at the z-axis. However, Equation (3.159) contains r in the denominator, and this fact may cause some confusion. In order to prove that $A_\varphi(0, z) = 0$, we have to expend the radicals in Equation (3.159) in a series. It gives

$$A_\varphi(r,z) = \frac{\mu_0 M_r}{4\pi} r\left[1 - \frac{1}{2}\frac{r^2}{(z-z_1)^2} - 1 + \frac{1}{2}\frac{r^2}{(z-z_2)^2}\right]$$

that is, near the z-axis this function tends to zero and disappears when $r=0$. By definition, the components of the magnetic field are defined as

$$\boldsymbol{B} = \frac{1}{r}\begin{vmatrix} \boldsymbol{i}_r & \boldsymbol{i}_\varphi & \boldsymbol{i}_z \\ \dfrac{\partial}{\partial r} & \dfrac{\partial}{\partial \varphi} & \dfrac{\partial}{\partial z} \\ 0 & rA_\varphi & 0 \end{vmatrix} \tag{3.160}$$

and we have

$$B_r(r,z) = -\frac{1}{r}\frac{\partial rA_\varphi}{\partial z}, \quad B_\varphi = 0, \quad B_z(r,z) = \frac{1}{r}\frac{\partial rA_\varphi}{\partial r}$$

Performing a differentiation of Equation (3.159), we obtain

$$B_r(r,z) = -\frac{\mu_0 M_r}{4\pi r}\left[\frac{1}{L_{q_1p}} - \frac{1}{L_{q_2p}} - \frac{(z-z_1)^2}{L_{q_1p}^3} + \frac{(z-z_2)^2}{L_{q_2p}^3}\right]$$

or

$$B_r(r,z) = \frac{\mu_0 M_r r}{4\pi}\left[\frac{1}{L_{q_2p}^3} - \frac{1}{L_{q_1p}^3}\right]$$

and

$$B_z(r,z) = \frac{\mu_0 M_r}{4\pi}\left[\frac{z-z_2}{L_{q_2p}^3} - \frac{z-z_1}{L_{q_1p}^3}\right] \tag{3.161}$$

Here

$$L_{q_1p} = [(z-z_1)^2 + r^2]^{1/2}, \quad L_{q_2p} = [(z-z_2)^2 + r^2]^{1/2}$$

For instance, at the z-axis, we have

$$B_r = 0$$

and

$$B_z = \frac{\mu_0 M_r}{4\pi}\left[\frac{1}{(z-z_2)^2} - \frac{1}{(z-z_1)^2}\right] = \frac{\mu_0 M_r}{4\pi}\frac{2z(z_2-z_1)+z_1^2-z_2^2}{(z-z_1)^2(z-z_2)^2}$$

If a point of observation is located far away from the magnet, we obtain the known expression for the field of the magnetic dipole at its axis:

$$B_r = 0 \quad \text{and} \quad B_z(0,z) = \frac{\mu_0 M}{2\pi z^3}$$

Finally, suppose that the origin of coordinates is located at the middle plane, $z_1 = -z_2$ and $z = 0$. Then we have

$$B_r = 0 \quad \text{and} \quad B_z = -\frac{\mu_0 M}{4\pi L^3_{q_1 p}}$$

Now let us make one comment about the magnetic field inside the magnet. By definition, we have:

$$\boldsymbol{B} = \mu \boldsymbol{H} + \mu_0 \boldsymbol{P}_r$$

Inasmuch as the sources of the field $\boldsymbol{H}$ are fictitious charges, situated at magnet ends, this field decreases with an increase of the distance from them and, correspondingly, at the central part of an elongated magnet we have an almost uniform magnetic field:

$$\boldsymbol{B} = \mu_0 \boldsymbol{P}_r$$

3.8.3. Scalar potential

Now we derive Equations (3.161) in a different way and demonstrate that the same magnetic field can be determined with the help of the scalar potential. Let us again start from the concept of the magnetic dipole of an elementary volume. As was shown in Chapter 1, its scalar potential is equal to

$$dU(p) = \frac{\mu_0}{4\pi}\frac{\boldsymbol{P}_r \cdot \boldsymbol{L}_{qp}}{L^3_{qp}}\,dV$$

By analogy with the previous case, the scalar potential caused by a thin cylindrical magnet with a constant magnetization is

$$U(p) = \frac{\mu_0 M_r}{4\pi} \int_{z_1}^{z_2} \frac{z - z_q}{[(z - z_q)^2 + r^2]^{3/2}} dz_q$$

Performing the same replacement of variables as before, we have

$$U(p) = -\frac{\mu_0 M_r}{4\pi r} \int_{\alpha_{11}}^{\alpha_2} \sin\varphi \, d\varphi = \frac{\mu_0 M_r}{4\pi r}(\cos\alpha_2 - \cos\alpha_1)$$

or

$$U(p) = \frac{\mu_0 M_r}{4\pi} \left\{ \frac{1}{[(z - z_2)^2 + r^2]^{1/2}} - \frac{1}{[(z - z_1)^2 + r^2]^{1/2}} \right\} \tag{3.162}$$

As in the case of the scalar potential of the electric field, let us assume that at each end of the thin magnet there is either a positive or a negative magnetic charge:

$$m = m_+ = M_0 = P_r S \quad \text{and} \quad m_- = -P_r S \tag{3.163}$$

that is, the fictitious charge is equal to product of the normal component of vector of magnetization and the area of a cross-section of the magnet. If the vector $\boldsymbol{P}_r$ is directed out of the magnet we imply the presence of a positive charge, while at the opposite end the negative charge "appears". Sometimes these places are called poles of the magnet. It is important to emphasize again that charges do not exist in reality. Correspondingly, Equation (3.162) can be rewritten as

$$U(p) = \frac{\mu_0}{4\pi} m \left(\frac{1}{L_{q_2 p}} - \frac{1}{L_{q_1 p}} \right) \tag{3.164}$$

For instance, the potential due to the positive elementary charge is

$$U(p) = \frac{\mu_0}{4\pi} \frac{m}{L_{q_2 p}} \tag{3.165}$$

By definition, $\boldsymbol{B} = -\text{grad}\, U$, that is

$$B_r = -\frac{\partial U}{\partial r} \quad \text{and} \quad B_z = -\frac{\partial U}{\partial z}$$

Performing differentiation of Equation (3.162), we obtain

$$B_r = \frac{\mu_0 m}{4\pi} r \left[\frac{1}{L_{q_2 p}^3} - \frac{1}{L_{q_1 p}^3} \right], \quad B_z = \frac{\mu_0 m}{4\pi} \left[\frac{z - z_2}{L_{q_2 p}^3} - \frac{z - z_1}{L_{q_1 p}^3} \right]$$

that of course coincides with Equations (3.161). For instance, if we deal with a semi-infinite thin magnet, then

$$B_r = \frac{\mu_0 m}{4\pi}\frac{r}{L_{qp}^3} \quad \text{and} \quad B_z = \frac{\mu_0 m}{4\pi}\frac{z - z_q}{L_{qp}^3} \tag{3.166}$$

By analogy with the gravitational and electric fields, it is a great temptation to think that the magnetic field, given by these formulas, is caused by a point magnetic charge located at the point q, but in reality this field is generated by magnetic dipoles distributed uniformly inside the magnet. Applying the principle of superposition, it is very convenient to use Equations (3.166) in order to find the magnetic field caused by a single magnet or a system of them. Now we consider a behavior of a magnet in the presence of an external magnetic field.

3.9. THE MAGNET IN A UNIFORM MAGNETIC FIELD

As follows from Ampere's and Biot–Savart laws, elements of linear as well as surface and volume currents, placed in a magnetic field $\boldsymbol{B}$, are subjected to the action of a force which is

$$\boldsymbol{F} = I\, d\boldsymbol{l} \times \boldsymbol{B}, \quad \boldsymbol{F} = (\boldsymbol{i} \times \boldsymbol{B})dS, \quad \boldsymbol{F} = (\boldsymbol{j} \times \boldsymbol{B})dV \tag{3.167}$$

Proceeding from these equations, we investigate several cases.

3.9.1. Force acting on a free charge

First, consider an elementary volume dV where an ordered motion of charges takes place.

By definition, the current density $\boldsymbol{j}$ of charges with the same sign can be represented as

$$\boldsymbol{j} = \delta \boldsymbol{W} = en\boldsymbol{W}$$

Here e is a charge of an electron and n a number of them in a unit volume, but δ and $\boldsymbol{W}$ are the volume density and velocity of the ordered motion of these charges, respectively. Therefore, the force of the magnetic field $\boldsymbol{B}$ acting on all electrons in the volume dV is

$$\boldsymbol{F} = ne(\boldsymbol{W} \times \boldsymbol{B})dV$$

and, correspondingly, every electron is subjected to a force equal to

$$\boldsymbol{F} = e\boldsymbol{W} \times \boldsymbol{B} \tag{3.168}$$

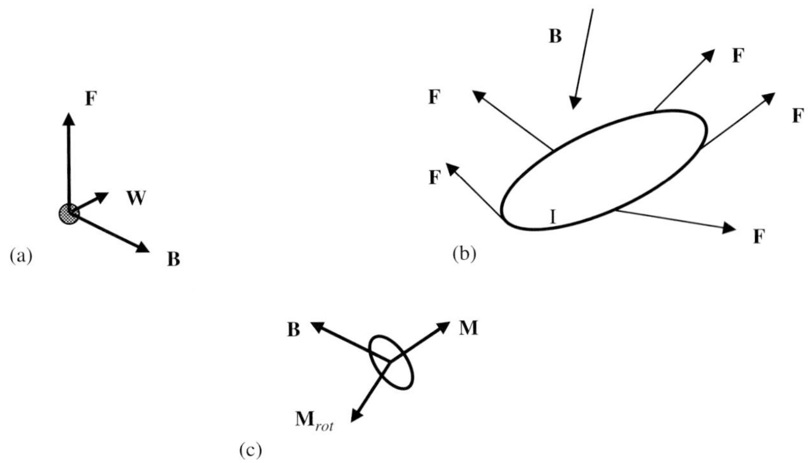

Fig. 3.6. (a) Magnetic force. (b) Forces acting on the closed circuit with current I. (c) Moment of rotation of a small loop.

This force is called the magnetic force, and its orientation is shown in Fig. 3.6(a) ($e<0$).

3.9.2. Linear current circuit in a uniform magnetic field B

Let us take an arbitrary linear circuit with current I, placed in the magnetic field $\boldsymbol{B}$, as is shown in Fig. 3.6(b). In accordance with Equation (3.167), the mechanical force acting on the contour L placed in the magnetic field $\boldsymbol{B}$ can be represented as

$$\boldsymbol{F} = \oint_L d\boldsymbol{F} = I \oint_L d\boldsymbol{l}(q) \times \boldsymbol{B}(q) \tag{3.169}$$

Here $d\boldsymbol{l}(q)$ is the circuit element. The magnetic field $\boldsymbol{B}(q)$ represents the superposition of fields caused by all conduction and magnetization currents except the current I. In accordance with the third Newton's law, its magnetic field does not influence on the integral in Equation (3.169). The resultant force $\boldsymbol{F}$ is a sum (integral) of forces applied at different points of the circuit. Assuming that the latter is rigid, this force can cause only a translation and rotation. As is well known from classical mechanics, an action of any system of forces can be replaced by the resultant force $\boldsymbol{F}$ applied at some point and a moment of rotation, $\boldsymbol{M}_{\rm rot}$.

3.9.3. Resultant force

Bearing in mind the fact that the field $\boldsymbol{B}$ is uniform in the vicinity of the circuit L, Equation (3.169) is simplified and we obtain

$$\boldsymbol{F} = -I \oint_L \boldsymbol{B} \times d\boldsymbol{l} = -I\boldsymbol{B} \times \oint_L d\boldsymbol{l}$$

It may be proper to emphasize here that we assume uniformity of the field only at points of the current circuit, but it may change at other places. Inasmuch as the integral is a sum of vectors $d\boldsymbol{l}$, which form a closed polygon L, we have

$$\oint_L d\boldsymbol{l} = 0$$

Thus, the total force $\boldsymbol{F}$ acting on the current contour equals zero:

$$\boldsymbol{F} = 0 \tag{3.170}$$

if the field $\boldsymbol{B}$ is the same at all points of the contour. This means that the translation is absent, and a circuit can be only involved in rotation. Such a result is not difficult to predict for a circuit of a simple shape, such as a square or circle, where always there are two elementary currents having opposite directions. Forces acting on them are equal by a magnitude but have opposite directions and correspondingly the resultant force vanishes. For a current circuit of an arbitrary shape, this result is not so obvious. Also it is proper to notice that in reality the field cannot be absolutely uniform and, correspondingly, resultant force is not equal to zero and it depends on the rate of change of the magnetic field in the vicinity of the current circuit.

3.9.4. Moment of rotation

Next we will find an expression for the moment $\boldsymbol{M}_{\text{rot}}$ which produces the rotation of the current circuit. With respect to any arbitrary point 0, the resultant moment is

$$\boldsymbol{M}_{\text{rot}} = \oint_L \boldsymbol{L}_{oq} \times d\boldsymbol{F} = I \oint_L \boldsymbol{L}_{oq} \times (d\boldsymbol{l} \times \boldsymbol{B}) \tag{3.171}$$

Here $\boldsymbol{L}_{oq} \times d\boldsymbol{F}(q)$ is the moment of force acting on the current element $I\,d\boldsymbol{l}$. To simplify Equation (3.171), we make use of an identity:

$$\boldsymbol{a} \times (\boldsymbol{b} \times \boldsymbol{c}) = (\boldsymbol{a} \cdot \boldsymbol{c})\boldsymbol{b} - \boldsymbol{c}(\boldsymbol{a} \cdot \boldsymbol{b})$$

This gives

$$\boldsymbol{M}_{\text{rot}} = I \oint_L (\boldsymbol{L}_{oq} \cdot \boldsymbol{B}) d\boldsymbol{l} - I \oint_L \boldsymbol{B}(\boldsymbol{L}_{oq} \cdot d\boldsymbol{l}) \tag{3.172}$$

The latter does not contain cross products and this certainly simplifies calculation of the moment even in general case when the field $\boldsymbol{B}$ may change from point to point. In the case of a uniform magnetic field in the vicinity of the contour L, the second integral in Equation (3.172) can be represented as

$$\oint_L \boldsymbol{B}(\boldsymbol{L}_{oq} \cdot d\boldsymbol{l}) = \boldsymbol{B} \oint_L \boldsymbol{L}_{oq} \cdot d\boldsymbol{l}$$

Applying Stokes' theorem, we have

$$\oint_L \boldsymbol{L}_{oq} \cdot d\boldsymbol{l} = \int_S \overset{q}{\text{curl}}\ \boldsymbol{L}_{oq} \cdot d\boldsymbol{S}$$

where S is the area surrounded by the contour L. Performing the calculation of $\overset{q}{\text{curl}}\ \boldsymbol{L}_{oq}$, for instance, in Cartesian system of coordinates, we find that

$$\overset{q}{\text{curl}}\ \boldsymbol{L}_{oq} = 0$$

and, therefore, instead of Equation (3.172) we have

$$\boldsymbol{M}_{\text{rot}} = I \oint_L (\boldsymbol{L}_{oq} \cdot \boldsymbol{B}) d\boldsymbol{l} \tag{3.173}$$

In order to simplify this equation, we will make use of the known equality from the vector analysis

$$\oint_L T\, d\boldsymbol{l} = \int_S d\boldsymbol{S} \times \overset{q}{\nabla} T$$

Letting $T = \boldsymbol{L}_{0q} \cdot \boldsymbol{B}$, we obtain

$$\boldsymbol{M}_{\text{rot}} = I \int_S d\boldsymbol{S} \times \overset{q}{\nabla}(\boldsymbol{L}_{0q} \cdot \boldsymbol{B})$$

Taking into account the fact that the magnetic field is uniform and it is a function of point p, the integrand is greatly simplified. By definition

$$\boldsymbol{L}_{0q} \cdot \boldsymbol{B} = x_q B_x + y_q B_y + z_q B_z$$

and performing a differentiation, we obtain

$$\overset{q}{\nabla}(\boldsymbol{L}_{0q} \cdot \boldsymbol{B}) = B_x\boldsymbol{i} + B_y\boldsymbol{j} + B_z\boldsymbol{k} = \boldsymbol{B}$$

This gives

$$\boldsymbol{M}_{\text{rot}} = I \int_S d\boldsymbol{S} \times \boldsymbol{B} = -I\boldsymbol{B} \times \int_S d\boldsymbol{S}$$

and we have arrived at a very elegant expression for the moment of rotation which is much simpler than Equation (3.171)

$$\boldsymbol{M}_{\text{rot}} = \boldsymbol{M} \times \boldsymbol{B} \tag{3.174}$$

Thus, the moment of rotation $\boldsymbol{M}_{\text{rot}}$ is equal to the cross-product of the magnetic moment of the current circuit $\boldsymbol{M}$ and the magnetic field $\boldsymbol{B}$, where:

$$\boldsymbol{M} = I \int_S d\boldsymbol{S} = I \int_S \boldsymbol{n}\, dS \quad \text{and} \quad M_{\text{rot}} = MB \sin(\boldsymbol{M}, \boldsymbol{B}) \tag{3.175}$$

that is, the magnitude reaches a maximum when the field $\boldsymbol{B}$ is perpendicular to the moment $\boldsymbol{M}$, and it is equal to zero if they are parallel to each other. The mutual position of vectors $\boldsymbol{B}$, $\boldsymbol{M}$, and $\boldsymbol{M}_{\text{rot}}$ is shown in Fig. 3.6(c). The simplicity of Equation (3.174) vividly illustrates that the use of the vector analysis is well justified and it demonstrates one fundamental result, namely, the moment of rotation is independent of the choice of the point 0, that is, the distance L_{oq} does not influence the moment of rotation. If the current contour is located in the plane, then the moment $\boldsymbol{M}$ is directed perpendicular to this plane and we have

$$\boldsymbol{M}_{\text{rot}} = IS\boldsymbol{n} \times \boldsymbol{B} \tag{3.176}$$

where S is the area surrounded by the current circuit. At the same time, the moment of rotation $\boldsymbol{M}_{\text{rot}}$ is parallel to the current plane. It may be proper to point out that the vector $\boldsymbol{M}_{\text{rot}}$ represents a superposition of elementary moments due to magnetic forces acting at each element $I\,\boldsymbol{dl}$ of the current circuit. For illustration, suppose that a small current loop is suspended and its position is shown in Fig. 3.7(a). It is clear that the moment of rotation is located in the plane of the current loop and its direction defines the axis around which the loop moves until the vector $\boldsymbol{M}$ becomes parallel to $\boldsymbol{B}$.

3.9.5. Thin and elongated magnet in a uniform magnetic field

Now we assume that a thin horizontal bar (compass) is installed on the vertical axis and the magnetic field of the earth is uniform at its vicinity (Fig. 3.7(b)). Taking

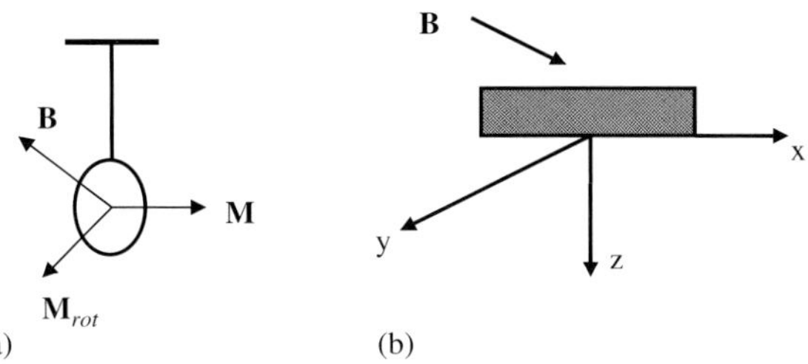

Fig. 3.7. (a) Current loop. (b) Rotation of a horizontal magnet.

into account this fact, the force acting on each elementary current of the magnet is zero and, therefore, the resultant force applied to the body is also zero. At the same time the moment of rotation for each elementary current is equal to

$$d\boldsymbol{M}_{\text{rot}} = d\boldsymbol{M} \times \boldsymbol{B} \quad \text{and} \quad d\boldsymbol{M} = \boldsymbol{P}_r S\, dz \tag{3.177}$$

Here S is the cross-section of the magnet. Thus, each element of the magnet is subjected to an action of the same moment; that is, they have the same magnitude and direction and their superposition gives

$$\boldsymbol{M}_{\text{rot}} = \boldsymbol{M} \times \boldsymbol{B} \quad \text{and} \quad \boldsymbol{M} = \boldsymbol{P}_r Sl \tag{3.178}$$

Here $\boldsymbol{M}$ is the total magnetic moment of the magnet. Assuming again that there are magnetic charges at its ends and taking into account Equation (3.163), we can write

$$\boldsymbol{M}_{\text{rot}} = (\boldsymbol{l} \times \boldsymbol{F}) \tag{3.179}$$

Here

$$\boldsymbol{F} = m\boldsymbol{B}, \quad \boldsymbol{M} = m\boldsymbol{l}, \quad m = P_r S \tag{3.180}$$

and $\boldsymbol{l}$ is the vector directed along the vector of magnetization and its magnitude is equal to the magnet's length. Equation (3.179) allows us to visualize the action of the rotation moment as if forces $\boldsymbol{F}$ and $-\boldsymbol{F}$ were applied to positive and negative fictitious charges located at the distance $l/2$ from the axis of rotation, and they produce a movement of the magnet around this axis. It is important to emphasize that in reality there are no charges and there are no such forces, but the torque produces exactly the same effect as that of the moment of rotation, $\boldsymbol{M}_{\text{rot}}$ (Equation (3.179)). Also, it is much simpler to think by analogy with the electric field that there is a force equal to the product of the magnetic charge m and the field $\boldsymbol{B}$. Since both forces produce equal torques, the resultant torque is defined by Equation (3.179). In other words, such concepts as magnetic charge, forces applied to them and torque are used in order to simplify calculations, but the physical foundation of magnet rotation is an interaction of the magnetic field with the dipole moments of the magnet. In order to understand better the effect produced by this moment of rotation, let us introduce a Cartesian system of coordinates x, y, z (Fig. 3.7(b)), and assume that the field $\boldsymbol{B}$ is located in the plane $X0Z$, that is, $B_y = 0$. Then, as follows from Equation (3.178)

$$\boldsymbol{M}_{\text{rot}} = \begin{vmatrix} \boldsymbol{i} & \boldsymbol{j} & \boldsymbol{k} \\ M_x & M_y & 0 \\ B_x & 0 & B_z \end{vmatrix} \tag{3.181}$$

Taking into account the fact that a compass may rotate only around the z-axis, we have

$$M^z_{\text{rot}} = -M_y B_x = -ml \sin \varphi B_x \tag{3.182}$$

Here φ is the angle between vertical planes where the compass and magnetic field of the earth are located.

Because of this moment, the compass rotates around the z-axis and the angle becomes smaller. As soon as the magnet and the field $\boldsymbol{B}$ are located at the same plane, the former stops and this is a purpose of the compass. Besides, Equation (3.182) demonstrates that such system can be used for measuring the horizontal component of the magnetic field.

3.10. INTERACTION BETWEEN TWO MAGNETS

Earlier we found out that concept of magnetic charges is very convenient to calculate the magnetic field due to magnetization currents, located on the lateral surface of the magnet. These charges "exist" at the front and back sides of the magnet and their strength is defined by the vector of magnetization and the cross-section of the magnet. In particular, the field caused by one charge is equal to

$$\boldsymbol{B}(p) = \frac{\mu_0}{4\pi} \frac{m}{L^3_{qp}} \boldsymbol{L}_{qp} \quad \text{and} \quad m_+ = m = P_r S, \quad m_- = -P_r S \tag{3.183}$$

Also, we have shown that the charges allow us to determine the moment of rotation of the magnet, placed in a uniform magnetic field. Next, we demonstrate that they are also useful to describe the force of interaction between two thin magnets which are arbitrarily oriented with respect to each other. At the beginning, consider the following case.

3.10.1. Two magnets are placed along the same line (Fig. 3.8)

Inasmuch as we deal with very thin magnets, it is natural to assume that within the cross-section of each of them the magnetic field is uniform. However, this means

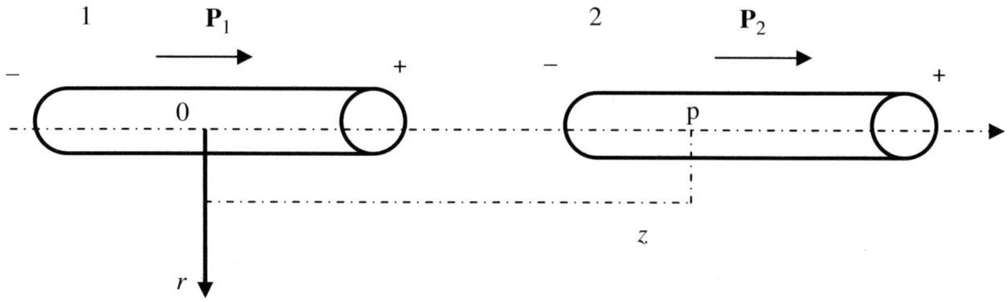

Fig. 3.8. Interaction of two magnets placed along the z-axis.

that the force acting at every elementary cross-section is equal to zero and therefore there is no interaction between magnets. Certainly, this contradicts experiments and, as was pointed out earlier, indicates that we have to take into account the change of the field within each cross-section regardless of how small it is. Now we investigate this question in some detail. First, consider the circular elementary current of the second magnet, located at the distance z from the origin of coordinates (Fig. 3.8). As follows from Ampere's law, the force acting at the vicinity of its element $i_2\,d\boldsymbol{l}_2$ is

$$d\boldsymbol{F}(p) = i_2\,d\boldsymbol{l}_2 \times \boldsymbol{B}_1(p)dz$$

Here $\boldsymbol{B}_1(p)$ is the field caused by the first magnet and $i_2\,d\boldsymbol{l}_2 = \boldsymbol{i}_\varphi r\,d\varphi$, where $\boldsymbol{i}_\varphi$ is the surface density of the current. Correspondingly, we have

$$d\boldsymbol{F}(p) = r\,d\varphi(\boldsymbol{i}_\varphi \times \boldsymbol{B}_1)dz \tag{3.184}$$

Because of symmetry, the magnetic field is independent on coordinate φ and it has only two components: $\boldsymbol{B}_{1r}$ and $\boldsymbol{B}_{1z}$:

$$\boldsymbol{B}_1 = B_{1r}\boldsymbol{i}_r + B_{1z}\boldsymbol{i}_z$$

We focus on the z-component of the force, since it characterizes attraction or repulsion of magnets. In accordance with Equation (3.184), we have

$$d\boldsymbol{F}_z(p) = r\,d\varphi(\boldsymbol{i}_\varphi \times \boldsymbol{B}_{1r}) = -i_\varphi r\,d\varphi B_{1r}(p)dz\boldsymbol{i}_z$$

Correspondingly, the force acting on an elementary ring is

$$dF_z(r,z) = -i_\varphi r\,dz \oint_L B_{1r}(r,\varphi,z)d\varphi \tag{3.185}$$

In our case, the magnitude of the radial component of the field is the same at all points of the current circuit but the direction varies, regardless of how small the current loop is. Moreover, there are always two elements of the circuit with opposite direction of the current, where the radial component of the magnetic field has also opposite directions. For this reason, these forces acting on such elements do not cancel each other, but they have equal magnitudes and the same direction. As we already know, the magnetic field caused by magnetic dipoles of the first magnet can be represented as a sum of fields generated by positive and negative charges. At the beginning, consider the effect due to the field caused by a positive magnetic charge of the first magnet. Its radial component is

$$B_{1r} = -\frac{\mu_0 m_1}{4\pi}\frac{\partial}{\partial r}\frac{1}{[(z-(l_1/2))^2+r^2]^{1/2}}$$

Substitution of this equation into Equation (3.185) and integration with respect to φ gives

$$dF_z(r,z) = \frac{\mu_0 m_1 i_\varphi r\, dz}{2}\frac{\partial}{\partial r}\frac{1}{[(z-(l_1/2))^2+r^2]^{1/2}}$$

or

$$dF_z(r,z) = -\frac{\mu_0 m_1 m_2\, dz}{2\pi}\frac{1}{[(z-(l_1/2))^2+r^2]^{3/2}} \tag{3.186}$$

Here m_2 is the product of the vector of magnetization of the second magnet and its cross-section, S_2. Finally, the total component, F_z, "caused" by both charges of the first magnet is

$$dF_z(r,z) = -\frac{\mu_0 m_1 m_2\, dz}{2\pi}\left\{\frac{1}{[(z-(l_1/2))^2+r^2]^{3/2}} - \frac{1}{[(z+(l_1/2))^2+r^2]^{3/2}}\right\} \tag{3.187}$$

For illustration, we consider the case when $(z_1-(l_1/2)) \gg r$. Then

$$dF_z(z) = -\frac{\mu_0 m_1 m_2}{2\pi}dz\left[\frac{1}{(z-(l_1/2))^3} - \frac{1}{(z+(l_1/2))^3}\right]$$

Performing integration we find that the resultant force acting on the second magnet, that is the force of interaction between both magnets, is

$$F_z = -\frac{\mu_0 m_1 m_2}{4\pi}\left[\frac{1}{(z_1-(l_1/2))^2} - \frac{1}{(z_2-(l_1/2))^2} + \frac{1}{(z_2+(l_1/2))^2} - \frac{1}{(z_1+(l_1/2))^2}\right] \tag{3.188}$$

Here z_1 and z_2 are coordinates faces of the second magnet: $z_2 = z_1 + l_2$, and l_2 its length.

Suppose that the length of both magnets is much greater than the distance between the front of the first magnet and the back of the second one. In this case, we have

$$F_z = -\frac{\mu_0 m_1 m_2}{4\pi(z_1 - (l_1/2))^2} \tag{3.189}$$

The latter represents Coulomb's law for two elementary fictitious magnetic charges, and it shows how an unreal force, applied to one end of the magnet, characterizes a superposition of real forces acting between magnetic dipoles inside of two semi-infinite magnets. Inasmuch as the vector of magnetization in both magnets has the same direction, we observe attraction; on the contrary, if they are opposite to each other, repulsion takes place. It is obvious that in our case the moment of rotation is equal to zero and there is only the force of interaction (Equation (3.188)). Let us make two comments: (1) We arrived at Coulomb's law from Biot–Savart's law. (2) In the case of a very thin magnet, fictitious charges are located at its opposite faces, but for real magnets poles are situated somewhere inside at some small distances from these faces.

3.10.2. *Current circuit in the magnetic field B*

Now we consider a more general case and suppose that in the vicinity of the current circuit the field $\boldsymbol{B}$ does not have axial symmetry as in the previous case. To simplify derivations, we assume that the circuit has a rectangular shape and the origin of a Cartesian system of coordinates coincides with its center (Fig. 3.9). The z-axis is normal to the circuit, and its direction and the current obey the right-hand rule. First, we focus our attention on the normal component of the magnetic force acting on the circuit and assume that its change within a circuit is so small that it can be described by a linear function. Since the vertical component of the field $\boldsymbol{B}$ does not influence the same component of the force, we consider only its tangential component:

$$\boldsymbol{B}_t = B_x\boldsymbol{i} + B_y\boldsymbol{j}$$

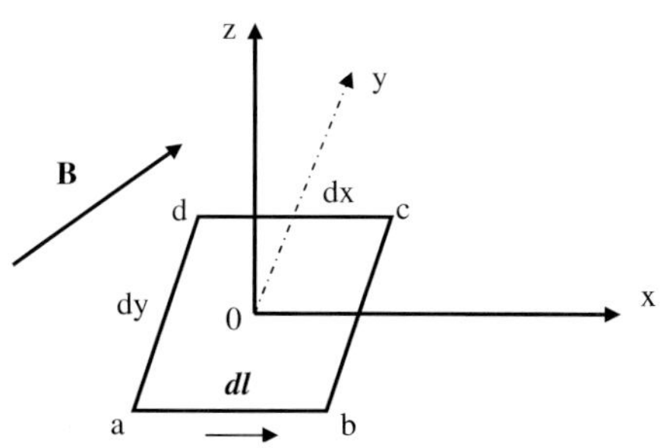

Fig. 3.9. Current circuit in a nonuniform magnetic field.

Expanding this in power series and discarding all terms except the first two ones, we have for components at points of coordinate axes

$$B_x(x,0) = B_x(0) + \frac{\partial B_x}{\partial x}dx, \quad B_y(0,y) = B_y(0) + \frac{\partial B_y}{\partial y}dy$$

By definition, the z-components of the forces acting on the element ab and cd are

$$\boldsymbol{F}_z = I(\boldsymbol{i}\times\boldsymbol{j})\left(B_y - \frac{\partial B_y}{\partial y}\frac{dy}{2}\right)dx \quad \text{and} \quad \boldsymbol{F}_z = -I(\boldsymbol{i}\times\boldsymbol{j})\left(B_y + \frac{\partial B_y}{\partial y}\frac{dy}{2}\right)dx$$

By analogy, the normal components of forces applied to elements bc and da are

$$\boldsymbol{F}_z = I(\boldsymbol{j}\times\boldsymbol{i})\left(B_x + \frac{\partial B_x}{\partial x}\frac{dx}{2}\right)dy \quad \text{and} \quad \boldsymbol{F}_z = -I(\boldsymbol{j}\times\boldsymbol{i})\left(B_x - \frac{\partial B_x}{\partial x}\frac{dx}{2}\right)dy$$

Thus, the resultant force acting along the z-axis is

$$F_z = -\left(\frac{\partial B_y}{\partial y} + \frac{\partial B_x}{\partial x}\right)IS$$

Here $S = dx\,dy$ is the area surrounded by the circuit. Taking into account the fact that $\operatorname{div}\boldsymbol{B} = 0$, we have

$$F_z = IS\frac{\partial B_z}{\partial z} \tag{3.190}$$

Next, we find horizontal components of the force and start from F_x which acts on the sides bc and da. As is seen from Fig. 3.9

$$\boldsymbol{F}_x = I(\boldsymbol{j}\times\boldsymbol{k})\left[B_z(0) + \frac{\partial B_z}{\partial x}\frac{dx}{2}\right]dy \quad \text{and} \quad \boldsymbol{F}_x = -I(\boldsymbol{j}\times\boldsymbol{k})\left[B_z(0) - \frac{\partial B_z}{\partial x}\frac{dx}{2}\right]dy$$

Therefore, the x-component of the force is

$$F_x = IS\frac{\partial B_z}{\partial z} \tag{3.191}$$

By analogy, for forces directed along the y-axis and acting on sides: ab and cd, we have

$$\boldsymbol{F}_y = I(\boldsymbol{i}\times\boldsymbol{k})\left[B_z(0) - \frac{\partial B_z}{\partial y}\frac{dy}{2}\right]dx, \quad \boldsymbol{F}_y = -I(\boldsymbol{i}\times\boldsymbol{k})\left[B_z(0) + \frac{\partial B_z}{\partial y}\frac{dy}{2}\right]dx$$

Their sum gives

$$F_y = IS\frac{\partial B_z}{dy} \tag{3.192}$$

Thus, we have expressed the force acting on the current circuit in terms of derivatives of the component B_z:

$$\boldsymbol{F} = IS\left(\frac{\partial B_z}{\partial x}\boldsymbol{i} + \frac{\partial B_z}{\partial y}\boldsymbol{j} + \frac{\partial B_z}{\partial z}\boldsymbol{k}\right) \tag{3.193}$$

or

$$\boldsymbol{F} = IS \text{ grad } B_z \tag{3.194}$$

Under the action of this force, the current circuit experiences a translation. At the same time, the moment of rotation is still defined by

$$\boldsymbol{M}_r = \boldsymbol{M} x \boldsymbol{B}$$

because for a current loop with a small radius, the contribution of the field change is negligible and $\boldsymbol{M} = IS\boldsymbol{n}$.

3.10.3. Magnet in a field of a point magnetic charge

Suppose that a thin magnet with vector magnetization $\boldsymbol{P}_2$ is directed parallel to the z-axis and it is placed into a magnetic field, "caused" by a point charge m_1 (Fig. 3.10):

$$\boldsymbol{B}(p) = \frac{\mu_0 m_1}{4\pi}\frac{\boldsymbol{L}_{qp}}{L_{qp}^3}$$

or

$$B_x(p) = \frac{\mu_0 m_1}{4\pi}\frac{x_p - x_q}{L_{qp}^3}, \quad B_y(p) = \frac{\mu_0 m_1}{4\pi}\frac{y_p - y_q}{L_{qp}^3}, \quad B_z(p) = \frac{\mu_0 m_1}{4\pi}\frac{z_p - z_q}{L_{qp}^3}$$

and

$$L_{qp} = [(x_p - x_q)^2 + (y_p - y_q)^2 + (z_p - z_q)^2]^{1/2} \tag{3.195}$$

3.10.4. Magnetic force

Let us imagine the surface currents of the magnet as a system of current rings with thickness dz. Each of them is subjected to a force and, in accordance with

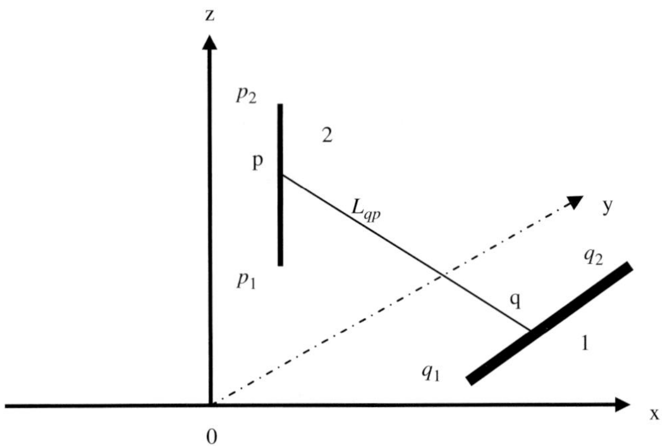

Fig. 3.10. Magnet in the field of a point charge.

Equation (3.194), this force is equal to

$$d\boldsymbol{F}(z) = m_2\, dz \text{ grad } B_z$$

Performing integration, we obtain for the resultant force acting on the whole magnet

$$\boldsymbol{F}(z) = m_2 \int_{z_1}^{z_2} \text{grad } B_z\, dz \tag{3.196}$$

First, we will find the vertical component of the resultant force which is equal to

$$F_z = m_2 \int_{z_1}^{z_2} \frac{\partial B_z}{\partial z}\, dz = m_2[B_z(p_2) - B_z(p_1)]$$

or

$$F_z = \frac{\mu_0 m_2 m_1}{4\pi}\left[\frac{z_2 - z_q}{L_{qz_2}^3} - \frac{z_1 - z_q}{L_{qz_1}^3}\right] \tag{3.197}$$

Here $p_2(x_2, y_2, z_2)$ and $p_1(x_1, y_1, z_1)$ are points where positive and negative charges of the magnet are located, respectively. For the x-component, we have

$$F_x = \frac{\mu_0 m_1 m_2}{4\pi}\frac{\partial}{\partial x}\int_{z_1}^{z_2} \frac{z_p - z_q}{L_{qp}^3}\, dz = \frac{\mu_0 m_1 m_2}{4\pi}\frac{\partial}{\partial x}\left[\frac{1}{L_{qz_1}} - \frac{1}{L_{qz_2}}\right]$$

or

$$F_x = \frac{\mu_0 m_1 m_2}{4\pi}\left[\frac{x_2 - x_q}{L^3_{qz_2}} - \frac{x_1 - x_q}{L^3_{qz_1}}\right]$$

or

$$F_x = m_2[B_x(p_2) - B_x(p_1)] \tag{3.198}$$

By analogy

$$F_y = m_2[B_y(p_2) - B_y(p_1)] \tag{3.199}$$

Combining Equations (3.197)–(3.199), we obtain

$$\boldsymbol{F} = m_2[\boldsymbol{B}(p_2) - \boldsymbol{B}(p_1)] \tag{3.200}$$

and this describes a superposition of forces acting on elementary currents, that is, forces applied at different points of the magnet. In other words, $\boldsymbol{F}$ is the resultant force acting on the magnet and is caused by a single magnetic charge located at some point q. We can imagine that this force is applied at any point, for instance, at the center of mass of this body and causes its translation. It is essential that in order to calculate this force we can use Coulomb's law for the magnetic charges: one of them is "located" at the point q, while others are mentally placed at end of the magnet (Fig. 3.10). In other words, we have determined the force of interaction between the single charge of the first magnet and the second magnet. Applying the principle of superposition, it is a simple matter to take into account the presence of the charge at the opposite end of the first magnet, and then the force of interaction of both magnets is

$$\boldsymbol{F} = m_2\{[\boldsymbol{B}(p_2, q_2) - \boldsymbol{B}(p_1, q_2)] - [\boldsymbol{B}(p_2, q_1) - \boldsymbol{B}(p_1, q_1)]\} \tag{3.201}$$

Here q_1 and q_2 are terminals points of the first magnet, where the positive and negative charges are placed, respectively. For instance, $\boldsymbol{B}(q_1, q_2)$ is the magnetic field caused by the positive charge of the first magnet at the point of the second magnet where the negative charge is "located". Note that the first magnet with terminal points q_1 and q_2 can be oriented arbitrarily with respect to the coordinate axes (Fig. 3.10).

3.10.5. Moment of rotation

By definition, the moment of rotation of an elementary current ring of the magnet with the thickness dz is

$$d\boldsymbol{M}_r = (\boldsymbol{M} \times \boldsymbol{B})dz$$

Here $\boldsymbol{M}$ is the magnetic moment directed along the z-axis and it is equal to

$$\boldsymbol{M} = i_2 S\boldsymbol{k} = P_2 S\boldsymbol{k} = m_2\boldsymbol{k}$$

and $\boldsymbol{B}$ is a uniform magnetic field in the vicinity of the current loop. In reality, this field changes from point to point of the circuit, but with a decrease in its dimensions these variations are negligible, and, as before, we can suppose that the field $\boldsymbol{B}$ is uniform at each cross-section of the magnet. Performing integration between the terminal points of the magnet, we obtain an expression for the rotation moment:

$$\boldsymbol{M}_r = \boldsymbol{M} \times \int_{z_1}^{z_2} \boldsymbol{B}(x, y, z)dz = m_2\boldsymbol{k} \times \int_{z_1}^{z_2} \boldsymbol{B}(x, y, z)dz$$

or

$$\boldsymbol{M}_r = M_x\boldsymbol{i} + M_y\boldsymbol{j}$$

Here

$$M_x = -m_2 \int_{z_1}^{z_2} B_y \, dz \quad \text{and} \quad M_y = m_2 \int_{z_1}^{z_2} B_x \, dz \tag{3.202}$$

3.11. ENERGY OF MAGNETIC DIPOLE IN THE PRESENCE OF THE MAGNETIC FIELD

Earlier we studied forces caused by the external magnetic field $\boldsymbol{B}$ and acting on the small current loop (magnetic dipole). Now we find the magnetic energy of this dipole and with this purpose in mind let us represent the dipole as a pair of two magnetic charges of equal magnitude but opposite sign, located very close to each other (Fig. 3.11). In order to bring the magnetic charge m_+ at the point q_2, it is necessary to perform the work equal to

$$\int_{\infty}^{q_2} m_+\boldsymbol{B} \, d\boldsymbol{l} = -m_+ \int_{\infty}^{q_2} \text{grad } U \, d\boldsymbol{l}$$

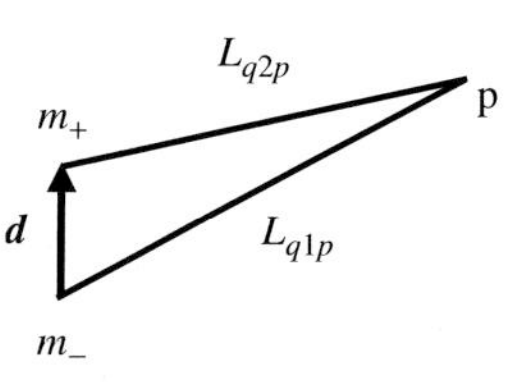

Fig. 3.11. Magnetic dipole as system of two magnetic charges.

Inasmuch as the integral is path independent, we have

$$m_+ \int_\infty^{q_2} \frac{\partial U}{\partial l} dl = m_+ U(q_2)$$

By analogy, the similar work for the negative charge is

$$m_- \int_\infty^{q_1} \frac{\partial U}{\partial l} dl = m_- U(q_1)$$

Thus, the total work which represents the magnetic energy of the dipole is

$$E = m[U(q_2) - U(q_1)]$$

since $m = m_+ = -m_-$. By definition of gradient, the latter can be written as

$$E = m\boldsymbol{d} \cdot \nabla U$$

Here $\boldsymbol{d}$ is the vector directed toward the positive charge and its magnitude is equal to the distance between magnetic charges. Thus, we finally obtain an expression for the magnetic energy of the magnetic dipole in the presence of the field $\boldsymbol{B}$:

$$E = -\boldsymbol{M} \cdot \boldsymbol{B} = -MB \cos \theta \tag{3.203}$$

Here $\boldsymbol{M} = m\boldsymbol{d}$ is the magnetic moment of the dipole. Note that this energy reaches a minimum when the external magnetic field is parallel to the magnetic moment and later (Chapter 6) we will use Equation (3.203) in studying magnetic properties of different substances.

3.12. PERMANENT MAGNET AND MEASUREMENTS OF THE MAGNETIC FIELD

During several centuries a permanent magnet was the main part of devices measuring the inclination and declination magnetic field of the earth, but first measurements of the field magnitude, based on a study of a period of oscillations of the magnet, were started at the end of 18th century. Imagine that the horizontal magnet is suspended by a vertical thread (Fig. 3.12(a)), and it is located in the plane of the magnetic meridian. It is clear that an action of the vertical component of the field does not cause a motion and the moment of rotation due to the horizontal component B_h

$$\boldsymbol{M}_r = \boldsymbol{M} \times \boldsymbol{B}_h \quad \text{and} \quad M_r = MB_h \sin \theta \tag{3.204}$$

is equal to zero, since $\theta = 0$. Here $\boldsymbol{M}$ is the magnetic moment of the magnet. In terms of forces acting on magnetic charges, we may say that these forces are directed along

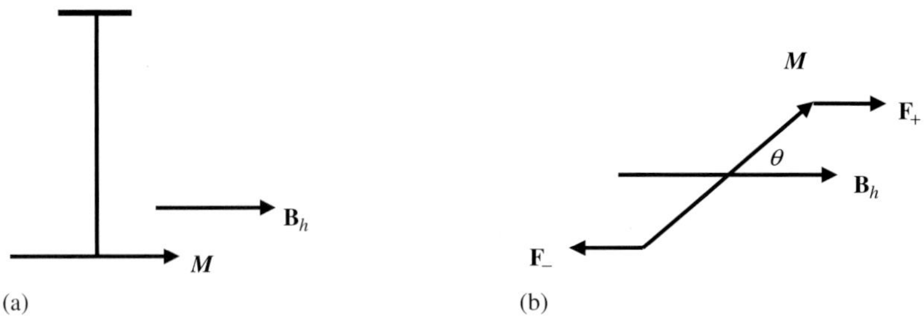

Fig. 3.12. (a) Magnet suspended by thread. (b) Movement of magnet toward the magnetic meridian.

the magnet in the opposite directions, and it is at rest. As soon as the magnet is moved from the equilibrium, the moment of rotation arises and increases with the angle θ (Fig. 3.12(b)). It is essential that regardless of the sign of the angle θ, the moment of rotation tends to return the magnet back to the plane of the magnetic meridian, $\theta = 0$, that is, this is a position of stable equilibrium. Suppose we placed the magnet at some position: $\theta = \theta_0$. Then under an action of the moment of rotation, it starts to move with the acceleration and the angle θ decreases. When the magnet is in plane of the magnetic meridian, the moment of rotation is equal to zero, but by inertia the body continues to move and the angle θ becomes negative. Inasmuch as the moment of rotation increases and tends to return a magnet at the equilibrium, the angular velocity decreases and the magnet stops. At this moment, the torque of forces acting on magnetic charges reaches maximum and the magnet begins to move back. Thus, we observe a periodic motion. Of course, there is always a friction and finally the mechanical energy will be transformed into heat and magnet stops. However, if this factor is very small, it is possible to observe periodic motion during time which is much greater than the period of oscillations. Bearing in mind that a deviation from an equilibrium is very small ($\theta \ll 1$), we may also neglect a contribution of the moment of rotation caused a torsion of thread. In such case, in accordance with the second Newton's law, an equation of motion of the magnet is

$$I\ddot{\theta} = -MB_{\mathrm{h}} \sin \theta \tag{3.205}$$

Here I is moment of inertia of the magnet and $\ddot{\theta}$ its angular acceleration. The minus sign indicates that the moment of rotation tends to decrease the angle θ. Taking into account that this angle is usually very small and $\sin\theta \approx \theta$, in place of Equation (3.205), we can write

$$\frac{d^2\theta}{dt^2} + \omega^2\theta = 0 \tag{3.206}$$

This is an equation of the harmonic oscillations and

$$\theta = \theta_0 \sin(\omega t + \phi)$$

where

$$\omega = \left(\frac{MB_{\mathrm{h}}}{I}\right)^{1/2} \tag{3.207}$$

Correspondingly, the period T of oscillations is equal to

$$T = 2\pi\left(\frac{I}{MB_{\mathrm{h}}}\right)^{1/2} \tag{3.208}$$

It is obvious that with increase of the moment inertia, the period increases too. At the same time, it becomes smaller with an increase of the moment of magnetization and the external field. In other words, the returning torque becomes bigger. Note that the magnet can be installed on the pivot, and if it is horizontal and perpendicular to the magnetic meridian, the magnet is at rest when both vectors: $\mathbf{M}$ and $\mathbf{B}$ have the same direction, and the period of oscillations depends on the total magnetic field $\mathbf{B}$.

3.12.1. Deflection method of measurements

At the beginning this method of measuring the period oscillations allowed one to study only relative changes of the horizontal component of the field B_{h} at the different points of the earth, since both parameters of the magnet I and M were unknown. Measurements of this component became possible when Gauss introduced the so-called deflection approach which was used during many years. Before we describe it, let us notice the following. The magnet in this device does not usually have a simple shape and it is related with the presence of attached lens or mirror to observe a motion. Correspondingly, the moment of inertia cannot be calculated even when its mass and size are given but this is determined experimentally. With this purpose in mind, the additional bar of very simple shape with the known value I_1 is connected with the magnet, so that the total moment of inertia becomes $I+I_1$. In accordance with Equation (3.208), the period of oscillations of the system: magnet and nonmagnet bar is

$$T_1 = 2\pi\left(\frac{I+I_1}{MH_{\mathrm{h}}}\right)^{1/2} \tag{3.209}$$

From measurements of periods with the bar and without it, we find that

$$\frac{T_1^2}{T^2} = \frac{I+I_1}{I}$$

Since I_1 is known, it is possible to evaluate the moment of inertia of the magnet I but we still need to take into account an influence of the magnetic moment M which

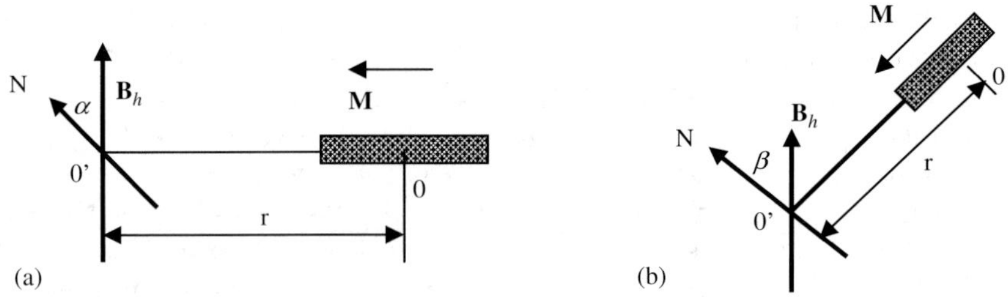

Fig. 3.13. (a) Gauss's method. (b) Lamont's method.

is unknown. As was pointed out, to overcome this problem, Gauss suggested deflection experiment, which is shown in Fig. 3.13(a).

After determination of the period T of free oscillations of the magnet and its moment of inertia I, consider an action of this magnet on the motion of a small magnetic needle located at the same plane. The needle is subjected to action of two moments of rotation. One of them is caused by the magnetic field of the earth and it is equal to

$$\boldsymbol{M}_r^{(1)} = \boldsymbol{M}_{\mathrm{N}} \times \boldsymbol{B}_{\mathrm{h}}$$

Here $\boldsymbol{M}_{\mathrm{N}}$ is the magnetic moment of the needle. The second moment of rotation is due to the field of the magnet and we have

$$\boldsymbol{M}_r^{(2)} = \boldsymbol{M}_{\mathrm{N}} \times \boldsymbol{B}_{\mathrm{M}}$$

where $\boldsymbol{B}_{\mathrm{M}}$ is the horizontal component of the magnetic field at the point $0'$ caused by the magnet. Unlike the first one, this moment of rotation tries to move the needle away from equilibrium that is, it produces a deflection of the needle from the position of rest. Correspondingly, the needle stops at some angle α when magnitudes of moments of rotation are equal. This gives

$$B_{\mathrm{h}} \sin \alpha = B_{\mathrm{M}} \cos \alpha \quad \text{or} \quad \tan \alpha = \frac{B_{\mathrm{M}}}{B_{\mathrm{h}}} \tag{3.210}$$

If the distance r is sufficiently large, the field of the magnet is practically equal to that of the magnetic dipole:

$$B_{\mathrm{M}} = \frac{\mu_0 M}{2\pi r^3} \tag{3.211}$$

and in place of Equation (3.210) we obtain

$$\tan \alpha = \frac{\mu_0 M}{2\pi r^3 B_{\mathrm{h}}} \tag{3.212}$$

Since we measure the period of oscillations T and the angle α, as well as the moment of inertia I and the distance r, Equations (3.208) and (3.212) allow us to calculate the horizontal component of the magnetic field of the earth B_h. Also, of course, this system gives us a value of the magnetic moment of the magnet. Let us make two comments:

1. If the distance r is not sufficiently large, the field of the magnet differs from that of the magnetic dipole, and then the correction coefficient is introduced into Equation (3.211), which is defined from measurements at different positions of the magnet.
2. As was already pointed out, the deflection method suggested by Gauss allowed one to measure the absolute value of the field and later Lamont improved an accuracy of measurements with different orientation of the needle and magnet (Fig. 3.13(b)).

3.12.2. Theory of the vertical magnetometer

As next example of an application of the magnet, we consider the Schmidt type of the magnetic balance for measuring variations of the vertical component of the magnetic field. During very long time, this instrument was used widely in applied geophysics and now we describe some of its features (Fig. 3.14). Its main part consists of two magnets balanced on a horizontal knife-edge which are perpendicular to their magnetic axis. These bars connected together carry a mirror and some weight so that the center of mass of the moving system is displaced horizontally and vertically with respect to the axis of rotation (knife-edge). This is an essential feature of the device. The magnet is oriented in an east–west direction so that the horizontal component of the magnetic field does not produce a torque regardless of the position of the magnet.

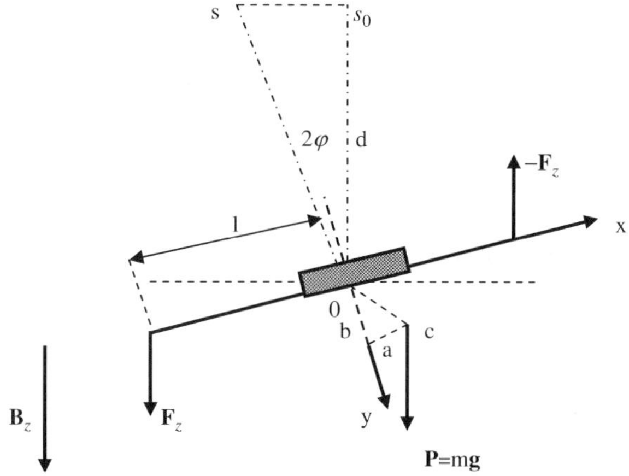

Fig. 3.14. Principle of the vertical magnetometer.

The moving system is subjected to an action of two moments of rotation. One of them is caused by the vertical component of the field, $\boldsymbol{B}_z$, and it is equal to

$$\boldsymbol{M}_r^{(1)} = \boldsymbol{M} \times \boldsymbol{B}_z \tag{3.213}$$

This moment tends to rotate the system counterclockwise. Here $\boldsymbol{M}$ is the magnetic moment of the magnet. For illustration, we have shown fictitious forces $\pm\boldsymbol{F}_z$ applied to magnetic charges which produce the same effect. The second moment of rotation is due to the gravitational force and it is applied at the center of mass (point c):

$$\boldsymbol{M}_r^{(2)} = \boldsymbol{r} \times \boldsymbol{P} \tag{3.214}$$

Here $\boldsymbol{r}$ is the radius-vector of the point c and $\boldsymbol{P}$ the weight: $\boldsymbol{P} = m\boldsymbol{g}$. The position of mass attached to the magnet is chosen in such a way that the second torque causes a rotation clockwise. As is seen from Fig. 3.14

$$M_r^{(1)} = MB_z \cos\varphi \tag{3.215}$$

and with an increase of the angle, this moment becomes smaller. Also we have

$$\boldsymbol{r} = a\boldsymbol{i} + b\boldsymbol{j}, \quad \boldsymbol{P} = mg(-\boldsymbol{i}\sin\varphi + \boldsymbol{j}\cos\varphi)$$

Thus, the second torque is

$$\boldsymbol{M}_r^{(2)} = mg\begin{vmatrix} \boldsymbol{i} & \boldsymbol{j} & \boldsymbol{k} \\ a & b & 0 \\ -\sin\varphi & \cos\varphi & 0 \end{vmatrix}$$

and its magnitude is

$$M_r^{(2)} = mg(a\cos\varphi + b\sin\varphi) \tag{3.216}$$

It is clear that the magnetic system stops when both rotation moments are equal by magnitude and it gives

$$MB_z\cos\varphi = mg(a\cos\varphi + b\sin\varphi)$$

or

$$\tan\varphi = \frac{MB_z - mga}{mgb} \tag{3.217}$$

Therefore, the latter establishes the relationship between the vertical component of the field B_z, and the angle φ which can be measured. Suppose that when the magnet is horizontal the scale value is s_0, but at equilibrium it is s (Fig. 3.14). Then

$$s - s_0 = d \tan 2\varphi$$

Since the angle is usually very small, we can write

$$\tan 2\varphi = 2 \tan \varphi$$

and it gives

$$\tan \varphi = \frac{s - s_0}{2d} \tag{3.218}$$

Its substitution into Equation (3.217) yields a relation between the field and the scale:

$$\frac{s - s_0}{2d} = \frac{MB_z - mga}{mgb} \tag{3.219}$$

As was mentioned earlier, Schmidt magnetometer was mainly applied in exploration geophysics for relative measurements. For illustration, consider a measurement at different point where we can write

$$\frac{s_1 - s_0}{2d} = \frac{MB_{1z} - mga}{mgb} \tag{3.220}$$

From the last two equations, we obtain

$$s - s_1 = \frac{2dM}{mgb}(B_z - B_{1z}) \quad \text{or} \quad B_z - B_{1z} = K(s - s_1) \tag{3.221}$$

where

$$K = \frac{mgb}{2dM} \tag{3.222}$$

is the scale constant of the instrument. Its value is determined by the scale reading produced by the known magnetic field caused by either the conduction current or magnet. As experience has shown, relative measurements could be performed with the precision of about 1γ.

Chapter 4

Main Magnetic Field of the Earth

4.1. ELEMENTS OF THE MAGNETIC FIELD OF THE EARTH

For several centuries it has been known that the magnetic field is present inside and outside of the earth as well as on its surface. This fact is a result of direct observations with a thin magnetic needle, called a compass. Let us imagine that such a needle is suspended and that it can freely rotate around its center of mass. More than thousand years ago people discovered that at any point of the earth's surface this needle tends to take a certain position around some axis of rotation but does not experience a noticeable displacement. Such a behavior indicates that there is a magnetic field, and, as was shown in Chapter 3, this field is almost uniform in the vicinity of observation point but it varies on the earth's surface. It may be proper to notice that while the magnetic field is almost directly observed, the presence of the gravity required a genius guess by Newton.

Inasmuch as the magnetic field is vector field, it is characterized by its magnitude and direction or its components along the coordinate axes. A study of the magnetic field of the earth can be done in different systems of coordinates.

For instance, in a Cartesian system we have

$$\boldsymbol{B} = B_x \boldsymbol{i} + B_y \boldsymbol{j} + B_z \boldsymbol{k}$$

where the x-axis is oriented along the geographical meridian and the direction to north is positive, the y-axis along the parallel with positive direction toward east and the z-axis is directed downward. The observation point 0 is the origin of coordinates. The vector $\boldsymbol{B}$ occupies some position with respect to the coordinate axes, and the following notations are sometimes also used:

$$X = B_x, \quad Y = B_y, \quad Z = B_z \tag{4.1}$$

These are called the north, east and vertical components, respectively. The projection of the magnetic field on the horizontal plane, H, is called the horizontal component of the field $\boldsymbol{B}$. Of course, it does not have any relation to the fictitious field $\boldsymbol{H}$. It is obvious that

$$H = (B_x^2 + B_y^2)^{1/2} \quad \text{and} \quad B = (B_z^2 + H^2)^{1/2}$$

Also, the vertical plane, where the vector $\boldsymbol{B}$ is located, is called the plane of magnetic meridian. Now we will describe the field in a system of coordinates with the same origin which only slightly differs from a spherical one. With this purpose in mind, we introduce two angles. The angle D between the plane of the magnetic meridian and coordinate plane $X0Z$ is called the declination, while the angle I between the horizontal plane and the vector $\boldsymbol{B}$ is called the inclination. It is clear that three parameters: the field magnitude B, the declination D and inclination I, define at each point the magnetic field in the same way as field components in the Cartesian system of coordinates. As seen from Fig. 4.1, they are related by

$$I = \tan^{-1}\frac{B_z}{(B_x^2 + B_y^2)^{1/2}} \quad \text{and} \quad D = \sin^{-1}\frac{B_y}{(B_x^2 + B_y^2)^{1/2}} \tag{4.2}$$

Also we have

$$B_x = H \cos D, \quad B_y = H \sin D, \quad B_z = H \tan I$$

and

$$\tan D = \frac{B_y}{B_x} \tag{4.3}$$

In this Cartesian system of coordinates (Fig. 4.1), the declination D is positive when the vector $\boldsymbol{B}$ is turned from north to east, and it is negative if it is turned in direction of west. Also we can see that in the northern hemisphere the inclination I is positive, since the field $\boldsymbol{B}$ is directed downward with respect to the earth's surface, and it is negative in the south hemisphere because the field is directed upward.

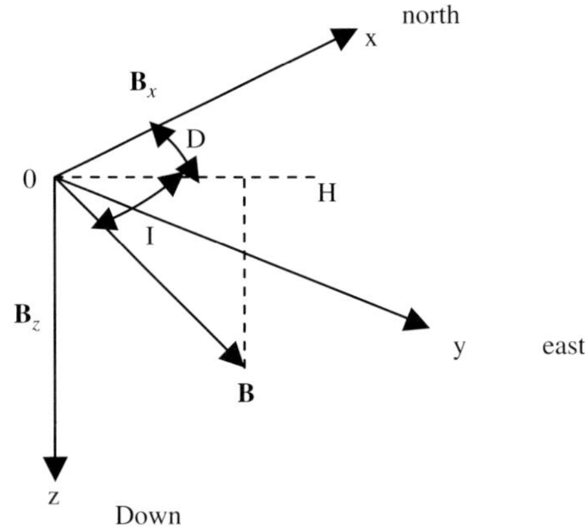

Fig. 4.1. The elements of magnetic field of the earth.

4.2. HISTORY OF THE EARTH MAGNETISM STUDY

4.2.1. *The discovery of the magnetic compass*

It is impossible to overestimate the importance of the invention of the magnetic compass, which explains a great interest in the origin of this amazing device. A thousand years ago people already knew two minerals, amber and magnetic iron ore, which possess remarkable properties. One of them, amber, when robed attracts light bodies; the other, magnetic iron ore, has the ability to attract or repeal other iron. Chinese mythology indicates that the directional properties of iron were known several thousand years ago and were even used for military operations. Also it is known that around 1000 years ago one Chinese inventor took a bowl with water and placed a lodestone on a small platform (boat) which can freely move on the surface of the water. However, due to the iron this tiny ship always rotated to face south. Certainly, this was one of the first magnetic compasses. There are claims that this invention was made even much earlier, though at this time the Chinese hardly knew how to use this device for navigation. Also, they noticed the effect of induced magnetization, and as was pointed out by one scholar "fortunate tellers rub the point of a needle with the stone of a magnet in order to make it properly indicate south". There is a strong indication that the Chinese, despite knowing directional properties of the magnet, did not use it for purposes of navigation until the end of the 13th century. Certainly, the ancient Greeks knew about the remarkable features of iron ore, and it is possible that the name "magnet" is related to the fact that lodestone was found near city of Magnesia in Asia Minor (Turkey).

For a long time it has been an opinion that this discovery in China was brought by Arabs to the Mediterranean and was used the Crusaders. At the same time, there are indications that the compass was independently invented in north-western Europe, probably in England, earlier than elsewhere. For instance, in the year 1186, the monk Alexander Neckham mentioned the compass as if it is already a well-known device.

4.2.2. *Pierre de Maricourt (Petrus Peregrinus)*

It is natural that the remarkable properties of the magnet attracted attention of scholars, and in 1269 Petri Peregrini de Maricourt (native of Picardy) in his "*Epistola Petri Pertegrini de Maricourt*" described the results of experiments, which can be treated as beginning of the study of the earth's magnetism. First of all, he used a magnet, perhaps a lodestone of spherical shape like the earth. He laid a needle at some point of the magnet and marked its orientation. Then, the needle was placed at a neighboring point and the same procedure was performed, allowing him to trace a line along which the needle was directed. In the same manner, measurements were performed on the whole surface and a system of lines was obtained. They covered the surface of magnet exactly in the same manner as meridians on the earth's surface; in particular there were two points at opposite ends of the stone where all lines merge, and by analogy with the north and south poles Peregrinus suggested calling these points the poles of the magnet. As we know, this terminology became conventional. In essence, Peregrinus plotted the vector lines of the magnetic field caused by a spherical magnet. Besides, he was perhaps the first who studied the interaction between permanent magnets and demonstrated that the attraction and repulsion are dependent on the mutual position of poles of different magnets. Without any doubt his experiments were an important contribution to the study of geomagnetism, even though at that time and much later this phenomenon was not understood and it remained a mystery. For instance, for several centuries the directional property of the compass was explained by the action of the stars which exerted a special magical force and in accordance with some reports steersman on British ships were forbidden to eat garlic because of the belief that its smell can destroy the magnetic power of a compass.

4.2.3. Magnetic compass and navigation

Nevertheless, the magnetic compass found a broad application in navigation, and it is difficult to imagine the great sea voyages of De Gamba, Columbus, and Magellan without the use of this device. These and other sea travels allowed one to learn much more about the behavior of the magnetic needle. First of all, it was found that there are different types of iron. For instance, wrought iron, which contains less than 0.3% of carbon, is "magnetically soft", and its magnetization disappears when lodestone is removed, while high carbon steel preserves its magnetization. At that time the traditional way to make a compass was the following. First, a flat steel needle was balanced horizontally on a pivot. Then it was rubbed gently by a lodestone. Very soon it was discovered that after the needle become magnetic, its north-pointing end was inclined down as if it gained weight. This phenomenon was noticed by George Hartmann in 1544 and investigated by British compass maker Robert Norman. He demonstrated that to the north of equator a needle end is always slanted downwards into earth, and this angle is now called the dip or inclination. Moreover, he suggested a dip circle allowing the measurement of the inclination with the help of compass needle pivoted to rotate freely in the north–south plane (inclinometer). Also, numerous observations during sea voyages have shown that the magnetic needle is not directed exactly northward, but its direction usually differs from that of the geographical meridian, and the angle (declination) between them depends on a position of the observation point. Measurements of this angle were performed by different types of declinometers. Thus, at the end of the 16th century, the magnetic compass played an extremely important role for sea navigation and also was the single instrument to study the behavior of the magnetic field of the earth even though at that time the concept of the magnetic field did not exist.

4.2.4. William Gilbert (1540–1603)

Gilbert was born at Colchester and obtained his education at Cambridge. After that he had a practice in London and became President of the Royal College of

Physicians. As a distinguished doctor, Gilbert was appointed physician to Queen Elizabeth. Somewhere around 1581 he decided to study magnetism, and for this purpose he performed numerous experiments, including those which were already well known. This was a new approach, fundamentally different from the conventional approach used by most scholars at that time. First of all, Gilbert confirmed the conclusion of Peregrinus about the existence of poles and discovered the main properties of permanent magnet such as:

(a) when a magnet is broken there are still two poles at its opposite ends;
(b) with an increase of temperature an iron bar loses its magnetism;
(c) when a hot iron bar is aligned along a meridian and its temperature decreases the bar again behaves as a magnet due to the magnetic field of the earth;
(d) when an iron is placed near a magnet the former becomes a magnet too.

In essence, Gilbert described the main features of inductive and permanent magnetization, as well as the influence of temperature, which are extremely important for understanding of geomagnetism. Then, Gilbert came to the conclusion that the earth causes the compass needle to orient toward the north. In other words, the earth itself is a giant magnet. Without any doubts it was a fundamental discovery and the beginning of a new science, geomagnetism. He built a model of a spherical earth, and, as Peregrinus, studied the behavior of the compass and its surface, in particular, its orientation northward and the existence of inclination (dip). These results, as well as others, were published in 1600 in the book "*De Magnete*". It may be proper to notice that Gilbert also studied electric properties of materials. For more than 2000 years there was an opinion that only amber and may be a couple of other materials have an attractive power. William Gilbert performed numerous experiments and demonstrated that due to friction many different bodies, such as glass, sulfur and others, display the same power, which was called by him the electric force. It is natural that Gilbert's book inspired scholars to find the cause of the magnetism; for instance, Descartes tried to explain this phenomenon with help of vortices. He thought that there is an interaction between a magnet and a fluid of vortices around each magnet. These vortices were thought to enter a material through one pole and leave by the other.

4.2.5. Edmond Halley (1656–1742)

At the beginning of the 17th century, there was a strong conviction that magnetism is caused by permanent magnets and that their parameters are independent of time. For instance, W. Gilbert thought that the earth is a permanent magnet. However, in 1634 Henry Gellibrand studied the declination near London and demonstrated that it varies with time. Numerous experiments were performed in other places and they also showed that the magnetic field of the earth varies. In order to explain this fact, one of the most famous scholars of that time, Edmond Halley, assumed that the earth consists of a system of concentric spherical shells with different parameters of magnetization and that they rotate differently with respect to each other. It is interesting to notice that an idea about a rotation of different layers of the earth with different velocity found, with some essential changes, its application in the modern theories of the origin of the magnetic field of the earth. Also, under the guidance of Halley the earliest magnetic survey was performed (1698), and as a result the first magnetic map of the Atlantic was prepared. Halley plotted lines connecting points with equal values of declination, and these lines were known as "Halleyan lines" or geomagnetic charts. With time the use of contour lines to describe the behavior of different parameters of the magnetic field as well as other quantities became conventional. Halley, using observations obtained by him and others, constructed and improved geomagnetic charts of declination (isogonics) and published them. It is very natural to appreciate his great contribution to the study of the magnetic field of the earth. Of course, we know Halley comet, because he predicted its return. Besides, as one of the leading members of "Royal Society", by his constant help and encouragement he played an outstanding role in publishing Newton's "*Principia*". Certainly, Edmond Halley was a great scientist and outstanding human being.

4.2.6. Charles Coulomb (1736–1806)

With an improvement of compass measurements, it became possible to see that different characteristics of magnetic field change. For instance, careful observations of a long compass needle allowed a London clockmaker George Graham to discover in 1722 that a needle changed its direction during 24 h and returned to its original position. Also in 1741 he and Anders Celsius from Sweden observed simultaneously perturbations due to polar aurora. As concerns these "diurnal" magnetic variations, they are very small but clearly indicate that the earth does not behave as the permanent magnet. In order to study this phenomenon and improve knowledge about the magnetic field of the earth, the Paris Academy of Sciences offered in 1773 a prize for "the best manner of constructing magnetic needles, of suspending them, of making sure that they are in true magnetic meridian, and finally, of accounting for their diurnal variations". The Academy announced this prize three times and finally in 1777 it was won by a French military engineer, Charles Coulomb. He was born in 1736 in the south of France and studied science and mathematics. Later he began his career as a military engineer and for several years supervised the construction of fortifications. In 1776 he settled in Paris and was involved in science; perhaps the next 13 years were the most productive and his accomplishments advanced enormously electricity and magnetism.

The principle of his device is called the "torsion balance" and for almost 200 years it was widely used for measuring the magnetic and gravitational fields. In Coulomb's instrument a magnetic needle is suspended on a wire with such parameters that relatively small torque acting on the needle produces a noticeable twist of the wire, which can be measured. The latter is measured with a help of a small mirror attached to the wire near the needle. Notice that the wire–needle system stops rotating when the moment generated by the magnetic force is compensated by the torque of the elastic force caused by the wire twist. An observation of a shift of the light spot reflected from mirror allows one to see very small movement of the needle. Coulomb very quickly realized that his instrument is an extremely sensitive device and correspondingly it is important to remove the influence of different types of noise, such as air flow and the action of static electric charges. In particular, Coulomb placed the instrument into a glass container in order to reduce the influence of motion stray of air. As was pointed out, Coulomb's invention played an extremely important role in developing methods of measuring magnetic and gravitational fields, that is, far away from the initial purpose of the instrument, namely, accurate measurements of small movement of the compass needle. For instance, almost 20 years later Henry Cavendish, applying the same approach, measured the much weaker force of interaction of two spherical masses of different mass and determined the gravitational constant. The importance of this experiment is difficult to overestimate, since knowledge of this constant made it possible to calculate the force of attraction of masses and, in particular, to evaluate the mass of the earth. It is possible that John Mitchell (England) invented the torsion balance even earlier but certainly both Coulomb and Mitchell made their discovery independently. It is known that Mitchell suggested to Cavendish to perform measurements of the gravitational constant using this device.

Making use of his instrument Coulomb investigated the interaction of poles of two different permanent magnets. Near a pole of the suspended needle he placed a pole of another magnet. Observations have shown that if poles are of the same kind their repulsion takes place but if the poles are of a different type they attract each other. Moreover, Coulomb discovered that as in the case of masses the force of interaction of poles is inversely proportional to the square of distance between them. This result is called Coulomb's law for poles, and it is useful for calculation of forces between permanent magnets. Then, Coulomb's experiments allowed him to discover the physical law, which plays the fundamental role in the theory of a constant electric field. He replaced the compass needle by a small straw covered by wax, carrying a pith ball at one end, while the other end has also some object to compensate the ball weight and keep the straw in the horizontal plane. Besides, he had an insulating stand where the same ball with a charge was placed. When the latter touched the first ball, both of them became charged and a movable ball was placed back on the stand. The force of interaction between these balls caused a twist of wire and its measurements for different distances between balls allowed one to establish the dependence of separation between charges. Coulomb was able to change the value of charges at each ball, and as a result of these experiments, he established the law for interaction of elementary charges: one of the greatest foundations of electromagnetism. Apart from the laws for the electric and magnetic forces, he made other significant contributions to electricity; for instance, Coulomb investigated the distribution of charges on a conductor surface and demonstrated that it follows from the law of interaction between them.

4.2.7. Oersted (1777–1851)

After discoveries by Coulomb there was long period, almost 40 years, when scientists cherished the amazing fact that so different features of nature such as the attraction of masses, interaction of electric charges and forces acting between permanent magnets obey almost the same law: such forces are directly proportional to the product of masses or charges or strength of poles and inversely proportional to the square of the distance between them. It is not difficult to imagine the satisfaction of scholars that so different and complicated phenomena of nature can be described by so simple and identical relationships which were discovered by two scientists: Newton and Coulomb. At the same time, the cause of the magnetic forces acting between permanent magnets remained unknown and in this sense the vortices suggested by Descartes could not satisfy the scientific community. Certainly, during this period nobody suspected that there is a link between the magnetic force and the ordered movement of charges (current).

The next step in understanding of magnetism was made by Hans Christian Oersted, who was born in a small town in the southern Denmark. In 1793 he became a student of University of Copenhagen. In the beginning his interests were not related to physics, but rather literature and law and later traveling. But in 1806 he joined his alma-matter as a regular professor and demonstrated interest in electricity and, especially, in relatively new subject, the electric battery. In the spring of 1820 he discovered, perhaps by chance (that by any means makes it less important), a fundamental fact of nature. According to different accounts he invited friends and some students to his home and gave a lecture about electricity and magnetism. The purpose of one of his experiments was to show the heating of a thin metal wire by electric current. It happened that near the battery and wire there was also a compass. Its presence may suggest that at that time Oersted perhaps thought about a relationship between the magnetic force and a current. Anyway, at this famous evening each time when current appeared in the wire the magnetic needle moved, but when the current vanished, the needle returned to its original position. It is vital that Oersted noticed this amazing fact, and later during several months he performed numerous experiments in order to understand this phenomenon but without success. First of all, he saw that in the presence of the current the needle turns at right angle to the current wire. Certainly, experiments did not indicate that there is an attraction or repulsion of the needle to the current wire. Also, each time when the direction of current was changed the compass needle was also reversed. Finally, without explanation Oersted published his experimental studies in July 21, 1820, which clearly demonstrated a connection between electricity and magnetism and from this moment the new direction in physics, electromagnetism, was started.

4.2.8. Andre-Marie Ampere (1777–1836)

This publication reached other scientists very quickly and on September 11 it was discussed at a meeting, where Ampere was present. During the next week he found an explanation of Oersted's measurements and then, performing series of brilliant experiments, created a completely new theory of magnetism. First of all, he discovered that magnetism or magnetic force is caused by currents, as electric force or force of attraction is generated by either charges or masses. This was one of the most important discoveries. Correspondingly, magnetic poles do not have any relation to the appearance of the magnetic force. Moreover, Ampere suggested that a permanent magnet generates magnetism because of presence of small currents in atoms which are lined up in such a way that they reinforce each other. As was shown in Chapter 1, this great physicist not only discovered the origin of the magnetic force but was also able to find the law of interaction of currents, which can be either conduction or magnetization ones.

4.2.9. Carl Gauss (1777–1855)

The influence of the revolutionary discoveries by Oersted and Ampere was very strong and in 1828 the naturalist Alexander von Humboldt suggested to the famous mathematician, the professor of mathematics of Göttingen University, Carl Gauss, to be involved in a study of magnetism. Expectations by Humboldt were well justified since Gauss was one of the sharpest minds in Europe. He worked in all areas of pure and applied mathematics: in number theory, algebra, function theory, differential geometry, probability theory, mechanics, geodesy, hydrostatic, mechanics, electrostatics, optics, and so on. Also, Gauss together with his assistant Wilhelm Weber, who later became one of the outstanding physicists of his time, made a great contribution to our knowledge of the magnetic field of the earth. At that time instruments did not allow one to measure the intensity of the magnetic field, but only inclination and declination. Gauss and Weber invented a very simple method of measuring the magnitude of the magnetic force. It was a great step in the study of the magnetic field. Then Gauss was actively involved in creating a global network of magnetic observatories where all components of the magnetic field were measured. The second outstanding contribution is related to the use of the mathematical method allowing one to represent analytically the magnetic field of the earth. This is the spherical harmonic analysis, which is described in the next sections, and it was introduced to geomagnetism by S. Poisson. Making use of results of measurements at different points of the earth, Carl Gauss described this magnetic field with a help of a series, and its coefficients became the conventional characteristic of the field on the earth. In 1834 Gauss and Weber created an international network of observations and with the help of Humboldt and other scientific organizations, and, especially, the British Royal Society. This was beginning of a systematic study of the magnetic field of the earth as well as other phenomena around the earth, associated with its magnetic field. Invention of the magnetometer, performance of the spherical analysis of data of measured field, and organization of network of magnetic stations around the word is an amazingly broad list of achievements by Gauss and Weber for development of geomagnetism. The first spherical analysis, performed by Gauss, was based on a very limited number of stations, which were not uniformly distributed over the earth surface, yet they gave very important information. For instance, it was proved that field mainly behaves in an extremely simple manner, namely, as a magnetic dipole. Perhaps, without the spherical analysis it would be hardly possible to deduce this remarkable fact from a comparison of measurements at separate stations. Later, the spherical analysis became a conventional approach in the study of the magnetic field of the earth and was regularly performed many times to investigate important features of this field; some of them will be briefly discussed in this monograph. Before we describe the spherical analysis of the magnetic field of the earth, it is useful to focus on the solution of Laplace equation in a spherical system of coordinates and its solutions.

4.3. SOLUTION OF THE LAPLACE EQUATION

It is obvious that measurements at each observatory first of all characterize the magnetic field at a given point of the earth surface. Now we will represent the field in the form of a series where every term characterizes the behavior of this field everywhere, and with this purpose in mind let us recall that the potential of the magnetic field outside of currents obeys Laplace's equation (Chapter 1). Taking into account the fact that the earth's surface is almost spherical and measurements are performed at this surface, it is natural to use a spherical system of coordinates with its origin at the earth's center. Our first goal is to find an expression for the potential; in other words, we have to solve the Laplace's equation, which in this system of coordinates has the form

$$\frac{1}{R}\left(R^2\frac{\partial U}{\partial R}\right)+\frac{1}{\sin\theta}\frac{\partial}{\partial\theta}\left(\sin\theta\frac{\partial U}{\partial\theta}\right)+\frac{1}{\sin^2\theta}\frac{\partial^2 U}{\partial\varphi^2}=0 \tag{4.4}$$

Here R, θ, and φ are the coordinates of any point and $U(R, \theta, \varphi)$ the potential of the magnetic field:

$$\boldsymbol{B} = -\text{grad } U \tag{4.5}$$

Of course, it is much more convenient to deal with the scalar function U than to operate with the vector $\boldsymbol{B}$. Applying the method of separation of variables, we represent the potential as a product of three functions:

$$U = \mathrm{T}\Omega\Phi = TS \tag{4.6}$$

where T is a function of R only, Ω a function of θ, and Φ depends only on φ. The function

$$S = \Omega\Phi$$

is called a surface spherical harmonic, and it is a function of two angles but is independent of the distance R. At the beginning we consider a product: $U = TS$, and its substitution into Equation (4.4) and division by TS gives

$$\frac{1}{T}\frac{\partial}{\partial R}\left(R^2\frac{\partial T}{\partial R}\right)+\frac{1}{S\sin\theta}\frac{\partial}{\partial\theta}\left(\sin\theta\frac{\partial S}{\partial\theta}\right)+\frac{1}{S\sin^2\theta}\frac{\partial^2 S}{\partial\varphi^2}=0 \tag{4.7}$$

The first term of this equation depends only on R, but the other two are functions of angles. This means that the left-hand side is equal to zero at each point, if the functions T and S satisfy equations:

$$\frac{1}{T}\frac{d}{dR}\left(R^2\frac{dT}{dR}\right) = K \tag{4.8}$$

and

$$\frac{1}{S\,\sin\theta}\frac{\partial}{\partial\theta}\left(\sin\theta\frac{\partial S}{\partial\theta}\right) + \frac{1}{S\,\sin^2\theta}\frac{\partial^2 S}{\partial\varphi^2} = -K \tag{4.9}$$

Here K is some constant. Thus, instead of Laplace's equation we have obtained one ordinary differential equation for the function T and one partial differential equation for the surface spherical harmonic, S. As is well known, differential equations are the most natural source of information about functions. For instance, a solution of Equation (4.8) has the form

$$T_n(R) = A_n R^n + B_n R^{-n-1} \tag{4.10}$$

where A_n and B_n are arbitrary constants, that is, they are independent of R. In solving Equation (4.8), we have also found an expression for the constant of separation K:

$$K = n(n+1) \tag{4.11}$$

since for other values of K this equation does not have a solution. Correspondingly, in order to satisfy Laplace's equation, the right-hand side of Equation (4.9) has to be equal to

$$-K = -n(n+1)$$

For each value of n we obtain solutions of Equations (4.8) and (4.9) and, therefore, particular solutions of Laplace's equation can be written as

$$U_n(R,\theta,\varphi) = (A_n R^n + B_n R^{-n-1})S_n(\theta,\varphi) \tag{4.12}$$

In order to find the general solution of this equation, we have to perform either summation or integration with respect to n. The choice of these operations depends on the problem. Let us notice that we did not consider the solution of Equation (4.8) for the case $n = 0$. As will be demonstrated later, this particular solution does not give any contribution to the magnetic field of the earth. We have already determined from the differential equation one function, $T(R)$, which describes the potential behavior in the radial direction. Next, we focus on the surface spherical harmonics, S_n, which are solutions of the partial differential equation of second order (Equation (4.9)):

$$\frac{1}{\sin\theta}\frac{\partial}{\partial\theta}\left(\sin\theta\frac{\partial S}{\partial\theta}\right)+\frac{1}{\sin^2\theta}\frac{\partial^2 S}{\partial\varphi^2}+n(n+1)S=0 \tag{4.13}$$

Before we solve this equation and find the function S, let us demonstrate that it possesses one very important feature and with this purpose notice the following. As follows from Equation (4.12), the function $R^n S_n$ is a solution of Laplace's equation for any value of n.

4.4. ORTHOGONALITY OF FUNCTIONS S_n

Consider two different solutions of Laplace's equation:

$$\Psi_n = R^n S_n \quad \text{and} \quad \Psi_m = R^m S_m \tag{4.14}$$

Then applying Green's formula (Chapter 3), we have

$$\int_V (\Psi_n \Delta\Psi_m - \Psi_m \Delta\Psi_n)dV = \oint_A \left(\psi_n \frac{\partial\Psi_m}{\partial n} - \Psi_m \frac{\partial\Psi_n}{\partial n}\right)dA \tag{4.15}$$

Here V is the volume and A the surface surrounding this volume. Inasmuch as the functions Ψ_n and Ψ_m obey Laplace's equation, Equation (4.15) yields

$$\oint_A \left(\Psi_n \frac{\partial\Psi_m}{\partial n} - \Psi_m \frac{\partial\Psi_n}{\partial n}\right)dA = 0 \tag{4.16}$$

Here n is a variable along the normal to the surface and it is directed outward. Suppose that A is a spherical surface of radius R, for instance, the surface of the earth. Then, by definition

$$dA = R^2\, d\omega \tag{4.17}$$

where $d\omega$ is the solid angle of the elementary surface dA under which it is seen from the origin. At the same time, we have

$$\frac{\partial\Psi_n}{\partial n} = \frac{\partial(R^n S_n)}{\partial R} = nR^{n-1}S_n$$

By analogy

$$\frac{\partial\Psi_m}{\partial n} = mR^{m-1}S_m$$

Correspondingly, Equation (4.16) can be written as

$$\oint R^2(mR^n R^{m-1} - nR^{n-1} R^m) S_n S_m \, d\omega = 0$$

or (4.18)

$$R^{n+m+1}(m-n) \oint S_m S_n \, d\omega = 0$$

We have arrived at the fundamental feature of the surface spherical functions, namely, if $m \neq n$, then the integral:

$$\oint S_m S_n \, d\omega = 0 \tag{4.19}$$

This is an extremely important result, since it will allow us to perform a spherical analysis. The equality (4.19) shows that the surface spherical harmonics, in the same manner as sinusoidal functions, are orthogonal. Later we will consider the special case when $m = n$.

4.5. SOLUTION OF EQUATION (4.13) FOR THE FUNCTIONS *S*

In accordance with Equation (4.6), the function

$$S = \Omega\Phi \tag{4.20}$$

depends on two arguments and obeys the partial differential equation (4.13). In order to find its solution we will again apply the method of separation of variables. Substituting Equation (4.20) into Equation (4.13) and dividing by the ratio $\Omega\Phi/\sin^2\theta$, we obtain

$$\frac{\sin\theta}{\Omega}\frac{\partial}{\partial\theta}\left(\sin\theta\frac{\partial\Omega}{\partial\theta}\right) + \frac{1}{\Phi}\frac{\partial^2\Phi}{\partial\varphi^2} + n(n+1)\sin^2\theta = 0$$

This means that functions Ω and Φ are solutions of the following ordinary differential equations:

$$\frac{\sin\theta}{\Omega}\frac{d}{d\theta}\left(\sin\theta\frac{d\Omega}{d\theta}\right) + n(n+1)\sin^2\theta = m^2, \quad \frac{1}{\Phi}\frac{d^2\Phi}{d\varphi^2} = -m^2 \tag{4.21}$$

The particular solution of the last equation:

$$\frac{d^2\Phi}{d\varphi^2} + m^2\Phi = 0$$

is well known and it has the form:

$$\Phi_m(\varphi) = C_m \cos m\varphi + D_m \sin m\varphi$$

Here it is appropriate to make two comments: (a) In principle, m can be an arbitrary number, but taking into account the fact that in our case of the spherical surface of the earth the potential is a periodic function of the angle φ, we chose only integer values of m. (b) For the same reason, we assumed that the left-hand side of the last equation of the set (4.21) is negative; otherwise we would have

$$\frac{d^2\Phi}{d\varphi^2} - m^2\Phi = 0$$

and its solution is not a periodic function. Thus, after the last separation of variables the particular solution of Laplace's equation is presented as

$$U(R, \theta, \varphi, m, n) = (A_n R^n + B_n R^{-n-1})\Omega(\theta, m, n)(C_m \cos m\varphi + D_m \sin m\varphi) \quad (4.22)$$

where the function Ω is a solution of the first equation of the set (4.21). After multiplication by $\Omega/\sin^2\theta$, it becomes

$$\frac{1}{\sin\theta}\frac{d}{d\theta}\left(\sin\theta\frac{d\Omega}{d\theta}\right) + \left[n(n+1) - \frac{m^2}{\sin^2\theta}\right]\Omega = 0 \quad (4.23)$$

Our goal is to describe the solution of this equation, and it is convenient to begin from the simplest case when $m = 0$.

4.6. LEGENDRE'S EQUATION AND ZONAL HARMONICS

Suppose that the potential is independent of the coordinate φ; that is, the function Φ_m is constant. As follows from Equation (4.22), this means that $m = 0$ and Equation (4.23) is greatly simplified. Introducing new variable $\mu = \cos\theta$, we have

$$d\mu = -\sin\theta\, d\theta$$

Since

$$\frac{d}{d\theta} = \frac{d}{d\mu}\frac{d\mu}{d\theta} = -\sin\theta\frac{d}{d\mu}$$

Equation (4.23) can be represented as

$$\frac{d}{d\mu}\left[(1-\mu^2)\frac{d\Omega_n}{d\mu}\right] + n(n+1)\Omega_n = 0 \quad (4.24)$$

As is well known (Chapter 3), this is Legendre's equation, and its solutions are often called the zonal harmonics.

4.7. SOLUTION OF LEGENDRE'S EQUATION

In order to find a solution of this equation, we first represent the function Ω in the form of a power series:

$$\Omega(\mu) = \sum a_r \mu^r \tag{4.25}$$

where a_r are unknown coefficients. Substitution of Equation (4.25) into Equation (4.24) and performing a differentiation gives

$$\sum r(r-1)a_r\mu^{r-2} + \sum [n(n+1) - r(r+1)]a_r\mu^r = 0$$

Then introducing a new variable $x = r-2$ and using the original notation, we have

$$\sum \{(r+2)(r+1)a_{r+2} + [n(n+1) - r(r+1)]a_r\}\mu^r = 0$$

This equality has to hold for any μ and this happens if the coefficients for different powers of μ are separately equal to zero. Collecting terms with the same power of μ, we obtain

$$(r+2)(r+1)a_{r+2} + [n(n+1) - (r+1)r]a_r = 0$$

that is

$$\begin{gathered} a_r = -\frac{(r+1)(r+2)}{(n-r)(n+r+1)}a_{r+2} \\ \text{or} \\ a_{r+2} = -\frac{(n-r)(n+r+1)}{(r+1)(r+2)}a_r \end{gathered} \tag{4.26}$$

Also

$$a_{r-2} = -\frac{(r-1)r}{(n-r+2)(n+r-1)}a_r \tag{4.27}$$

Thus, the function $\Omega(\mu)$ is a solution of Equation (4.24), if coefficients obey Equations (4.26) and (4.27). These recursion formulas allow us to see some interesting features of these coefficients. First of all, it turns out that if $a_r = 0$, then

$$a_{r-2} = a_{r-4} = \cdots = 0$$

Second, as follows from Equation (4.27), for finite values of a_0 and a_1 we have

$$a_{-1} = a_{-2} = 0$$

From the last two equalities we see that the series $\Omega(\mu)$ does not contain negative powers of μ, and this is a very important feature of the function. It is conventional to consider two special cases of the series (4.25), so that their sum gives the function $\Omega(\mu)$, and since in both cases the coefficients satisfy Equation (4.26) these series also are solutions of Equation (4.24).

Case one: Series has only even powers of μ and $a_0 = 1$.

In accordance with Equation (4.26), this solution can be written as

$$p_n = 1 - \frac{n(n+1)}{2!}\mu^2 + \frac{n(n-2)(n+1)(n+3)}{4!}\mu^4 - \cdots \qquad (4.28)$$

Case two: Series has only odd powers of μ and $a_1 = 1$.

In this case we have only odd powers and the new solution has the form

$$q_n = \mu - \frac{(n-1)(n+2)}{3!}\mu^3 + \frac{(n-1)(n-3)(n+2)(n+4)}{5!}\mu^5 - \cdots \qquad (4.29)$$

Therefore, the general solution of Equation (4.24) for any value of n is

$$\Omega_n(\mu) = A_n p_n(\mu) + B_n q_n(\mu) \qquad (4.30)$$

Here the argument changes within the range

$$-1 < \mu < +1$$

and, as long as the power series converges, n can be either integer or fraction, complex or real number.

4.7.1. *Index n of functions p_n and q_n is positive*

Now we assume that n is positive and show a fundamental feature of both functions p_n and q_n. First, we consider a solution $p_n(u)$ when n is an even positive number and $a_0 = 1$. Inasmuch as n and r are even positive numbers, we let $n = 2k$ and $r = 2s$. Their substitution into Equation (4.26) gives

$$a_{r+2} = -\frac{(k-s)(2k+2s+1)}{(2s+1)(s+1)}a_r$$

and when $k = s$ the coefficient $a_{2(s+1)}$ and the following ones are equal to zero. Therefore, in place of the infinite series we have a finite sum of terms. In other words, in this case the function p_n is polynomial. For instance, as is seen from Equation (4.28):

$$p_0 = 1, \quad p_2(\mu) = 1 - 3\mu^2, \quad p_4(\mu) = 1 - 10\mu^2 + \frac{35}{3}\mu^4 - \cdots$$

It is obvious that number of terms which are different from zero is equal to

$$N_p = \frac{n+2}{2}$$

Next consider the function q_n when n is odd positive number; letting $n = 2k-1$ and $r = 2s-1$, we obtain

$$a_{2s-3} = -\frac{(k-s)(2k+2s-1)}{s(2s+1)} a_{2s-1}$$

and this means that when $k = s$ the terms of the series as well as the following are equal to zero; that is, the function q_n is a polynomial too. In accordance with Equation (4.29), we have

$$q_1(\mu) = \mu, \quad q_3(\mu) = \mu - \frac{5}{3}\mu^3, \quad q_5(\mu) = \mu - \frac{14}{3}\mu^3 + \frac{63}{15}\mu^5$$

and number of terms different from zero is still equal to

$$N_q = \frac{n+1}{2}$$

4.8. RECURSION FORMULAS FOR THE FUNCTIONS *P* AND *Q*

It turns out that there are recursion formulas for functions p_n and q_n, that is, relationships between them with different indices n. As follows from Equation (4.28)

$$p_{n-1} = 1 - \frac{(n-1)n}{2!}\mu^2 + \frac{(n-1)(n-3)n(n+2)}{4!}\mu^4 - \cdots$$

and

$$p_{n+1} = 1 - \frac{(n+1)(n+2)}{2!}\mu^2 + \frac{(n+1)(n-1)(n+2)(n+4)}{4!}\mu^4 - \cdots$$

Taking the difference of these functions, we obtain

$$p_{n-1} - p_{n+1} = \left[\frac{(n+1)(n+2) - n(n-1)}{2!}\mu^2 - \frac{(n+1)(n+4) - n(n-3)}{4}\frac{(n-1)(n+2)}{3!}\mu^4 + \cdots\right]$$
$$= (2n+1)\mu\left[\mu - \frac{(n-1)(n+2)}{3!}\mu^3 + \cdots\right]$$

or

$$p_{n-1} - p_{n+1} = (2n+1)\mu q_n \tag{4.31}$$

In the same manner, we have

$$(n+1)^2 q_{n+1} - n^2 q_{n-1} = (2n+1)\mu p_n \tag{4.32}$$

The last two equalities show that if we know one function with some values of n, the others can be calculated from it. Next, we derive another type of recursion relation. Making use of Equation (4.29) and performing a differentiation, we obtain:

$$nq'_{n-1} + (n+1)q'_{n-1} = \left[(2n+1 - \frac{(n-2+n+3)n(n+1)}{2!}\mu^2 + \cdots\right]$$
$$= (2n+1)\left[1 - \frac{n(n+1)}{2!} + \cdots\right] = (2n+1)p_n \tag{4.33}$$

By analogy

$$(n+1)p'_{n-1} + np'_{n+1} = -n(n+1)(2n+1)q_n \tag{4.34}$$

4.9. LEGENDRE POLYNOMIALS

Next, we introduce the functions which play a vital role in the spherical analysis of the magnetic field of the earth; they are related in a very simple manner to the functions p_n and q_n. First, suppose that n is a positive and even number. Then, Equation (4.28) can be written in the form

$$p_n = (-1)^{n/2}2^n\frac{[(n/2)!]^2}{n!}\sum_{r=0}^{n/2}(-1)^{r-(n/2)}\frac{(n+2r)!\mu^{2r}}{2^n[(n-2r)/2]![(n+2r)/2]!(2r)!}$$

and the Legendre polynomials are defined as

$$P_n(\mu) = \frac{(-1)^{n/2} n!}{2^n [(n/2)!]^2} p_n \tag{4.35}$$

If n is a positive and odd number, then the series (4.29) has $(n+1)/2$ terms, and it can be written in the form

$$q_n = (-1)^{(n-1)/2} 2^{n-1} \frac{[((n-1)/2)!]^2}{n!} \sum_{r=0}^{(n-1)/2} (-1)^{r-[(n-1)/2]} \times \frac{(n+2r+1)\mu^{2r+1}}{2^n [(n-2r-1)/2]! [(n+2r+1)/2]! (2r+1)!}$$

In this case, the Legendre polynomials are defined as

$$P_n(\mu) = \frac{(-1)^{(n-1)/2} n!}{2^{n-1} \{[(n-1)/2]!\}^2} q_n \tag{4.36}$$

If n is an integer and positive number, then in place of Equations (4.35) and (4.36) we can write one series for the Legendre polynomials. Letting $s = (n/2) - r$ in Equation (4.35) and $s = [(n-1)/2] - r$ in Equation (4.36), we obtain

$$P_n(\mu) = \sum_{s=0}^{m} (-1)^s \frac{(2n-2s)!}{2^n (s!)(n-s)!(n-2s)!} \mu^{n-2s} \tag{4.37}$$

Here m is equal to either $n/2$ or $(n-1)/2$ and this depends on which of them is integer. Thus, we have represented the solution of Equation (4.23) (Legendre polynomials), in the form of a series with finite number of terms. As follows from Equation (4.37)

$$P_n(\mu) = \frac{1}{2^n n!} \sum_{s=0}^{m} (-1)^s \frac{n!}{s!(n-s)!} \frac{(2n-2s)!}{(n-2s)!} \mu^{n-2s}$$
$$= \frac{1}{2^n n!} \frac{d^n}{d\mu^n} \sum_{s=0}^{n} (-1)^s \frac{n!}{s!(n-s)!} \mu^{2n-2s}$$

This sum is an expansion of the function $(\mu^2-1)^n$, and therefore we arrive at the Rodrigue's formula

$$P_n(\mu) = \frac{1}{2^n n!} \frac{d^n}{d\mu^n} (\mu^2 - 1)^n \tag{4.38}$$

The last two formulas describe a solution of Legendre equation for any values of μ, for instance, for harmonics of elongated spheroid (Chapter 3), and the argument varies in the interval: $0<\mu<\infty$. In this case, Equation (4.37) gives an asymptotic expression for these functions. Since for very large values of μ the term with the highest degree of μ plays the dominant role, we obtain

$$P_n(\mu) \to \frac{(2n)!}{2^n (n!)^2} \mu^n \tag{4.39}$$

4.9.1. Recursion formulas for Legendre's polynomials

Now we demonstrate that as in the case of the functions p_n and q_n, there are recursion relationships between Legendre polynomials. First, suppose that n is an odd integer number. Then, substitution of the functions p_{n+1}, p_{n-1}, and q_n from Equations (4.35) and (4.36) into Equation (4.31) gives

$$nP_{n-1} + (n+1)P_{n+1} = (2n+1)\mu P_n \tag{4.40}$$

which remains valid for even values of n, also. Applying the same approach, we have for any integer n a different recursive relation:

$$P'_{n+1} - P'_{n-1} = (2n+1)P_n \tag{4.41}$$

The latter allows us to find the integral of Legendre function. In fact, performing integration of Equation (4.41) we have:

$$\int P_n(\mu)d\mu = \frac{P_{n+1} - P_{n+1}}{2n+1} \tag{4.42}$$

4.10. INTEGRAL FROM A PRODUCT OF LEGENDRE POLYNOMIALS

This subject is very important for many applications since it turns out that Legendre polynomials, as sinusoidal functions and many other special functions, are orthogonal functions. In accordance with Equation (4.19), we see that the integral from the product of polynomials with different indices within the interval: $-1<\mu+1$ or $0<\theta<\pi$ is equal to zero:

$$\int_{-1}^{+1} P_n(\mu)P_m(\mu)d\mu = 0, \quad \text{if } m \neq n \tag{4.43}$$

Now consider the case when $m = n$; that is, the value of the integral:

$$\int_{-1}^{+1} P_n^2(\mu)d\mu$$

Making use of Rodrigue's formula, we have

$$\int_{-1}^{+1} P_n^2(\mu)d\mu = \frac{1}{2^n n!}\int_{-1}^{+1} P_n(\mu)\frac{d^n}{d\mu^n}(\mu^2 - 1)^n d\mu$$

Subsequent integration by parts, where

$$u = P_n(\mu), \quad v = (\mu^2 - 1)^n$$

gives

$$\int_{-1}^{+1} P_n^2(\mu)d\mu = \frac{(-1)^n}{2^n n!}\int_{-1}^{+1} \frac{d^n P_n(\mu)}{d\mu^n}(\mu^2 - 1)^n d\mu$$

Applying again the Rodrigue's formula, we can show that

$$\frac{dP_n(\mu)}{d\mu^n} = \frac{(2n)!n!}{2^n n!n!} = \frac{(2n)!}{2^n n!}$$

whence

$$\int_{-1}^{+1} P_n^2(\mu)d\mu = \frac{(2n)!}{2^{2n}(n!)^2}\int_{-1}^{+1}(1 - \mu^2)^n d\mu = \frac{(2n-1)!!}{(2n)!!}\int_0^{\pi} \sin^{2n+1}\theta \, d\theta$$

The last integral is well known and finally we obtain

$$\int_{-1}^{+1} P_n^2(\mu)d\mu = \frac{2}{2n+1} \tag{4.44}$$

Certainly, this is an amazingly simple expression for the integral of such a complicated function.

4.11. EXPANSION OF FUNCTIONS BY LEGENDRE POLYNOMIALS

Assuming that the argument varies within the interval $(-1 < \mu + 1)$, let us express a function $f(\mu)$ in the following series:

$$f(\mu) = a_0 P_0(\mu) + a_1 P_1(\mu) + a_2 P_2(\mu) + \cdots + a_n P_n(\mu) + \cdots \tag{4.45}$$

Here $f(\mu)$ is a continuous function, except at a finite number of points where the function may have a discontinuity and the latter has a limited value. As in the case of Fourier's series, the sum of the series (Equation (4.45)) is equal to the mean value of the function at the opposite sides of a point of a discontinuity. Multiplying both sides of Equation (4.45) by $P_m(\mu)$ and integrating, we obtain

$$a_m = \frac{2m+1}{2}\int_{-1}^{+1} f(\mu)P_m(\mu)d\mu \tag{4.46}$$

Of course, in deriving the latter we used Equations (4.44) and (4.45).

4.11.1. *Expressions for Legendre polynomials*

For illustration we show the expressions for some of polynomials:

$$P_0(\mu) = 1, \quad P_1(\mu) = \mu, \quad P_2(\mu) = \frac{1}{2}(3\mu^2 - 1), \quad P_3(\mu) = \frac{1}{2}(5\mu^3 - 3\mu)$$

$$P_4(\mu) = \frac{35\mu^4 - 30\mu^2 + 3}{8}, \quad P_5(\mu) = \frac{63\mu^5 - 70\mu^3 + 15\mu}{8}$$

$$P_6(\mu) = \frac{231\mu^6 - 315\mu^4 + 105\mu^2 - 5}{16}, \quad P_7(\mu) = \frac{429\mu^7 - 693\mu^5 + 315\mu^5 - 35\mu}{16}$$

and

$$P_8(\mu) = \frac{6435\mu^8 - 12,012\mu^6 + 5930\mu^4 - 1260\mu^2 + 35}{128}$$

As is seen from the theory, Legendre polynomials are alternating power series with finite numbers of terms; their values vary between -1 and $+1$ and the number of zero values corresponds to the index of the polynomial. In conclusion, it is proper to notice that for each n, Equation (4.24) has another solution: zonal harmonics of the second type, $Q_n(\mu)$, related to functions p_n and q_n.

4.12. SPHERICAL ANALYSIS OF THE EARTH'S MAGNETIC FIELD WHEN THE POTENTIAL IS INDEPENDENT OF LONGITUDE

For illustration of the main concepts of spherical analysis at the beginning suppose that the potential of the magnetic field on the earth's surface depends on the angle θ only. In such a case, performing a summation of partial solutions, we obtain

$$U(R,\theta) = \sum_{n=1}^{\infty}(A_n R^n + B_n R^{-n-1})P_n(\cos\theta) \tag{4.47}$$

It is essential to note that the right-hand side of Equation (4.47) is a sum of two terms. One of them decreases with an increase in distance from the earth's center

$$\sum_{n=1}^{\infty} B_n R^{-n-1} P_n(\cos\theta)$$

the other

$$\sum_{n=1}^{\infty} A_n R^n P_n(\cos\theta)$$

becomes greater as R increases. For this reason it is natural to interpret the first and second sums as potentials of the magnetic fields caused by currents in the earth and ionosphere, respectively. By definition, we have

$$\boldsymbol{B} = -\text{grad}\, U$$

that is

$$B_R = -\frac{\partial U}{\partial R}, \quad B_\theta = -\frac{1}{R}\frac{\partial U}{\partial \theta}, \quad B_\varphi = 0$$

Differentiation in Equation (4.47) gives

$$\begin{aligned} B_R &= \sum_{n=1}^{\infty}[-nA_n R^{n-1} + (n+1)B_n R^{-n-2}]P_n(\cos\theta) \\ B_\theta &= -\sum_{n=1}^{\infty}[A_n R^{n-1} + B_n R^{-n-2}]\frac{\partial P_n(\cos\theta)}{\partial\theta} \\ B_\varphi &= 0 \end{aligned} \tag{4.48}$$

Before we continue let us notice that we discarded the term $n = 0$, which is equal to

$$B_R = B_0 R^{-2} \quad \text{and} \quad B_\theta = B_\varphi = 0$$

Such a field corresponds to that of a magnetic charge, the distribution of which is independent of the angles θ and φ. Inasmuch as magnetic charges do not exist, the sums are started from $n = 1$. Introducing notations used in geomagnetism

$$Z = -B_R \quad \text{and} \quad X = -B_\theta$$

we obtain for points located at the earth's surface

$$Z = \sum_{n=1}^{\infty}[nA_n R_0^{n-1} - (n+1)B_n R_0^{-n-2}]P_n(\cos\theta)$$

and

$$X = \sum_{n=1}^{\infty}[A_n R_0^{n-1} + B_n R_0^{-n-2}]\frac{\partial}{\partial\theta}P_n(\cos\theta) \tag{4.49}$$

where R_0 and θ are the earth's radius and the latitude of the observation point, respectively. Thus, we have represented the vertical and horizontal components of the magnetic field on the earth's surface as a combination of spherical harmonics, and each of them is a sum of two terms, characterizing the magnetic field caused by currents above and beneath the surface. This type of representation is vital for separating the total field into two parts, generated by the external and internal currents. It is proper to point out that the derivative P_n' is also expressed through orthogonal functions and this presentation will be considered later. First, suppose that we performed a spherical analysis of the measured values of Z and X at different observation points of the earth. In other words, applying Equation (4.46), we found coefficients of the series

$$Z = \sum_{n=1}^{\infty} Z_n P_n(\cos\theta) \quad \text{and} \quad X = \sum_{n=1}^{\infty} X_n \frac{\partial P_n(\cos\theta)}{\partial\theta} \tag{4.50}$$

Here Z_n and X_n are characteristics of the vertical and horizontal components of the magnetic field on the earth surface. Note that the same result can be obtained differently. Taking a number of terms of the series which is smaller than number of points of observations, we obtain for each set of coefficients a system of equations where the number of unknowns is less than the number of equations. Then, making use the least squares method, we determine Z_n and X_n. Next, comparing Equations (4.49) and (4.50) and taking into account again the orthogonality of spherical functions, we arrive at two linear equations with two unknowns

$$Z_n = nA_nR_0^{n-1} - (n+1)B_nR_0^{-n-2}, \quad X_n = A_nR_0^{n-1} + B_nR_0^{-n-2} \tag{4.51}$$

Solving this system, we obtain for the amplitudes of the spherical harmonics, describing the fields of the external and internal currents, the following expressions:

$$A_n = \frac{(n+1)X_n + Z_n}{(2n+1)R_0^{n-1}} \quad \text{and} \quad B_n = \frac{nX_n - Z_n}{2n+1} R_0^{n+2} \tag{4.52}$$

By definition, these coefficients are independent of the position of a point on the earth's surface; they characterize the magnetic field of the earth as a whole. Besides, their values describe a relative contribution of currents, located above and beneath the earth's surface. In other words, in principle we have performed a separation of the total field on the external and internal parts. At the same time, it is proper to note that the best estimate of the ionosphere field comes from direct satellite measurements of the currents.

4.13. THE PHYSICAL MEANING OF COEFFICIENTS B_n

Now let us discuss the internal part of the field and start from the term $n = 1$. As follows from Equations (4.49) we have

$$Z = -\frac{2B_1}{R^3}\cos\theta, \quad X = -\frac{B_1}{R^3}\sin\theta$$

or

$$Z = \frac{2\mu_0 M}{4\pi R^3}\cos\theta, \quad X = \frac{\mu_0 M}{4\pi R^3}\sin\theta \tag{4.53}$$

since $n = 1$ and

$$P_1(\cos) = \cos\theta, \quad \frac{\partial}{\partial\theta}P_1(\cos\theta) = -\sin\theta$$

and the moment $\boldsymbol{M}$ is directed from north to south along the rotation axis of the earth, since we have assumed that the field is independent of longitude. As is well known (Chapter 1), Equations (4.53) describe the field of a magnetic dipole. As was established by Gauss this is the main part of the magnetic field of the earth, caused by currents in the earth's core. In our approximation, $\partial/(\partial\varphi) = 0$, the poles are the points where the z-axis intersects the earth's surface and the magnetic field is approximately equal to 60×10^3 nT. In accordance with the first equation of the set (4.53), the moment magnitude is

$$M \approx 10^{23}\ \text{A m}^2$$

If we suppose that the radius of a current system is around 1000 km, then the total current is

$$I \approx 3 \times 10^{10}\ \text{A}$$

and this is really strong current.

As we know, the magnetic dipole is a pure mathematical concept of an infinitely small current loop (Chapter 1), but if an observation of a magnetic field takes place at distances, which are much greater than dimensions of the current system, generating this field, then it behaves almost as the field of a magnetic dipole. With an increase of the distance from the real system, this approximation becomes more accurate. Inasmuch as we study the field on the earth's surface, but the conduction currents are located inside the core, the distance is practically three times greater than the dimensions of this current system. Correspondingly, it is natural to expect that the main part of the field is defined by that of a magnetic dipole, but it does not mean that there is a mystical magnetic dipole at the earth's center. Of course, the next terms of the series (4.49) formally can be interpreted as multi-poles of higher order. For instance, the term with $n = 2$ characterizes the field of the quadrupole, which can be imagined as a system of two dipoles with opposite direction and equal moments, located again at the earth's origin. For instance, two small parallel loops with the same radius and opposite direction of currents of the equal magnitude behave at relatively large distances almost as a quadrupole. Perhaps, it is more proper to treat Equations (4.49) as only a mathematical representation of the magnetic field, caused by any system of currents which possesses axial symmetry with respect to the z-axis, and do not try to see physical meaning of each term of these series. At the same time, it is natural to expect that the relative contribution of each term depends on the shape of the current system and its distance to the observation point.

4.14. ASSOCIATED LEGENDRE FUNCTIONS

Next we consider a more general case of the field behavior and with this purpose in mind we return back to Equation (4.23). Earlier we have shown that a potential of the magnetic field, that is, a solution of Laplace's equation, can be represented as a product of three functions for each value of n and m, and two of them have the form

$$T_n = A_n R^n + B_n R^{-n-1}, \quad \Phi_m = C_m \cos m\varphi + D_m \sin m\varphi \tag{4.54}$$

but the last one, $\Omega(\theta)$, obeys Equation (4.23). Introducing the variable $\mu = \cos\theta$, it becomes

$$\frac{d}{d\mu}\left[(1-\mu^2)\frac{d\Omega}{d\mu}\right] + \left[n(n+1) - \frac{m^2}{1-\mu^2}\right]\Omega = 0 \tag{4.55}$$

At the beginning, in order to find its solution we let $m = 0$ in Equation (4.55). Then, differentiation of its first term gives

$$(1-\mu^2)\frac{d^2y}{d\mu^2} - 2\mu\frac{dy}{d\mu} + n(n+1)y = 0 \tag{4.56}$$

This equation does not differ from the Legendre equation and, correspondingly, its solutions are Legendre functions of the first and second kind, $y = P_n(\mu)$ and $y = Q_n(\mu)$. Now we demonstrate that the functions $\Omega(\mu)$ satisfying Equation (4.55) can be expressed through Legendre functions. Performing a differentiation of Equation (4.56) m times and using the notation: $v = (d^m y)/(d\mu^m)$, we obtain

$$(1-\mu^2)\frac{d^2 v}{d\mu^2} - 2\mu(m+1)\frac{dv}{d\mu} + (n-m)(n+m+1)v = 0$$

Introducing the notations $\omega = (1-\mu^2)^{m/2}v$ or $v = (1-\mu^2)^{-m/2}\omega$, the last equation becomes

$$(1-\mu^2)\frac{d^2\omega}{d\mu^2} - 2\mu\frac{d\omega}{d\mu} + \left[n(n+1) - \frac{m^2}{1-\mu^2}\right]\omega = 0 \qquad (4.57)$$

Performing a differentiation of the first term of Equation (4.55), we see that it coincides with Equation (4.57). Therefore, solution of this equation has the form

$$\Omega = \omega = (1-\mu^2)^{m/2}v = (1-\mu^2)^{m/2}\frac{d^m y}{d\mu^m}$$

This is a very important result, since we were able to express a solution of Equation (4.55) in terms of the known Legendre functions. It is natural to represent its solution as

$$\Omega = A' P_n^m(\mu) + B' Q_n^m(\mu) \qquad (4.58)$$

Within an interval $-1<\mu<+1$, the functions $P_n^m(\mu)$ and $Q_n^m(\mu)$ are defined as

$$P_n^m(\mu) = (1-\mu^2)^{m/2}\frac{d^m P_n(\mu)}{d\mu^n} \qquad (4.59)$$

and

$$Q_n^m(\mu) = (1-\mu^2)^{m/2}\frac{dQ_n(\mu)}{d\mu^n} \qquad (4.60)$$

are solutions of Equation (4.55).

In those cases when the argument is real or imaginary and its magnitude exceeds unity, the associated functions of Legendre are defined as

$$P_n^m(\mu) = (\mu^2-1)^{m/2}\frac{d^m P_n(\mu)}{d\mu^n} \qquad (4.61)$$

and

$$Q_n^m(\mu) = (\mu^2-1)^{m/2}\frac{d^m Q_n(\mu)}{d\mu^n} \qquad (4.62)$$

4.14.1. Examples of the associated Legendre functions ($\mu < 1$)

As follows from Equations (4.59) and (4.60), we have

$$P_1^1(\mu) = (1-\mu^2)^{1/2}, \quad P_2^1(\mu) = 3(1-\mu^2)^{1/2}\mu, \quad P_2^2(\mu) = 3(1-\mu^2)$$
$$P_3^1(\mu) = \frac{3}{2}(1-\mu^2)^{1/2}(5\mu^2-1), \quad P_3^2(\mu) = 15(1-\mu^2)\mu$$
$$P_3^3(\mu) = 15(1-\mu^2)^{3/2}, \quad P_4^1(\mu) = \frac{5}{2}(1-\mu^2)^{1/2}(7\mu^3-3\mu)$$
$$P_4^2(\mu) = \frac{15}{2}(1-\mu^2)(7\mu^2-1), \quad P_4^3(\mu) = 105(1-\mu^2)^{3/2}\mu$$
$$P_4^4(\mu) = 105(1-\mu^2)^2$$

and

$$Q_1^1(\mu) = (1-\mu^2)^{1/2}\left(\frac{1}{2}\ln\frac{1+\mu}{1-\mu} + \frac{\mu}{1-\mu^2}\right)$$
$$Q_2^1(\mu) = (1-\mu^2)^{1/2}\left(\frac{3}{2}\mu\ \ln\frac{1+\mu}{1-\mu} + \frac{3\mu^2-2}{1-\mu^2}\right) \tag{4.63}$$
$$Q_2^2(\mu) = (1-\mu^2)\left(\frac{3}{2}\ln\frac{1+\mu}{1-\mu} + \frac{5\mu-3\mu^3}{(1-\mu^2)^2}\right)$$

It is clear that the associated Legendre functions of the second kind have a singularity when $\theta = 0$, and for this reason they are not used in the spherical analysis of the magnetic field.

In order to solve the boundary-value problem when a spheroid is placed in a magnetic field (Chapter 3), we used the associated Legendre functions with the argument exceeding unity. In this case, formulas for these functions are obtained from Equations (4.63) by the following replacement:

$$(1-\mu^2)^{m/2} \to (\mu^2-1)^{1/2}$$

and in the logarithmic term of the function $Q_n^m(\mu)$:

$$1-\mu \to \mu - 1$$

Making use of Equations (4.61) and (4.62) and the formulas for Legendre polynomials, we have the formulas in the case of imaginary argument, $\mathrm{i}\xi$:

$$P_1(\mathrm{i}\xi) = \mathrm{i}\xi, \quad Q_1(\mathrm{i}\xi) = \xi\coth^{-1}\xi - 1$$

$$P_1^1(\mathrm{i}\xi) = \mathrm{i}(1+\xi^2)^{1/2}, \quad Q_1^1(\mathrm{i}\xi) = (1+\xi^2)\left(\coth^{-1}\xi - \frac{\xi}{1+\xi^2}\right)$$

$$P_2(i\xi) = -\frac{1}{2}(3\xi^2 + 1), \quad Q_2(i\xi) = \frac{i}{2}[(3\xi^2 + 1)\coth^{-1}\xi - 3\xi]$$

$$P_2^1(i\xi) = -3(1 + \xi^2)^{1/2}\xi, \quad Q_2^1(i\xi) = i(1 + \xi^2)^{1/2}\left[3\xi \coth^{-1}\xi - \frac{3\xi^2 + 2}{1 + \xi^2}\right]$$

$$P_2^2(i\xi) = -3(1 + \xi^2), \quad Q_2^2(i\xi) = i(1 + \xi^2)\left[3 \coth^{-1}\xi - \frac{5\xi + 3\xi^3}{(1 + \xi^2)^2}\right]$$

These functions are also used in solving boundary-value problems.

4.14.2. Integrals from a product of the associated Legendre functions

As follows from Equation (4.19), the integral

$$\int_{-1}^{+1}\int_0^{2\pi} [P_n^m(\mu)P_{n'}^{m'}(\mu)(A \cos m\varphi + B \sin m\varphi)(A' \cos m'\varphi + B' \sin m'\varphi)]d\mu d\varphi = 0 \tag{4.64}$$

Because of presence of trigonometric functions this integral is equal to zero, m is integer number, and $m \neq m'$ regardless of the values of n and n'. In order to find the value of the integral when $n = n'$, we will make use of Rodrigue's formula and Equation (4.59). Applying, as in the case of Legendre polynomials, integration by parts, we obtain

$$\int_{-1}^{+1} [P_n^m(\mu)]^2 d\mu = \int_{-1}^{+1} u\, dv = \frac{(-1)^m}{2^{2n}(n!)^2}\int_{-1}^{+1}\left[(\mu^2 - 1)^m \frac{d^{n+m}}{d\mu^{n+m}}(\mu^2 - 1)^n\right]$$

$$d\left[\frac{d^{n+m-1}}{d\mu^{n+m-1}}(\mu^2 - 1)^n\right] = -\frac{(-1)^m}{2^{2n}(n!)^2}\int_{-1}^{+1}\left\{\frac{d}{d\mu}\left[(\mu^2 - 1)^m \frac{d^{n+m}}{d\mu^{n+m}}(\mu^2 - 1)^n\right]\right\}$$

$$d\left[\frac{d^{n+m-2}}{d\mu^{n+m-2}}(\mu^2 - 1)^n\right]$$

Again we will perform integration by parts and let each time

$$u = \frac{d^{s-1}}{d\mu^{s-1}}\left[(\mu^2 - 1)^m \frac{d^{n+m}}{d\mu^{n+m}}(\mu^2 - 1)^n\right], \quad v = \frac{d^{n+m-s}}{d\mu^{m+n-s}}(\mu^2 - 1)^s$$

The product uv becomes equal to zero at terminal points of the interval of integration, since u contains term (μ^2-1) if $m \geq s$, and v has the same term if $m \leq s$. Correspondingly, after an $m+n$-fold integration by parts, we obtain

$$\int_{-1}^{+1} [P_n^m(\mu)]^2 d\mu = \int_{-1}^{+1} \frac{(1-\mu^2)^n}{2^{2n}(n!)^2} \left\{ \frac{d^{n+m}}{d\mu^{n+m}} \left[(\mu^2-1)^m \frac{d^{n+m}}{d\mu^{n+m}} (\mu^2-1)^n \right] \right\} d\mu \qquad (4.65)$$

Inasmuch as after differentiation the degree of μ decreases, the second term of the integrand becomes constant. Therefore, preserving the highest power of μ, this term finally gives

$$[2n(2n-1)(2n-2)\cdots(n-m+1)](n+m)! = \frac{(2n)!(n+m)!}{(n-m)!}$$

and the integral (4.65) becomes equal to

$$\int_{-1}^{+1} [P_n^m(\mu)]^2 d\mu = \frac{(n+m)!}{(n-m)!} \frac{(2n)!}{2^{2n}(n!)^2} \int_{-1}^{+1} (1-\mu^2)^n d\mu = \frac{2}{2n+1} \frac{(n+m)!}{(n-m)!} \qquad (4.66)$$

Note that by analogy with the Legendre functions there are recursive relations between the associated Legendre functions and some examples, if $\mu < 1$, are given as follows:

$$\begin{aligned} \Omega_{n+1}^{m+1} &= (m+n+1)[(1-\mu^2)^{1/2}\Omega_n^m + \mu\Omega_n^{m+1}], \\ \Omega_{n-1}^{m+1} &= (m-n)[(1-\mu^2)^{1/2}\Omega_n^m + \mu\Omega_n^{m+1}], \\ (1-\mu^2)^{1/2}\Omega_n^{m'} &= -m\mu(1-\mu^2)^{-1/2}\Omega_n^m + \Omega_n^{m+1} \end{aligned}$$

4.15. SPHERICAL HARMONIC ANALYSIS OF THE MAGNETIC FIELD OF THE EARTH

Earlier we described the spherical analysis of the magnetic field on the earth's surface provided that the field is independent of the longitude, but now let us consider the general case when $\boldsymbol{B}(R, \theta, \varphi)$. As before, we assume that conduction currents are absent in the vicinity of the earth's surface, in particular, the vertical component of the current is equal to zero. In other words, there is no current flow into the air; otherwise the circulation of the magnetic field on the earth's surface (Chapter 1):

$$\oint \boldsymbol{B} \cdot \boldsymbol{dl}$$

would differ from zero. Calculations performed independently and at different time show that this integral gives for the current density value about $10^{-8}\,\mathrm{A\,m^{-2}}$, but its distribution has a completely random behavior. It may suggest that nonzero value of the integral is due to errors of measuring and processing of data. This allows us to use the concept of the potential and in accordance with Equation (4.6) we have for points of the earth's surface as well as above and beneath it:

$$U(R,\theta,\varphi) = a\sum_{n=1}^{\infty}\sum_{m=0}^{n}\left(\frac{R}{a}\right)^{n}(b_n^m\cos m\varphi + c_n^m\sin m\varphi)P_n^m(\cos\theta)$$
$$+\,a\sum_{n=1}^{\infty}\sum_{m=0}^{n}\left(\frac{a}{R}\right)^{n+1}(g_n^m\cos m\varphi + h_n^m\sin m\varphi)P_n^m(\cos\theta) \qquad (4.67)$$

Here a is the earth's radius. The partial solution of Laplace's equation is often written as

$$[R^n(b_n^m\cos m\varphi + c_n^m\sin m\varphi) + R^{-n-1}(g_n^m\cos m\varphi + h_n^m\sin m\varphi)]P_n^m(\cos\theta)$$

Inasmuch as after a multiplication by a constant a function remains a solution of Laplace equation, Equation (4.67) is usually preferred, because in this case coefficients have the same dimension as the magnetic field, nT. In geomagnetism the functions $P_{n,m}$ are usually used and they are called Schmidt functions:

$$P_{n,m} = \left[\frac{2(n-m)!}{(n+m)!}\right]^{1/2}P_n^m$$

and this choice simplifies Equation (4.66).

As we already know, the magnetic field corresponding to $n = 0$ and caused by currents inside the earth is equal to zero and this is the reason why summation with respect to n is started from $n = 1$. First, consider an unrealistic case when the potential is known. In accordance with Equation (4.67), at points of the earth's surface we have

$$U(a,\theta,\varphi) = a\sum_{n=1}^{\infty}\sum_{m=0}^{n}[(b_n^m + g_n^m)\cos m\varphi + (c_n^m + h_n^m)\sin m\varphi)]P_n^m(\cos m\theta)$$

or

$$U(a,\theta,\varphi) = a\sum_{n=1}^{\infty}\sum_{m=0}^{n}(A_n^m\cos m\varphi + B_n^m\sin m\varphi)P_n^m(\cos m\varphi) \qquad (4.68)$$

where

$$A_n^m = b_n^m + g_n^m, \quad B_n^m = c_n^m + h_n^m$$

Thus, Equation (4.68) contains two sets of unknowns, A_n^m and B_n^m, and both of them depend on the external and internal magnetic fields. Let us illustrate how in principle these coefficients can be determined, and with this purpose in mind consider one term of the series, for example, $n = 2$:

$$U_2(\theta,\varphi) = a[A_2^0 P_2^0(\cos\,\theta) + (A_2^1\,\cos\,\varphi + B_2^1\,\sin\,\varphi)P_2^1(\cos\,\theta) \\ + (A_2^2\,\cos\,2\varphi + B_2^2\,\sin\,2\varphi)P_2^2]$$

We see that this term has five unknowns and with an increase of n their number also increases. Since we supposed that the potential U is known at each point of the earth's surface, a determination of these coefficients can be done in two steps. First of all, let us multiply both sides of Equation (4.68) by the associated Legendre function, for instance, $P_2^1(\cos\,\theta)$ and integrate with respect to the argument, μ, that is, along the longitude. Taking into account Equation (4.66), we obtain

$$U_2^1(\varphi) = a\int_{-1}^{+1} U(\theta,\varphi)P_2^1(\cos\,\theta)d\mu = a\frac{48}{5}(A_2^1\,\cos\,\varphi + B_2^1\,\sin\,\varphi) \qquad (4.69)$$

since due to orthogonality of the associated Legendre functions, integrals from other terms of the series (4.68) vanish. The second step is obvious. Multiplying the left- and right-hand sides of Equation (4.69) by either $\sin\varphi$ or $\cos\varphi$ and integrating from 0 to 2π, we find unknown coefficients A_2^1 and B_2^1. In the same manner, other coefficients can be determined. In essence, we have described a calculation of coefficients of the Fourier's series of a function, given on the spherical surface. Of course, the potential of the magnetic field on the earth's surface is unknown, and our goal is to perform the spherical harmonic analysis of components of the magnetic field which are observed on this surface. In order to solve this task we can use Equation (4.67), but let us represent the potential in a slightly different form:

$$U = a\sum_{n=1}^{\infty}\sum_{m=0}^{n}\left[c_n^m\left(\frac{R}{a}\right)^n + (1-c_n^m)\left(\frac{a}{R}\right)^{n+1}\right]A_n^m P_n^m(\cos\,\theta)\cos\,m\varphi \\ + \sum_{n=1}^{\infty}\sum_{m=0}^{n}\left[s_n^m\left(\frac{R}{a}\right)^n + (1-s_n^m)\left(\frac{a}{R}\right)^{n+1}\right]B_n^m P_n^m(\cos\,\theta)\sin\,m\varphi \qquad (4.70)$$

Here, as in the case of Equation (4.67), each term contains four unknowns, and c_n^m and s_n^m are numbers between 0 and 1. They characterize a contribution to the potential on the earth's surface, caused by currents located outside and inside the earth. For instance, if the influence of the external currents is absent, then

$$c_n^m = s_n^m = 0 \qquad (4.71)$$

Note that we use the same notation c_n^m in Equations (4.67) and (4.70), but they have very different meaning and value. Now we will find expressions for components of the field, which are known on the earth surface:

$$X = -B_\theta = \frac{1}{R}\frac{\partial U}{\partial \theta}, \quad Y = B_\varphi = -\frac{1}{R \sin \theta}\frac{\partial U}{\partial \varphi}, \quad Z = -B_R = \frac{\partial U}{\partial R} \tag{4.72}$$

First, let us find an expression for the vertical component of the field on the earth's surface. Performing a differentiation with respect to R and letting $R = a$, Equation (4.70) gives

$$\begin{aligned} Z(\theta, \varphi) = &\sum_{n=1}^{\infty}\sum_{m=0}^{n}[nc_n^m - (n+1)(1 - c_n^m)]A_n^m P_n^m(\cos\theta)\cos m\varphi \\ &+ \sum_{n=1}^{\infty}\sum_{m=1}^{n}[ns_n^m - (n+1)(1 - s_n^m)]B_n^m P_n^m(\cos\theta)\sin m\varphi \end{aligned} \tag{4.73}$$

As in the case of the potential, each term of the series contains four unknowns. Suppose that we have performed the spherical harmonic analysis of measured values of the vertical component and it is represented as

$$Z(R, \theta) = \sum_{n=1}^{\infty}\sum_{m=0}^{n}(\alpha_n^m \cos m\varphi + \beta_n^m \sin m\varphi)P_n^m(\cos\theta) \tag{4.74}$$

By definition, the coefficients α_n^m and β_n^m are known. Inasmuch as spherical harmonics are orthogonal functions, from the equality of sums in Equations (4.73) and (4.74) it follows that each term of them are also equal, and we have

$$[nc_n^m - (n+1)(1 - c_n^m)]A_n^m = \alpha_n^m$$

and

$$[ns_n^m - (n+1)(1 - s_n^m)]B_n^m = \beta_n^m \tag{4.75}$$

Thus, for each harmonic we have obtained two equations with four unknowns. Certainly, this result is a great simplification with respect to the equality of sums, which contain an infinite number of terms. At the same time, the set (4.75) shows that with the help of only a vertical component we cannot determine the amplitude of harmonics and it is necessary to use one of the two tangential components. For instance, a differentiation of Equation (4.70) with respect to φ gives for the component Y at the earth surface:

$$Y(\theta, \varphi) = \frac{1}{\sin\theta}\sum_{n=1}^{\infty}\sum_{m=0}^{n}(mA_n^m \sin m\varphi - mB_n^m \cos m\varphi)P_n^m(\cos\theta) \tag{4.76}$$

Carrying out the spherical analysis of the measured field $Y(\theta, \varphi)$, we simultaneously define both coefficients A_n^m and B_n^m. Thus, this tangential component allows us to determine two unknowns and their substitution into Equations (4.75) gives two other coefficients: c_n^m and s_n^m. Of course, in place of the component Y we can use results of measuring $X(\theta, \varphi)$ which can be written as

$$X(\theta, \varphi) = \sum_{n=1}^{\infty} \sum_{m=0}^{n} (A_n^m \cos m\varphi + B_n^m \sin m\varphi) \frac{\partial P_n^m(\cos \theta)}{\partial \theta} \tag{4.77}$$

Performing a spherical analysis we should again obtain the same coefficients A_n^m and B_n^m. If there is a discrepancy between two pairs of values of these coefficients, it may indicate that the field components cannot be derived from the potential. In other words, there is a vertical component of the current through the air. Making use of available data, Gauss performed calculations and demonstrated that values of A_n^m and B_n^m, obtained from each tangential component, coincide with each other. Also it was demonstrated that using these data, the coefficients c_n^m and s_n^m are equal to zero; that is, the magnetic field on the earth's surface is caused by currents inside the earth. The coefficients characterizing this field:

$$g_n^m = (1 - c_n^m) A_n^m, \quad h_n^m = (1 - s_n^m) B_n^m \tag{4.78}$$

are called Gauss coefficients. For instance, if the influence of the external field is absent, we have

$$g_n^m = A_n^m, \quad h_n^m = B_n^m \tag{4.79}$$

Correspondingly, the potential of the field, caused by currents inside the earth, can be written as

$$U(\theta, \varphi) = a \sum_{n=1}^{\infty} \sum_{m=0}^{n} \left(\frac{a}{R}\right)^{n+1} (g_n^m \cos m\varphi + h_n^m \sin m\varphi) P_n^m(\cos \theta) \tag{4.80}$$

For illustration, values of the Gauss coefficients for the first several harmonics are given in Table 4.1.

We see that main contribution is due to the first harmonic, which is independent of the angle φ. Let us notice that Gauss calculated coefficients up to $n = 4$, but later the number of terms was increased. Current calculations go to degrees over 100 but the contribution of the core field beyond degree 12 is a subject of controversy.

As was pointed out earlier, there are different methods to represent the measured magnetic field as a sum of the spherical harmonics; one of them is the solution of the system of equations, when their number exceeds number of unknowns. Such an approach, based on the use of the least squares method, was applied by Gauss. He demonstrated that the internal magnetic field practically behaves as that of the magnetic dipole, and it is defined by the first spherical harmonic, U_1. In accordance with Equation (4.80), it can be written as a sum of

Table 4.1. Main field (nT).

N	m	g_n^m	h_n^m
1	0	−29,682.0	0.0
1	1	−1789.0	5318
2	0	−2197.0	0.0
2	1	3074.0	−2356.0
2	2	1685.0	−425.0
3	0	1329.0	0.0
3	1	−2268	−263.0
3	2	1249	302.0
3	3	769.0	−406.0

three terms

$$U_1 = U_{11} + U_{12} + U_{13}$$

where

$$U_{11} = g_1^0 \frac{a^3}{R^2} P_1^0(\cos\theta) = \frac{4\pi g_1^0}{4\pi R^2} \cos\theta \tag{4.81}$$

$$U_{12} = g_1^1 \frac{a^3}{R^2} P_1^1(\cos\theta) \cos\varphi = \frac{4\pi g_1^1}{4\pi R^2} \sin\theta \cos\varphi \tag{4.82}$$

$$U_{13} = h_1^1 \frac{a^3}{R^2} P_1^1(\cos\theta) \sin\varphi = \frac{4\pi h_1^1}{4\pi R^2} \sin\theta \sin\varphi \tag{4.83}$$

It is obvious that the first term U_{11} describes the potential of the magnetic dipole located at the origin of coordinates and oriented along the z-axis, since g_1^0 is negative. In order to interpret meaning of the other two terms, let us consider a Cartesian system of coordinates and find the angle between the radius-vector $\boldsymbol{R}$ and the x- and y-axes. By definition

$$\boldsymbol{R} = x\boldsymbol{i} + y\boldsymbol{j} + z\boldsymbol{k} \quad \text{and} \quad \boldsymbol{R}\cdot\boldsymbol{i} = R\cos\psi_x = x, \quad \boldsymbol{R}\cdot\boldsymbol{j} = R\cos\psi_y = y$$

Making use of the known relationships between coordinates in both systems, we have

$$x = R\sin\theta\cos\varphi \quad \text{and} \quad y = R\sin\theta\sin\varphi$$

Thus

$$\cos \psi_x = \sin \theta \cos \varphi, \quad \cos \psi_y = \sin \theta \sin \varphi$$

This gives

$$U_{12} = \frac{4\pi g_1^1}{4\pi R^2} \cos \psi_x \quad \text{and} \quad U_{13} = \frac{4\pi h_1^1}{4\pi R^2} \cos \psi_y$$

that is, these terms describe the potential of the magnetic dipoles oriented along the coordinates x and y, respectively. Hence, the first spherical harmonic U_1 behaves as the potential of a magnetic dipole, located at the origin of coordinates and its moment is

$$\boldsymbol{M} = 4\pi a^3 (g_1^1 \boldsymbol{i} + h_1^1 \boldsymbol{j} + g_1^0 \boldsymbol{k}) \tag{4.84}$$

where

$$M = 4\pi a^3 [(g_1^1)^2 + (h_1^1)^2 + (g_1^0)^2]^{1/2} \tag{4.85}$$

while the angle α between the moment and the z-axis is equal to

$$\alpha = \cos^{-1} \frac{g_1^0}{M} \tag{4.86}$$

In the same manner, we may treat other terms of the series as the potential of multi-poles of different order but this is hardly productive, since the approach does not reflect a real distribution of currents. Certainly, the discovery that the magnetic field of the earth is mainly caused by currents inside the earth, and that it behaves as that of the field of the magnetic dipole, oriented almost along the axis of the earth rotation, was a great contribution made by Gauss. In conclusion of this section, let us make several comments:

1. As was well known for several centuries, all components of the magnetic field of the earth are functions of space and time. The change of the magnetic field with time is called variations. Their observations during a relatively small time of the order of 24 h show that they have a periodic character. The period of these variations changes from very small parts of a second to hundreds of thousands of seconds and they differ from each other in amplitude and phase. Observations during longer time intervals, for instance, several years, show that average annual values of the field also vary, but their change has rather monotonic character. However, they also demonstrate a periodic character, provided that we compare their values at much greater times. Thus, there are two types of the variations of the magnetic field: the rapid periodic variations and relatively slow, called secular, variations. It is essential that their origin is also different. As was pointed out above, the first one is caused by currents in

ionosphere, while the secular variations are generated by currents in deep parts of the earth.

2. In order to distinguish the latter, the measurements of the magnetic field during one year over a space of around $10^6\,km^2$ of the earth surface are averaged, and the result is called the main geomagnetic field. The procedure of averaging allows one to reduce the influence of currents in an ionosphere and the conduction and magnetization currents in the upper part of the earth, creating the local anomalies of the magnetic field.
3. The values of the main geomagnetic field are used for the spherical harmonic analysis and, making use of available data, Gauss performed the first such analysis and demonstrated that this field is caused by currents inside the earth and it behaves almost as the field of the magnetic dipole. In this light it may be proper to notice that measuring an inclination of the field on the surface of a spherical magnet, Gilbert suggested that the earth is a uniformly magnetized sphere. As we know (Chapter 3), outside this body, its magnetic field coincides with that of a magnetic dipole.
4. The main geomagnetic field on the earth surface is a relatively weak field, for instance, it can be hundreds or thousands of times smaller than the field of magnets used in a laboratory and industry.
5. In accordance with the spherical harmonic analysis, the main geomagnetic field is a sum of the dipole and nondipole fields. As was pointed out earlier, the dipole part plays the dominant role and as a function of time it behaves differently than the nondipole part. Numerous studies discovered the behavior of poles of the dipole and total fields, as well as a westward drift of the nondipole part and reversal of the magnetic fields and other interesting phenomena.

Chapter 5

Uniqueness and the Solution of the Forward and Inverse Problems

5.1. INTRODUCTION

In this chapter, we will discuss some features of solutions of the forward and inverse problem, when the magnetic methods are applied in exploration geophysics. As usual, we represent the magnetic field as a sum:

$$\boldsymbol{B} = \boldsymbol{B}_{\mathrm{N}} + \boldsymbol{B}_{\mathrm{a}} \tag{5.1}$$

where $\boldsymbol{B}_{\mathrm{N}}$ and $\boldsymbol{B}_{\mathrm{a}}$ are the normal and anomalous fields, respectively. As in the gravitational method, the normal field can be either the main field of the earth or of other fields, which change relatively slow within an area of a survey. There are different ways of introduction of the normal field, and all of them serve as a background allowing us to distinguish an anomaly. The main feature of an anomaly is a relatively high rate of change of the field components with respect to that of the normal field. For instance, in the case of the normal field it can be several nanoteslas per km., while the rate for anomalous field my be around tens and hundreds nanoteslas per km. If magnetic rocks or some other bodies are located very close to the earth's surface such changes can be observed within a very small area. As a rule, an intensity of the anomaly does not exceed 10% of that of the normal field; that is, it is around several thousand nanoteslas. Of course, there are exceptions, where the normal field is smaller the anomalous one; such behavior is observed over large areas, exceeding hundreds of kilometers. Regional anomalies can extend over hundreds of kilometers; dimensions of local anomalies may change from several meters to dozens of kilometers. By definition, the magnetic anomalies are caused by magnetization of rocks and their intensity as well as their shape depending on the geology. Consider several typical cases. In a basin with thick sedimentary formations anomalies may spread over dozens and hundreds of square kilometers, and they change rather slowly. This behavior is related to the fact that sediments are practically nonmagnetic and magnetic substances are located within a crystalline basement at a depth of several kilometers. These anomalies are sufficiently small and rather extended, and they are caused by large bodies inside of the basement or a system of relatively small ones. A different picture occurs when crystalline rocks are located near the earth's surface, covered by sediments with a small thickness. In this case magnetic anomalies may be very large and may reach many hundreds of nanoteslas and sometimes even thousands of nanoteslas. Also, the rate of a change of the field components may be very high, sometimes thousands of nanoteslas per 100 m. The area of such anomalies varies from a few square meters to a few hundred square kilometers. The classical example of a very large anomaly is the Kursk Magnetic Anomaly, caused by thick layers of iron quartz located at a depth of 100–600 m. Also, with an increase of sensitivity of measurements of the magnetic field it has become possible to search for anomalies caused by sediments with a relatively higher concentration of magnetic substances.

The anomalous field is caused by the magnetization of rocks, is, in general, a sum of two vectors

$$\boldsymbol{P} = \boldsymbol{P}_{\rm i} + \boldsymbol{P}_{\rm r} \tag{5.2}$$

where $\boldsymbol{P}_{\rm i}$ and $\boldsymbol{P}_{\rm r}$ are the induced and remanent magnetization, respectively. As a rule, the vector of the induced magnetization is directed along the normal field; however, sometimes there are exceptions caused by magnetic anisotropy. In reality, the vectors of magnetization in Equation (5.2) have different directions, but frequently an interpretation implies that they have the same orientation. In the case of the intrusive and metamorphic rocks the remanent magnetization is very often dominant; in other words, the anomalous field is primarily caused by the remanent magnetization. There are also cases when the induced magnetization prevails. It is conventional to think that the anomaly is positive, if the component of anomalous field has the same direction as the corresponding component of the normal field. If they have opposite orientations, the anomaly is negative. As a rule, the sign of magnetic anomalies caused by bodies of finite dimensions changes with a position of an observation point. Before we start to discuss some features of solutions of the forward and inverse problems let us remember the following.

Each elementary volume of the magnetic body can be treated as a magnetic dipole with the moment

$$d\boldsymbol{M}(q) = \boldsymbol{P}(q)dV \tag{5.3}$$

Here, q is any point inside the body, $\boldsymbol{P}$, the vector of magnetization, and it is the vector sum of the induced and remanent magnetizations. Correspondingly, the potential of the field due to this dipole at an observation point p is

$$dU(p) = \frac{\mu_0}{4\pi}\frac{(\boldsymbol{P}\cdot\boldsymbol{L}_{qp})}{L_{qp}^3}dV \tag{5.4}$$

and L_{qp} is the distance between points q and p. Applying the principle of superposition we obtain for the potential of the anomalous field

$$U(p) = \frac{\mu_0}{4\pi}\int_V \frac{\boldsymbol{P}(q)\cdot\boldsymbol{L}_{qp}}{L_{qp}^3}dV \tag{5.5}$$

and

$$\boldsymbol{B}(p) = -\text{grad}\, U(p) \tag{5.6}$$

Equation (5.5) has very simple meaning but it is rather cumbersome because it requires a calculation at each point of a dot product and integration over the volume of a body. For this reason, it is proper to simplify it. As was shown in Chapter 2

$$\frac{\boldsymbol{P}(q)\cdot \boldsymbol{L}_{qp}}{L_{qp}^3} = \boldsymbol{P}(q)\cdot \overset{q}{\nabla}\frac{1}{L_{qp}} \quad \text{and} \quad \overset{q}{\nabla}\frac{\boldsymbol{P}(q)}{L_{qp}} = \boldsymbol{P}(q)\cdot \overset{q}{\nabla}\frac{1}{L_{qp}} + \frac{1}{L_{qp}}\cdot \overset{q}{\nabla}\boldsymbol{P}(q)$$

Thus, Equation (5.5) becomes

$$U(p) = \frac{\mu_0}{4\pi}\int_V \text{div}\frac{\boldsymbol{P}(q)}{L_{qp}}dV - \frac{\mu_0}{4\pi}\int_V \frac{\text{div}\,\boldsymbol{P}(q)}{L_{qp}}dV \tag{5.7}$$

It is essential that in both integrals integration and differentiation are performed with respect to the point q, which belongs to the volume V. Making use of Gauss theorem, the first integral of Equation (5.7) can be transformed into the surface integral and then in place of Equation (5.7), we have

$$U(p) = \frac{\mu_0}{4\pi}\oint_S \frac{P_n dS}{L_{qp}} - \frac{\mu_0}{4\pi}\int_V \frac{\text{div}\,\boldsymbol{P}}{L_{qp}}dV \tag{5.8}$$

The latter is still complicated since it requires integration over the volume and differentiation of the magnetization vector. As practice shows, in most cases the vector of magnetization is practically constant within the magnetic body and this means that

$$\text{div}\,\boldsymbol{P} = 0 \tag{5.9}$$

Then, Equation (5.7) is drastically simplified and we obtain

$$U(p) = \frac{\mu_0}{4\pi}\oint_S \frac{\boldsymbol{P}\cdot d\boldsymbol{S}}{L_{qp}} = \frac{\mu_0}{4\pi}\oint_S \frac{P_n(q)dS}{L_{qp}} \tag{5.10}$$

This is the basic equation for the solution of the forward and inverse problems in the magnetic method of geophysics. Here q is a point on the surface of the magnetic body, S, and P_n is the normal component of the total vector of magnetization; the normal $\boldsymbol{n}$ is directed outward. Correspondingly, the component P_n can be either positive or negative, and it depends on the angle between the vector of magnetization and the normal $\boldsymbol{n}$. Of course, this equation is much simpler than

Equation (5.5) and this fact defines its important role in the theory of this method. In this light let us write down expressions for the potential for the electric field and that of attraction:

$$U_{\mathrm{e}}(p) = \frac{1}{4\pi\varepsilon_0} \oint\limits_S \frac{\sigma(q)}{L_{qp}} dS \quad \text{and} \quad U_{\mathrm{a}}(p) = k \oint\limits_V \frac{\delta(q)}{L_{qp}} dV \tag{5.11}$$

The analogy with the potential of the electric field is obvious: in both cases integration should be performed over the surface S; in the case of the electric field $\sigma(q)$ is the density of real surface charges, while P_n is the normal component of the vector magnetization, which characterizes a distribution of magnetization currents on the surface of a body. It is proper to notice that if the magnetization varies inside a body, the latter can be represented as a system of smaller bodies with constant but different vectors $\boldsymbol{P}$. Then, applying Equation (5.10) to each of them and performing a summation, in principle, we can find the potential and therefore the magnetic field for a more complicated model. As concerns the potentials of the magnetic and attraction fields, it turns out that they are related to each other and we will consider this question in the next section.

5.2. POISSON'S RELATIONSHIP BETWEEN POTENTIALS U AND U_{a}

Consider a magnetic body with a constant magnetization vector $\boldsymbol{P}$ and a constant mass density δ. This body creates both magnetic and gravitational fields and their potentials are

$$U(p) = \frac{\mu_0}{4\pi} \int_V \frac{\boldsymbol{P} \cdot \boldsymbol{L}_{qp}}{L_{qp}^3} dV \quad \text{and} \quad U_{\mathrm{a}}(p) = k \int_V \frac{\delta}{L_{qp}} dV \tag{5.12}$$

The first integral can be represented in the following way:

$$\int_V \boldsymbol{P} \cdot \overset{q}{\nabla} \frac{1}{L_{qp}} dV = \int_V \boldsymbol{i}_P P \cdot \overset{q}{\nabla} \frac{1}{L_{qp}} dV$$

Here $\boldsymbol{i}_P$ is the unit vector in the direction of the vector of magnetization. Taking into account the fact that

$$\overset{q}{\nabla} \frac{1}{L_{qp}} = -\overset{p}{\nabla} \frac{1}{L_{qp}} \quad \text{and} \quad \boldsymbol{P} = \text{constant}$$

we have

$$\int_V \boldsymbol{P} \cdot \overset{q}{\nabla} \frac{1}{L_{qp}} dV = -P\boldsymbol{i}_P \cdot \overset{p}{\nabla} \int_V \frac{1}{L_{qp}} dV \tag{5.13}$$

Such a change of the order of integration and differentiation is possible, because they are performed with respect to different points. By definition of the gradient the scalar product in Equation (5.13) is the directional derivative along the vector of magnetization. Thus,

$$\int_V \boldsymbol{P} \cdot \overset{q}{\nabla} \frac{1}{L_{qp}} dV = -P \frac{\partial}{\partial l} \int_V \frac{1}{L_{qp}} dV \tag{5.14}$$

Here ∂l is displacement along the vector of magnetization. Comparing Equation (5.14) with the second equation of the set (5.11) and bearing in mind that $\delta = \text{constant}$, we obtain Poisson's relation

$$U(p) = -\frac{P}{k\delta} \frac{\partial}{\partial l} U_{\mathrm{a}}(p) \tag{5.15}$$

The latter states that up to a constant the potential of the magnetic field is the derivative of the potential of the field of attraction along the vector $\boldsymbol{P}$, provided that both parameters $\boldsymbol{P}$ and δ are constant. The last equation can be also written as

$$U(p) = -\frac{P}{k\delta} g_l \tag{5.16}$$

that is, the potential of the magnetic field and the component of the attraction field differ from each other by a constant. Consider one special and important case when the vector of magnetization is directed along the z-axis: $\boldsymbol{P} = P_z \boldsymbol{i}_z$. Then,

$$U(p) = -\frac{P}{k\delta} g_z, \quad B_x(p) = -\frac{P}{k\delta} \frac{\partial g_z}{\partial x}, \quad B_y(p) = -\frac{P}{k\delta} \frac{\partial g_z}{\partial y} \tag{5.17}$$

Thus, knowing the potential of the attraction field we can determine components of the magnetic field. Certainly, this is useful when the potential of attraction field is given in the form which is convenient for differentiation. There many relatively simple shapes of bodies with a constant density where the potential of the field of attraction is expressed in terms of elementary functions that allow one to derive expressions for the magnetic field performing only differentiation, which is usually much simpler than integration. Of course, in the general case of a magnetic body with an arbitrary shape it is natural to perform a numerical integration using directly Equation (5.10). Correspondingly, for components of the field we have

$$B_x(p) = \frac{\mu_0}{4\pi} \oint_S \frac{P_n(x_p - x_q)}{L_{qp}^3} dS, \quad B_y(p) = \frac{\mu_0}{4\pi} \oint_S \frac{P_n(y_p - y_q)}{L_{qp}^3} dS \tag{5.18}$$

and

$$B_z(p) = \frac{\mu_0}{4\pi} \oint_S \frac{P_n(z_p - z_q)}{L_{qp}^3} dS$$

5.3. SOLUTION OF THE FORWARD PROBLEM WHEN THE INTERACTION BETWEEN MAGNETIZATION CURRENTS IS NEGLIGIBLE

Now we will describe one approximate but very important method of solution of the forward problem. With this purpose in mind consider an arbitrary body located in the magnetic field $\boldsymbol{B}_0$. At each point inside this body, magnetization currents are oriented in some direction and the induced magnetization is characterized by the vector $\boldsymbol{P}_\mathrm{i}$. At the beginning, we assume that a remanent magnetization is absent, $\boldsymbol{P}_\mathrm{r}=0$. Then for the induced magnetization, we have

$$\boldsymbol{P}_\mathrm{i} = \chi \frac{\boldsymbol{B}}{\mu} \tag{5.19}$$

where χ, μ are magnetic parameters of the body, and the total field $\boldsymbol{B}$ is a sum of the primary and secondary fields:

$$\boldsymbol{B}(q) = \boldsymbol{B}_0(q) + \boldsymbol{B}_\mathrm{s}(q) \tag{5.20}$$

Here $\boldsymbol{B}_0$ and $\boldsymbol{B}_\mathrm{s}$ are the primary (for instance, earth field) and secondary fields, respectively. It is proper to emphasize that $\boldsymbol{B}_\mathrm{s}$ is a superposition of fields caused by magnetization currents in all parts of a body, and in general it can be comparable with the primary field. In other words, magnetization takes place due to the action of the resultant field. However, we assume that inside the body the secondary field is much smaller than the primary one:

$$\boldsymbol{B}_0(q) \gg \boldsymbol{B}_\mathrm{s}(q) \tag{5.21}$$

that is, magnetization occurs due to the primary field only. This assumption can be expressed differently, namely, we neglect the interaction of magnetization currents. Inasmuch as the ratio in Equation (5.19): $\boldsymbol{B}/\mu$ is equal to

$$\frac{\boldsymbol{B}}{\mu} = \boldsymbol{H} = \boldsymbol{H}_0 + \boldsymbol{H}_\mathrm{s} = \frac{\boldsymbol{B}_0}{\mu_0} + \frac{\boldsymbol{B}_\mathrm{s}}{\mu}$$

in this approximation:

$$\frac{\boldsymbol{B}}{\mu} \approx \frac{\boldsymbol{B}_0}{\mu_0}$$

and therefore

$$\boldsymbol{P}_{\mathrm{i}n} = \chi \frac{\boldsymbol{B}_{0n}}{\mu_0} \tag{5.22}$$

that is, the magnetization of every elementary volume is known. Correspondingly, as follows from Equation (5.10), an expression for the potential of the secondary field outside the body is

$$U_s(p) = \frac{\chi}{4\pi} \oint\limits_S \frac{B_{0n}}{L_{qp}} dS \tag{5.23}$$

since it is assumed that the field $\boldsymbol{B}_0$ is uniform inside the body. Of course, the normal component of this field B_{0n} varies on the surface S. If in addition, there is a residual (remanent) magnetization, we have

$$U_s(p) = \frac{\chi}{4\pi} \oint\limits_S \frac{B_{0n}(q)}{L_{qp}} dS + \frac{\mu_0}{4\pi} \oint\limits_S \frac{P_{rn}(q)}{L_{qp}} dS$$

since

$$\boldsymbol{P} = \frac{\chi}{\mu_0} \boldsymbol{B}_0 + \mu_0 \boldsymbol{P}_0$$

It is a simple matter to derive expressions for the field components. For instance, in the case when $\boldsymbol{P}_r = 0$, we have for components in a Cartesian system of coordinates

$$B_{sx}(p) = \frac{\chi}{4\pi} \oint\limits_S \frac{B_{0n}(q)(x_p - x_q)}{L_{qp}^3} dS$$

$$B_{sy}(p) = \frac{\chi}{4\pi} \oint\limits_S \frac{B_{0n}(q)(y_p - y_q)}{L_{qp}^3} dS \tag{5.24}$$

$$B_{sz}(p) = \frac{\chi}{4\pi} \oint\limits_S \frac{B_{0n}(q)(z_p - z_q)}{L_{qp}^3} dS$$

Similar expressions can be written for the field due to the remanent magnetization, only instead of the product χB_{0n} we have to write $\mu_0 P_{0n}$. For illustration of Equation (5.24) consider a model shown in Fig. 5.1(a). Since the normal component of the primary (earth's) field is equal to zero on the lateral surface of the body, integration is performed over its top and bottom only. It is proper to notice that the normal component B_{0n} has different signs on these surfaces; in particular, it is negative on the upper side.

We have described the method of solution of the forward problem, which is used in practice in geophysics, and from the physical point of view we can imagine two

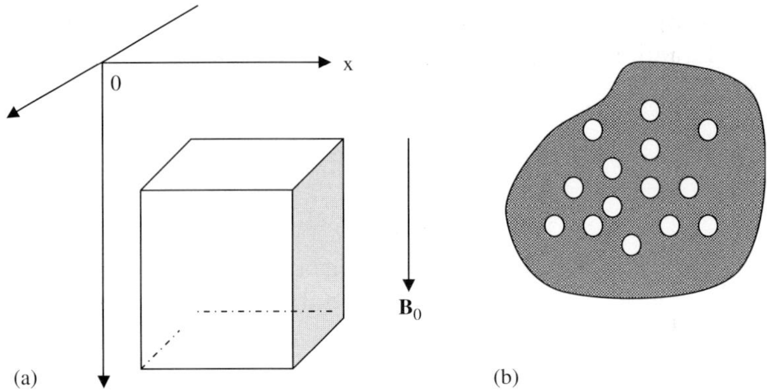

Fig. 5.1. (a) Illustration of Equation (5.24) and (b) disseminated medium.

cases where this approach has a sufficient accuracy. First of all, suppose that a body has a very small susceptibility χ. Then the magnitude of the magnetization currents is also small, and it is natural to neglect the interaction between them. The second case is a disseminated body which consists of many elementary ferromagnetic particles so that the distance between them is much greater than the particle size, but the rest part of a body is not ferromagnetic. For this reason we may expect that the field of each magnetization current decreases very rapidly and in the vicinity of neighboring particles it can be neglected with respect to the field of the earth (normal field). Of course, it is proper to use this approximation if the concentration of particles is relatively low. Let us notice that unlike the previous case the magnitude of magnetization currents can be large. In performing field calculations we mentally replace the real nonuniform medium by a uniform one that produces the same secondary field. To illustrate this procedure let us assume that small spherical particles with magnetic permeability μ and radius a are distributed uniformly within the nonmagnetic medium (Fig. 5.1(b)). Consider inside the body a spherical volume with radius R_0, which contains N particles. Then, making use of results obtained earlier, we can represent the potential caused by the currents in every particle in the form

$$U_n(p) = \frac{\mu - \mu_0}{\mu + 2\mu_0} \frac{a^3}{R_n^2} B_0 \cos \theta_n$$

where R_n is the distance between the observation point p and the sphere center and θ_n, the angle formed by the primary field and the radius-vector $\boldsymbol{R}_n$. Assume that the distance R from the point p to the center of the spherical volume is much greater than its radius R_0; then the potential due to all N particles located within this volume is

$$U(p) = N \frac{\mu - \mu_0}{\mu + 2\mu_0} \frac{a^3}{R^2} B_0 \cos \theta \tag{5.25}$$

since $R_n \approx R$ and $\theta_n \approx \theta$; θ is the angle between vectors $\boldsymbol{R}$ and $\boldsymbol{B}_0$.

Now we will suppose that the spherical volume is filled by a uniform medium with magnetic permeability μ_*, such that its magnetization currents generate the same magnetic field as those of the original model. It is clear that the potential of this field is

$$U(p) = \frac{\mu_* - \mu_0}{\mu_* + 2\mu_0} \frac{R_0^3}{R^2} B_0 \cos\theta \tag{5.26}$$

By equating the right-hand side of last two equations we determine an equivalent magnetic permeability μ_*:

$$\frac{\mu_* - \mu_0}{\mu_* + 2\mu_o} R_0^3 = N \frac{\mu - \mu_0}{\mu + 2\mu_0} a^3 \tag{5.27}$$

It is convenient to introduce a new parameter

$$V = \frac{Na^3}{R_0^3} = \frac{(4\pi/3)a^3 N}{(4\pi/3)R_0^3}$$

which characterizes the volume of particles per unit volume of a body. Solving Equation (5.27), we obtain

$$\mu_* = \mu_0 \frac{1 + 2VK_{12}}{1 - VK_{12}} \tag{5.28}$$

Here

$$K_{12} = \frac{\mu - \mu_0}{\mu + 2\mu_0} = \frac{\chi}{\chi + 3}$$

Inasmuch as the term VK_{12} is usually very small, we can rewrite Equation (5.28) as

$$\mu_* \approx \mu_0(1 + 3VK_{12}) \quad \text{or} \quad \chi_* \approx \frac{V\chi}{1 + (\chi/3)} \tag{5.29}$$

For instance, if magnetite occupies 1% of the volume and its susceptibility equals one, then the equivalent uniform medium has susceptibility a little less than 10^{-2}. As follows from Equation (5.29), the parameter χ_* is often directly proportional to the susceptibility of ferromagnetic particles and the relative volume occupied by them if $\chi < 1$. We have derived Equation (5.29) provided that all particles are spherical and their interaction is absent. At the same time it is natural to expect that the susceptibility of the equivalent uniform medium χ_m depends on the shape, dimensions, and mutual orientation of particles, as well as their susceptibility. For instance, if elongated particles are not oriented in the same direction, then the induced magnetization is different for different particles.

Therefore, unlike the case of spheres, the secondary field represents a vector sum of fields caused by every particle. This shows that the susceptibility of an equivalent medium χ_* may be a function of the orientation of the primary field $\boldsymbol{B}_0$.

5.4. DEVELOPMENT OF A SOLUTION OF THE FORWARD AND INVERSE PROBLEMS

At the beginning, as in the gravitational method, the forward problem was solved for magnetic bodies of relatively simple form, where the field components, Equation (5.18), can be expressed through elementary functions. These cases are well known, and they include many important models of different local bodies and structures. It is impossible to overestimate the importance of these solutions at the time when computers were not available. Several generations of geophysicists over the world laid down a foundation of the magnetic and gravitational methods long before a computer revolution. In fact, a careful study of the field behavior in the presence of different magnetic bodies allowed one to develop several very effective methods of solution of the inverse problem, which have found broad application in practice. One such method is the method of characteristic points, which is based on relations between some geometrical parameters of a body and values of the field at some specific points of curves of the field components. These points are usually points of either maximum or minimum, as well as points where field is equal to zero. Also they can be points, where field magnitude is half of the maximum, etc. For example, in the case of an inclined and relatively thick layer the center of its top is located under the point of the curve B_z where the straight line connecting the points of maximum and minimum intersects this curve. The second method which also obtained the wide application is the method of tangents, used for estimation of body parameters and specially the depth to the body. Like the method of characteristic points, it was developed from a careful study of field behavior. There are different modifications of this approach, and one of them is the following. At the flexure point we define the angle which the tangent at this point forms with the x-axis and draw the bisector between the tangent and the horizontal line at this point. Then we draw two tangents to the curve of B_z component near a maximum and minimum which are parallel to the bisector. It turns out that the difference of the x-coordinates of the tangent points, Δx, is approximately related to the depth to the body as

$$\Delta x \approx 1.6h$$

and this result is valid for many bodies of layer type. It is clear that there must be an outstanding depth of understanding of the field behavior in order to realize that there is a relationship between the position of these points and the body depth. Besides several other methods of solution of the inverse problem were developed, for instance, a group of so-called integral methods, which are based on relationships between the area bounded by the curves or lines of equal values of the field component. This approach allows one, under certain conditions, to determine such parameters as the total magnetic moment, its orientation, and in some case the depth h to the upper part of a body.

With the application of computers the method of characteristic points and tangent are still used for estimation of body parameters. In particular, they are very useful as the first guess for computer methods of solution of the inverse problem for an arbitrary body. The better the first approximation the smaller number of iterations is required to solve inverse problem. There are different methods of solution of the forward problem for a three-dimensional body, where either a volume is replaced by a system of simpler bodies, like prisms, or a surface, surrounding this volume, is approximated by a polygon with a plane faces. In the first case, the field due to each prism is known and correspondingly integrals in Equation (5.18) are sums of fields caused by each prism. In the second case, the determination of the normal component of the vector of magnetization at each face of the polygon is a very simple procedure and the summation for each component is also rather simple operation. As in the gravitational method, solution of the forward problem or numerical integrations of Equation (5.18) is a well-developed procedure. Now we are prepared to discuss the main subject of this chapter, namely, uniqueness and ill- and well-posed problems. At the same time it may be instructive to first consider two examples illustrating an interesting difference in the behavior of magnetic and attraction fields.

5.4.1. Example 1: Uniform half space

First, suppose that a medium is a uniform half space with one plane boundary perpendicular to the z-axis, and a constant vector of magnetization has only a z-component. For simplicity, assume that the observation point is located on the z-axis; then in place of Equation (5.18), we have

$$B_x(p) = -\frac{\mu_0}{4\pi}\oint_S P_{\rm n}\frac{x_q}{L_{qp}^3}\,dS, \quad B_y(p) = -\frac{\mu_0}{4\pi}\oint_S P_{\rm n}\frac{y_q}{L_{qp}}\,dS \tag{5.30}$$

and

$$B_z(p) = \frac{\mu_0}{4\pi}\oint_S P_{\rm n}\frac{(z_p - z_q)}{L_{qp}^3}\,dS$$

Here, the point q belongs to a plane interface and P_n is the projection of the vector $\boldsymbol{P}$ on the normal to the plane. Inasmuch as we can always find two points on the plane symmetrically located with respect the z-axis $x(q_1) = -x(q_2)$, $y(q_1) = -y(q_2)$, the first two integrals are equal to zero and therefore the horizontal components of the magnetic field are absent:

$$B_x(p) = B_y(p) = 0$$

The z-component of the field can be represented as

$$B_z(p) = -\frac{\mu_0}{4\pi} P_{\rm n} \int_S \frac{z_q - z_p}{L_{qp}^3} dS$$

where P_n is a scalar component, which can be either positive or negative, and $z_p > z_q$. Correspondingly, we have

$$B_z(p) = -\frac{\mu_0}{4\pi} P_{\rm n} \int_S \frac{\mathbf{L}_{qp} \cdot d\mathbf{S}}{L_{qp}^3} = -\frac{\mu_0}{4\pi} P_{\rm n}\omega(p) \tag{5.31}$$

Here $\omega(p)$ is the solid angle under which we see a plane from point p, and as is well known it is equal to -2π. Thus, the vertical component of the magnetic field caused by a homogeneous magnetization of a half space is

$$B_z(p) = \frac{\mu_0 P_{\rm n}}{2} \tag{5.32}$$

For instance, if the vector $\boldsymbol{P}$ is directed downward, then P_n is negative and $B_z < 0$. Certainly, a uniform half space is very abstract model which, in particular, does not take into account the opposite interface located in this case at infinity. We see that the field above a magnetic half space with a constant magnetization is uniform and, in accordance with Equation (5.30), in the case of an arbitrary orientation of the vector $\boldsymbol{P}$ the field is caused by its vertical component only.

5.4.2. *Example 2: Layer of finite thickness*

Next, consider a horizontal layer of thickness h when the vector of magnetization is still constant. Inasmuch as at both boundaries the normal component P_n has opposite signs but the same magnitude from Equation (5.32) it follows that the field outside is equal to zero. If we take a point inside the layer and mentally divide the layer into two smaller layers: one above and other beneath we come to the same result, that is, the field inside also vanishes. Thus, we have discovered that a layer with a constant magnetization does not create a magnetic field:

$$\boldsymbol{B} = 0$$

Certainly, this is not obvious result, but it can be easily explained proceeding from Ampere's law. At the same time, such a layer with density δ creates a vertical component of the attraction field equal to

$$g_z = \pm 2\pi k\delta h$$

Here k is the gravitational constant, and as is well known, this equation describes one of the important corrections reducing the influence of topography. In the gravitational method, we also take into account the change of the normal field with elevation. In this light let us evaluate the change of the dipole part of the main magnetic field. Suppose the difference of elevations between two points is equal to h. Then the difference of the field's component, caused by a dipole, can be represented as

$$B(p)\left[1-\frac{R^3}{(R+h)^3}\right] \approx B(p)\left(1+3\frac{h}{R}\right)$$

that is, the correction factor per meter is

$$\frac{\Delta B}{h} \approx 3\frac{B(p)}{R} \quad \text{and} \quad \frac{\Delta B}{h} \approx 2\times 10^{-2}\gamma$$

if $R = 6.4 \times 10^6$ km, $B = 50{,}000\gamma$. Because the normal field of attraction is relatively large, the correction for elevation is stronger in the gravitational method.

5.5. CONCEPT OF UNIQUENESS AND THE SOLUTION OF THE INVERSE PROBLEM IN THE MAGNETIC METHOD

5.5.1. Main steps of interpretation

Suppose that we know the anomalous magnetic field, for instance, on the earth's surface and our goal is to determine the geometrical parameters of a magnetic body and its magnetization. In other words, we have to solve the inverse problem or perform an interpretation of results of measuring with a magnetometer. Undoubtedly, this is the most important element of any geophysical method. To outline this subject we begin with the simplest model and then consider more complicated ones. First, assume that measurements of the magnetic field are performed on the surface of a horizontally layered medium and, in general, the layers have different thicknesses and vectors of magnetization. Inasmuch as at each point of observation the magnetic field is equal to zero, we conclude that magnetic method is not able to perform soundings; that is, to determine the parameters of each layer. Thus, in order to obtain information about the magnetization beneath the earth's surface we have to have lateral changes. Further, we mainly consider a model of a magnetic body of finite dimensions, surrounded by a nonmagnetic medium, shown in Fig. 5.2(a), and suppose that at least one component of the anomalous field is known by measurements along a profile or a system of profiles. Then, the purpose of interpretation is to determine the location, shape, dimensions, and magnetization of the subsurface body. This task is often called the inverse problem of magnetic field theory, since it is necessary to find parameters of a body, when its field is known along some profile or in some area. It is essential that the field is not known in the volume of the magnetic body, since measurements are almost always performed at some distance from this body, and this is the main reason why interpretation becomes a rather complicated problem. First let us analyze the measured field. With this purpose in mind we will proceed from

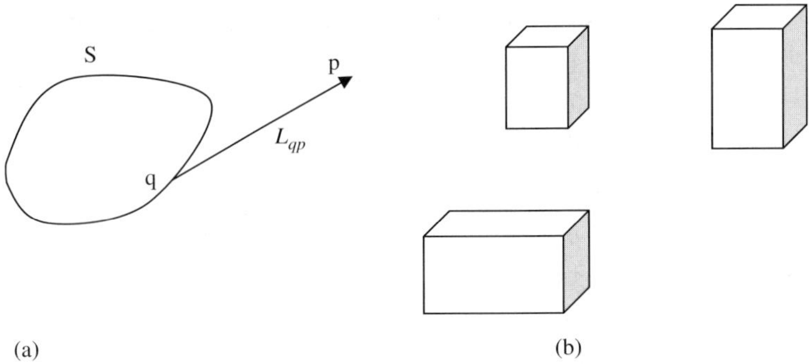

Fig. 5.2. (a) Illustration of Equation (5.33) and (b) system of different prisms.

Equation (5.10) and assume that the normal component of the vector of magnetization characterizes a surface density of fictitious sources of the magnetic field:

$$\sigma = P_{\mathrm{n}}$$

Of course, these charges do not exist, but the word "charge" is used to replace the less convenient "normal component of magnetization on the surface of a body." In particular, for the vertical component of the magnetic field, Equation (5.18), we have

$$B_z(p) = \frac{\mu_0}{4\pi} \oint_S \sigma(q) \frac{(z_p - z_q)}{L_{qp}^3} dS \tag{5.33}$$

In accordance with this equation the vertical component, as well as the two other components, can be represented at every observation point as a sum of fields caused by elementary "charges" on the surface of a body, and their contributions depend on the size, location, and orientation of these surfaces with respect to the observation point. For instance, those "charges" that are located far away from the observation point only slightly affect the field magnitude. On the contrary, "charges" situated closer to the point of observation have a stronger influence. Certainly, an orientation of the vector of magnetization with respect to the normal of an elementary surface is also important. Strictly speaking, at every observation point the field is subject to the influence of all parameters of the body, although to different extents, but each element of the surface, regardless of its position, makes a contribution to the measured field. Their relative effect varies from point to point of observation since they have different positions with respect to the body. In other words, the influence of different parts of a body changes with the position of the point of observation. Thus, the field measured at every observation point contains some information about the magnetization and geometry of a body. Taking into account this simple but fundamental fact let us formulate the main steps of interpretation.

1. First, we will make some assumptions about the magnitude and direction of the total vector of magnetization and ascribe values of parameters of a body which characterize its position with respect to the earth's surface and dimensions. Such a step is usually called the first guess or the first approximation. It is mainly based on some geological information and data obtained from other geophysical methods, and they approximately define the number of parameters and their numerical values. For example, if measurements of the magnetic field are performed over a known mining deposit, there is usually some information about a shape and dimensions of ore bodies. Of course, the difference between the first approximation and the factual values of body parameters can vary significantly depending on our knowledge of the geology.
2. The second step of interpretation consists of calculating synthetic values for the field along the profile, using the first approximation and comparing the measured and calculated fields. A reasonable coincidence of these fields may indicate that the chosen parameters of the model are close to the real ones.
3. If there is a difference between the measured and calculated fields, all parameters of the first approximation or some of them are changed in such a way that a better fit to these fields is achieved. Thus, we obtain a second approximation for parameters of a body. Of course, in those cases when even the new set of parameters does not provide a satisfactory match of these fields, this process of calculation has to be continued. As we see from this procedure, every step of interpretation requires application of Equation (5.10); that is, a solution of the forward problem.

Later we will add one more element, caused by the fact that the field containing information about the parameters of a body (useful signal) is never known exactly. Inasmuch as in the process of interpretation every step is reasonably well defined, we may arrive at the impression that the solution of the inverse problem is straightforward and does not contain any complications. Unfortunately, in reality this is not true and if there are exceptions, then they have purely theoretical interest. In order to realize some difficulties of solution of the inverse problem it is useful to start from an unrealistic situation.

5.5.2. Uniqueness and its application

Suppose that both the calculated and measured magnetic fields are known with infinitely high accuracy. Of course, it is impossible to know the value of the field without any error. This means that any digit after the decimal point, describing the field, is known, regardless of how small its contribution. At the same time, any computer or magnetometer provides a value of the field, for which we know only its first digits. In spite of the fact that we are going to consider this unrealistic case, it is very useful to discuss this subject for understanding of principles of interpretation in the magnetic method. Thus, assume that we know the fields exactly and perform all steps of interpretation described above. Suppose that sequentially repeating the solution of the forward problem at each step and comparing calculated and measured values of the field, we obtain a set of parameters such that a difference between these fields is infinitely small. Then the following question arises. Does this mean that by providing an unrealistic ideal fit between the measured and calculated fields, it is always possible to determine with an infinitely small error the shape, dimensions, location as well as magnetic parameters of a body that creates the given field? In general the answer is negative and the solution of the inverse problem sometimes is not unique; that is, different distribution of surface "charges" may create exactly the same field along a profile or a system of profiles. In other words, in general, but not always, different magnetic bodies can generate a field that provides an exact match to the measured field. The simplest example of such nonuniqueness is the very well known case in which the field is caused by different spheres with the same total magnetization, $\boldsymbol{M}$, and common center but different magnitude of the vector of magnetization, $\boldsymbol{P}$, and radii. At the same time, if we look more carefully at this subject, then it becomes clear that the phenomenon of nonuniqueness is hardly obvious. In fact, Equation (5.10) tells us that a change of body geometry should result in a change of the magnetic field. However, from nonuniqueness it follows that different magnetic bodies can create outside them exactly the same field, even if we know the fields without error. In other words, it is impossible to detect the difference between fields generated by such bodies. It is difficult to get rid of the impression that nonuniqueness is an amazing, unexpected fact which is more natural to treat as a paradox rather than an obvious consequence of the behavior of magnetic fields. In this light let us imagine for a moment that nonuniqueness is always present. Then it is clear that in such a case the interpretation of magnetic data would always be impossible. In fact, having determined parameters of a body that generates a given field, we have to also assume that due to nonuniqueness there are always other magnetic bodies which create exactly the same field. Certainly, we can say that such an ambiguity would be a disaster for the application of the magnetic method. Fortunately, this whole subject of uniqueness is not relevant in practice, because we never know the fields exactly and for this reason it would be natural to avoid the discussion of this topic. At the same time it is worthwhile to clarify some aspects of uniqueness in solving the inverse problem. First of all, as the theory of the potential shows, our treatment of nonuniqueness as a paradox is often correct. There are at least two classes of bodies for which a solution of the inverse problem is unique. One of them is a prism or a system of them (Fig. 5.2(b)). M. Brodsky proved that if we know the magnetic field exactly caused by some prism then there is only one prism which generates this field. Other prisms cause different fields. In other words, the solution of the inverse problem is unique. For illustration, consider a single prism, characterized by a vector of magnetization, dimensions of a cross-section, and distances from the top and bottom to the plane of observation. Regardless how small the prism is with respect to the distances to the observation points a solution of inverse problem will give exact values of the parameters of the prism. To emphasize this fact we can imagine that the distance between the prism and observation points is comparable with that from the earth to the moon, but the dimensions of the prism are around few centimeters. At the same time, different magnetic prisms cause different fields;

that is, a solution of the inverse problem for bodies which belong to the class of prisms is unique. The same result holds if instead of a single prism we have a system of prisms of different sizes and magnetization (Fig. 5.2(b)). It is interesting to notice that an arbitrary body can be usually represented as a system of different prisms. Another class of magnetic bodies for which the solution of the inverse problem is unique is the so-called star-shaped bodies, which are characterized by a more general shape. By definition, every ray drawn from any point of the body intersects the body surface only once. Unlike prisms, in this class of bodies uniqueness requires knowledge of the vector of magnetization. This theorem was proved by P. Novikov. The simplest example of a star-shaped body is a spherical one. Of course, prisms are also star-shaped bodies but due to their special form, that causes field singularities at corners, the inverse problem is unique even without knowledge of the vector of magnetization. It is obvious that these two classes of bodies include a wide range of magnetic bodies; besides it is very possible there are other classes of bodies for which the solution of the inverse problem is also unique. It seems that this information is already sufficient to think that nonuniqueness is not obvious but looks like a paradox.

Assuming that the measured and calculated fields, caused by a magnetic body, are known exactly it is a simple matter to outline the main steps of interpretation and, as was pointed out earlier, it is a straightforward task. Suppose that we deal with a class of bodies for which uniqueness holds. Then, the main steps of interpretation were already formulated above and they are:

1. Proceeding from the observed field and making use of additional information we approximately define the parameters of a magnetic body (first guess).
2. Substituting values of these parameters into Equation (5.10) we solve the forward problem and compare the measured and calculated fields.
3. This process of comparison allows us to determine how the parameters of the first guess have to be changed in order to decrease the difference between the measured and calculated fields.

Performing a solution of the forward problem with the new parameters we again compare fields, and this process can continue until the accuracy of determination of the parameters satisfies our requirements. It is essential that in performing the solution of the inverse problem we can in principle reduce the error in evaluating the parameters of a body to zero. Note that if the first guess contains parameters which do not characterize a body and they are introduced by error, this procedure of solution of the inverse problem allows us to detect and eliminate them. Of course, there are classes of bodies for which the solution of the inverse problem in the magnetic method is not unique. The same is usually true if bodies of different classes are considered. This fact is not surprising, and is also observed in other geophysical methods. For this reason, it is difficult to understand why the geophysical literature often emphasizes the fact that a solution of the inverse problem in so-called potential methods (gravitational and magnetic methods) is not unique without any reference to other methods. In this light it may be proper to notice that a concept of potential is used in all geophysical methods, so a division on the potential and nonpotential methods is hardly proper. At the same time, theorems of uniqueness are proven for certain well-specified class of bodies, such as prisms, star-shaped bodies, etc. For instance, a solution of the inverse problem for prisms does not mean that we cannot find some body different from a prism, which creates exactly the same field as that of the prism. As was pointed out earlier, uniqueness of a solution of an inverse problem has purely mathematical interest because it implies that the calculated and measured fields are known exactly. Since in reality this does not hold this subject will be put aside.

5.6. SOLUTION OF THE INVERSE PROBLEM AND INFLUENCE OF NOISE

Now we are ready to make one step forward and discuss some aspects of interpretation for real conditions when the magnetic field is measured with some error; that is, the numbers that describe the field are accurate to only some decimal places. This is the fundamental difference from the previous case where we assumed that the field generated by fictitious "charges" is known exactly. The presence of error is caused by two following factors:

1. Measurements are always accompanied by errors, which depend on the design of the instrument as well as external factors such as variation of temperature, etc.
2. The measured field consists of two parts: one of them B_U, caused by the magnetic body, which has to be found (anomaly); this is usually called the useful signal. The other part B_N, is due to the surrounding medium. In the same manner as in the gravitational method, there are different means in the magnetic method, such as various types of filters and upward analytical continuation, which allow one to perform a separation between these signals, provided that they behave differently as functions of coordinates. All of them are in detail described in the literature, and this subject is beyond the scope of our monograph.

Thus, the measured field B_a is a sum:

$$B_a = B_U + B_N \tag{5.34}$$

The latter is applied to any component of the field, and it can be measured with a relatively high accuracy. However, the error of determination of the useful signal:

$$B_U = B_a - B_N \tag{5.35}$$

is also dependent on the contribution of noise. Note that reduction of this noise is one of the most important elements of interpretation of any method including, of course, the magnetic one, since the determination of parameters of a body is based on a comparison of a calculated field (solution of forward problem) not with the measured field but the useful signal, which only contains information about the body. Bearing in mind that at any observation point we never know the value of

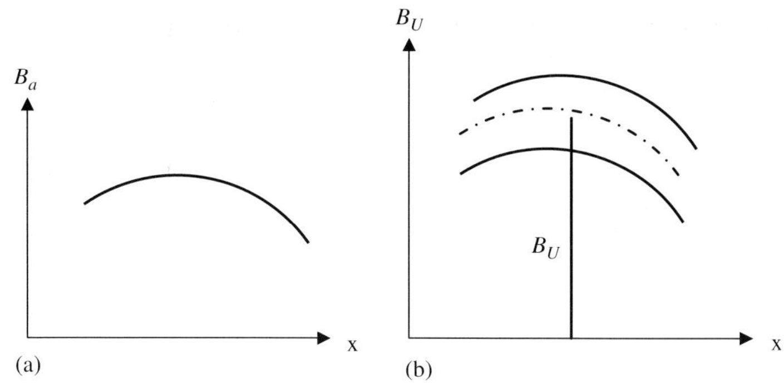

Fig. 5.3. (a) The measured signal; (b) interval of a change of the useful signal.

the noise signal but rather an interval of its change, it is appropriate to speak also about an interval of variation of the useful signal. To emphasize this fact consider the curves in Fig. 5.3(a and b). On the left side of Fig. 5.3 we show a graph of the measured signal, while in Fig. 5.3(b) the interval of a change of the useful signal is shown, as well as the graph of B_U (dotted line). Along the horizontal axis we plot the coordinate x of the observation point.

Because of the influence of noise, we only know that the value of B_U is located somewhere inside of the interval (bar). Its boundary is obtained from Equation (5.35), assuming a certain level of noise. If at an observation point the measured field and that due to noise have different signs, then we have $B_U > B_a$. In contrary, when B_a and B_N have the same sign, the value of the useful signal is smaller than the measured field. Notice that the boundary of a bar, where the useful signal is located, is defined approximately and based on some additional information about accuracy of measurements and the influence of the surrounding medium. From this consideration it is clear that the accuracy of field calculation (forward problem) can be practically the same as that of the measured field, and because of this there is always a difference between these fields. For this reason any attempt to achieve fitting of the calculated field and useful signal with an accuracy exceeding that of their determination has no meaning.

Taking into account the fact that the useful signal is known with some error which sometimes reaches several percent or more, let us consider the influence of this factor on the interpretation. First, as was pointed out, any component of the field can be represented at every point as a sum of fields caused by "charges" located at different parts of a body, and their contribution depends essentially on the location and distance of these "charges" from the observation point. In particular, "charges" located closer to the observation point give larger contributions, while remote parts of the surface produce smaller effects. It is obvious that there are always "charges" on the surface such that their contribution to the useful signal is so small that within a given accuracy of its measurement it cannot be detected. For instance, we can imagine such changes of shape, dimensions, and location of a body, as well as magnetization that the useful signal would still remain somewhere inside an interval (bar). In other words, due to the presence of noise there can be an unlimited number of different magnetic bodies that generate practically the same useful signal. For instance, this may happen when the bodies belong to the class of prisms, the same prisms for which a solution of the inverse problem is unique when the field is known exactly. Inasmuch as the secondary (anomalous) field is caused by all surface "charges" – that is, an integrated effect is measured – some changes of "charges" and the position of a surface in relatively remote parts of a body can be significant; but it turns out that their contribution to the field can be small. At the same time similar changes in those parts of a body closer to observation points will result in much large changes of the field. For this reason, in performing an interpretation it is natural to distinguish at least two groups of parameters describing a magnetic body, namely:

1. Parameters that have a sufficiently strong effect on the field, that is, relatively small changes in their values produce a change of the useful signal that can be detected.
2. Parameters that have a noticeable influence on the field only if their values are significantly changed. This simply means that they cannot be defined from the useful signal measured with some error.

Therefore, we can say that an interpretation or a solution of the inverse problem consists of determining the first group of parameters of the body even though they incompletely characterize its geometry and magnetization. It is clear that this so-called stable group of parameters describes a model of a body that differs to some extent from the actual one, but both of them have common parameters. For instance, these can be the depth to the top of the body or the product of its thickness and some component of the vector of magnetization, or others. Certainly, the most important factor which in essence defines all features of the interpretation, is the fact that the useful signal or the field caused by only a magnetic body is known with some error and because of this the error of evaluation of some parameters (unstable ones) can be unlimitedly large. In other words, these parameters cannot be practically determined. Such inverse problems are called ill-posed problems. In general, an inverse problem in the magnetic method, as well as in other geophysical methods, is ill posed. To illustrate this fact, let us write the following relation between the change of the useful signal $\Delta B_{\rm U}$ and that of a body parameter, Δp_i

$$\Delta p_i = k_i \Delta B_{\rm U} \qquad (5.36)$$

Here k_i is the coefficient of proportionality for the ith parameter of the body, and $\Delta B_{\rm U}$ is the change of the useful field (an interval width) at an observation point; and this varies from point to point. The coefficient k_i is different for different parameters; if it is small, then for a given interval width the change of the parameter P_i is also relatively small. This means that by performing a solution of the inverse problem we can determine the value of this parameter with a sufficiently high accuracy. In contrast, when k_i is large and $\Delta B_{\rm U}$ has a finite value the range of a change of the parameter P_i can be great, and we cannot evaluate this parameter. In principle, an upper limit cannot be established for some coefficients k_i, and this is the most important feature of the ill-posed problem. In other words, even an unlimited change of some parameters of the body does not produce a noticeable variation of the field, exceeding an interval, and as a result of this, it is impossible to define these parameters. Thus, there are always two groups of parameters: they are called stable and unstable parameters of the body and our goal is to separate them and determine the stable ones. The latter characterize a new model of a real body and the coefficients k_i for its parameters are relatively small. In this case we arrive at well-posed problem, but are not able to determine some parameters, and this fact reflects a reality of the solution of the inverse problem in geophysics, in particular, in the magnetic method. The transition from an ill-posed problem to a well-posed one is called the regularization of the inverse problem, and it is of a great practical interest. It is obvious that the interpretation of magnetic data is useful if the parameters of a model, approximating a magnetic body, are defined within such a narrow range of values that is sufficient from a practical point of view. Usually a choice of this group of parameters is automatically selected making use of the corresponding algorithm of the inverse problem. Of course, with an increase in the number of model parameters the approximation of a real magnetic body can in principle be better. However, the error with which some of these parameters are determined also increases. As in the theoretical case when the field is known exactly we can expect that the interpretation of magnetic data is greatly facilitated by the presence of additional information about a body derived from geology and other geophysical methods.

Now let us formulate the main steps of a solution of the inverse problem, taking into account the fact that the useful signal is subjected to the influence of error. These steps are:

1. Making use of preliminary information about parameters of the body we formulate a first guess and completely repeat what was done in the case of uniqueness.
2. The second step is a solution of the forward problem, applying Equation (5.18).
3. As a result of this calculation we obtain a set of values of the field component which can be graphically represented as a curve $B_{\mathrm{U}}(x)$. Suppose that this curve is situated beyond the interval (Fig. 5.3(b)). Then, changing parameters of the body we again use Equation (5.18) and obtain a curve of a field which is closer to the interval of the useful signal. This process continues until a curve of the calculated field is located inside the interval (bar). So far the steps of interpretation are identical to those when we considered the case of uniqueness. Now we will observe a fundamental difference.
4. As soon as values of a calculated field are located inside the observation interval further improvement of matching between the measured and calculated fields does not have any meaning, because we do not know where inside the interval the useful signal is located. Therefore, we stop the process of fitting of fields and start a new procedure which also requires a solution of the forward problem. In the last stage of matching we obtained the set of parameters

$$p_1, p_2, p_3, \ldots, p_n$$

that places the calculated useful signal inside the interval. Our goal is to determine the range of change of each parameter so that the calculated useful signal remains inside the interval. This procedure is usually repeated several times and it is accompanied by a solution of the forward problem. Of course, every step causes a movement of the curve of the useful signal, and so long as its position is inside the interval it is equivalent to the previous one. As result of these steps we obtain for each parameter its range:

$$p_1^{\min} < p_1 < p_1^{\max}, \ldots, p_n^{\min} < p_n < p_n^{\max}$$

As was pointed out earlier, within these ranges the degree of "matching" with the measured useful signal is the same for any set of parameters. Certainly, knowledge of variation of these parameters is the most important step in solving the inverse problem, because this table allows us to separate the stable from unstable parameters, and correspondingly, perform a transformation from the ill-posed to a well-posed problem. Thus, the interpretation gives us a set of stable parameters of a body which causes the useful signal. It is natural to raise the following question. Is it possible that there are other magnetic bodies that produce the same field? Without any doubts the answer is positive, but due to additional information about geology and previous geophysical surveys in the same area or similar ones this ambiguity is often reduced to minimum. In this light it may be appropriate to notice that uniqueness and this ambiguity are not the same. For instance, if the useful signal is caused by a prism or a system of prisms, the solution of the inverse problem is unique; however, we still observe ambiguity as soon as the useful signal is defined with some error. From a review of the solution of the inverse problem it is clear that with a decrease of the interval width the range of each possible parameter of a body decreases too and the number of unstable parameters may become smaller. At the same time stable parameters can be determined with higher confidence. For this reason, reduction of different types of noise is very important subject. We outlined the main features of interpretation for classes of models where uniqueness takes place. It turns out that if models can be described by finite number of parameters, the solution of inversion will be stable and one can get inequalities for potential errors. As concerns an interpretation within class of models with no uniqueness situation is completely different. We can say that a solution of inverse problem in such case is hardly possible (speaking strictly it is senseless). In fact, even in an absence of a noise we always have an infinite number of models which create the same measured field. At the same time it may be possible to determine some generalized characteristics of a body such as the total vector of magnetization. Thus, uniqueness theorems which appear to be mainly of academic interest are actually very important in solving practical inverse problems.

Chapter 6

Paramagnetism, Diamagnetism, and Ferromagnetism

6.1. INTRODUCTION

In order to derive a system of equations of a magnetic field in a magnetic medium, we earlier introduced such concepts as magnetization, vector of magnetization, and magnetization currents and very briefly described different types of magnetic materials. From the classical point of view it follows that these currents are generators of the magnetic field but in reality this field is caused by moving particles inside atoms. This means that in macroscopic scale magnetization currents represent these elementary atomic currents. In other words, the vector of magnetization, characterizing the dipole moment of an elementary volume, is a sum of the atomic dipole moments. In this chapter, using principles of the classical and quantum physics, we will describe a relation between magnetic properties of different magnetic substances and motion of elementary particles within atoms and, first of all, focus on a motion of electrons. Taking into account extremely small dimensions of atoms, the current associated with a motion of an electron can be treated as a magnetic dipole. It is convenient to think that the magnetic moment of the dipole is a sum of two terms. One of them is due to an orbital motion of electron around the nucleus and it has obvious analogy in the classical electrodynamics. The second term is related to the quantum mechanical property of the electron and sometimes it is interpreted in terms of classical mechanics as a rotation of electron around itself (spin rotation). As is well known, in general there are different shells where electrons are located and they may have different moments. Correspondingly, the total magnetic moment of dipoles caused by a motion of electrons is a sum of dipoles due to each electron. Besides, there are magnetic moments associated with a motion of protons and neutrons of a nucleus, and they are several orders smaller than that due to electron motion, and yet these moments play the important role in the nuclear magnetic resonance (NMR) and in measuring the magnetic field of the earth.

Because of heat we may expect that in the absence of an external magnetic field, either all or a part of dipoles is involved in a random motion, that is accompanied by some type of collision between them, and at each instant the average value of the magnetic moments of these dipoles is zero. It may happen that moments due to orbital motion and spin rotation exactly cancel each other. This means that the resultant magnetic moment and, correspondingly, the vector of magnetization are equal to zero. When we place this kind of material into an external magnetic field, it seems that the secondary field does not appear. However, it is not true since when an external magnetic field changes with time, the electromagnetic induction causes a motion of an electron around the nucleus and a magnetic dipole of the inductive origin arises. As follows from the law of electromagnetic induction, a direction of its moment and the ambient magnetic field are opposite to each other. This means that susceptibility is negative and we deal with a diamagnetic material. As long as there is a motion of an electron and electromagnetic induction, there is an induced magnetic dipole. For this reason, diamagnetism is present in materials which are placed into a magnetic field. Taking into account the induced origin of currents, their orientation only slightly depends on a motion of atoms and, correspondingly, the effect of temperature on diamagnetism is relatively weak.

In the next group of substances, which are called paramagnetic materials, each atom has a permanent magnetic moment; that is, a sum of magnetic moments due to orbital motion of electrons and their spin rotation is not equal to zero. Thus, every atom has the magnetic moment caused by a motion of electrons and due to the thermal motion they are oriented randomly. At the same time, in the presence of the ambient magnetic field these dipoles are lined up with the magnetic field. Thus, magnetization takes place and the vector of magnetization is directed along the field. Correspondingly, the field increases inside the paramagnetic material and susceptibility is positive. As in the case of the diamagnetic materials, the secondary field caused by these dipoles is extremely small and in the absence of an external field it disappears.

As is well known, there is the third group of substances (iron, nickel, cobalt, and different alloys) where we observe ferromagnetism. One of the most important features of this phenomenon is a spontaneous magnetization when a material becomes a permanent magnet even in the absence of an external magnetic field. This occurs due to a strong interaction between atoms of material like iron, and it turns out that a force between them exceeds in thousand times the force which follows from Ampere's law. Now, following mainly "*The Feynman's Lectures on Physics*", we start to describe some features of magnetic materials as well as some methods of measuring the magnetic field based on a behavior of electrons and nucleus of atoms.

6.2. THE ANGULAR MOMENTUM AND MAGNETIC MOMENT OF AN ATOM

At the beginning consider the simplest model of a single atom in the absence of an external magnetic field when there is only one electron rotating around a nucleus in the plane perpendicular to the z-axis (Fig. 6.1(a)). First, we focus on this orbital motion. In accordance with the classical mechanics, let us assume that the radius of its orbit is r. By definition, a motion of an electron represents a current I along the orbit and, correspondingly, this small loop can be treated as the magnetic dipole

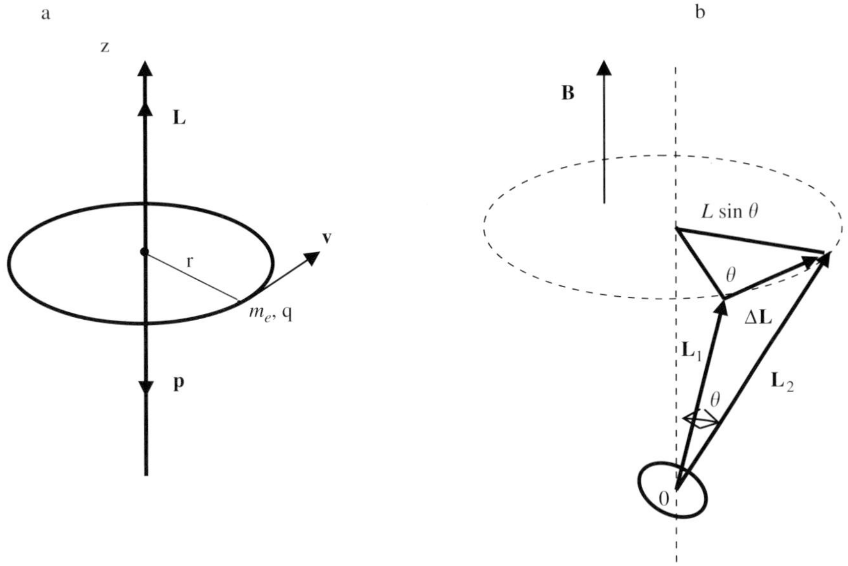

Fig. 6.1. (a) Relation between magnetic moment and angular momentum. (b) Precession of the magnetic dipole.

with the moment $\boldsymbol{p}$ directed along the z-axis:

$$\boldsymbol{p} = I\pi r^2 \boldsymbol{z}_0 \tag{6.1}$$

Here $\boldsymbol{z}_0$ is a unit vector directed along the z-axis and a direction of this vector and that of the electron velocity obeys the right-hand rule. Inasmuch as the current is amount of a charge passing through any point of the orbit per unit time, we have

$$I = \frac{q}{T} \tag{6.2}$$

where q is the electron charge and T the period of its rotation. Introducing the linear velocity of the electron v, we have

$$T = \frac{2\pi r}{v}$$

and therefore the current can be represented as

$$I = \frac{qv}{2\pi r}$$

Thus, the magnetic moment of the dipole is

$$\boldsymbol{p} = \frac{qvr}{2}\boldsymbol{z}_0 \tag{6.3}$$

and it is directly proportional to the negative charge q, its velocity, and a radius of the orbit. It is clear that this tiny current system creates its own magnetic field which is extremely small. The second parameter of an orbital motion of an electron is its angular momentum, $\boldsymbol{L}$. This concept appears in the classical mechanics when we use the second Newton's law to describe a rotation and its expression is

$$\boldsymbol{L} = m_e \boldsymbol{r} \times \boldsymbol{v}$$

Here m_e is a mass of the electron and $\boldsymbol{r}$ the vector directed away from the axis of rotation (Fig. 6.1(a)). Correspondingly, the angular momentum is directed along the z-axis and it is equal to

$$\boldsymbol{L} = m_e r v \boldsymbol{z}_0 \tag{6.4}$$

From Equations (6.3) and (6.4), we obtain an important relation between the magnetic moment and angular momentum:

$$\boldsymbol{p} = \frac{q}{2m_e} \boldsymbol{L} \tag{6.5}$$

Thus, the coefficient of proportionality is defined by the ratio between the electron charge and its mass, but it is independent of the velocity and radius of the orbit. It is essential that the magnetic moment and angular momentum are related with each other, and a change of one of them leads in general to a change of the magnitude and direction of the other. It is proper to emphasize that the same result (Equation (6.5)) follows from the quantum mechanics. Let us represent the last equation slightly differently

$$\boldsymbol{p} = -\frac{q_e}{2m_e} \boldsymbol{L} \tag{6.6}$$

and q_e is positive. Thus, in the case of an orbital motion of an electron, vectors $\boldsymbol{p}$ and $\boldsymbol{L}$ have opposite directions. In accordance with the quantum mechanics, the magnetic field also arises due to an electron rotation around its axis (spin rotation), and this motion is characterized by the magnetic moment and angular momentum too; in this case, in place of Equation (6.6), we have

$$\boldsymbol{p}_s = -\frac{q_e}{m_e} \boldsymbol{L}_s \tag{6.7}$$

Here $\boldsymbol{p}_s$ and $\boldsymbol{L}_s$ characterize a rotation of electron around itself. Note that in a general case when there are several electrons rotating along different orbits and each electron is involved in a spin rotation, we can perform a summation of the vectors $\boldsymbol{p}$ as well as $\boldsymbol{L}$. This gives us the total magnetic moment and the total angular momentum. Then a relation between them is written as

$$\boldsymbol{p} = -g \frac{q_e}{m_e} \boldsymbol{L} \tag{6.8}$$

where g-factor is of the order of unity. As we already know, in the case of the orbital motion of the single electron and of its spin rotation, this factor is equal to 1/2 and 1, respectively. We focus on a motion of electrons and the two parameters $\boldsymbol{p}$ and $\boldsymbol{L}$ associated with this motion. Next let us review a motion inside a nucleus which contains protons and neutrons. As follows from the quantum mechanics, it is also possible to speak about an orbital motion of these particles and their spin rotation. Besides, regardless how this sounds surprising, a motion of a neutron causes the magnetic moment too. It is remarkable that a relation between the resultant moment and resultant angular momentum, associated with motion of protons and neutrons, is similar to that for electrons and has a form:

$$\boldsymbol{p} = g\left(\frac{q_e}{2m_p}\right)\boldsymbol{L} \tag{6.9}$$

where m_p is a mass of the proton and the coefficient g is called the nuclear factor and its value is of the order of unity and depends on a nucleus. For instance, in the case of hydrogen

$$g = 5.58 \tag{6.10}$$

It is instructive to evaluate even approximately quantities which are used in Equations (6.8) and (6.9). First of all, as is well known

$$m_e = 9.1 \times 10^{-31}\ \text{kg}, \quad m_p = 1.67 \times 10^{-27}\ \text{kg}, \quad q_e = q_p = 1.6 \times 10^{-19}\ \text{C}$$

and

$$r \approx 5 \times 10^{-11}\ \text{m}, \quad v \approx 10^5\ \text{m s}^{-1}$$

Correspondingly, for an orbital motion of an electron, we have

$$L \approx 4.5 \times 10^{-36}\ \text{kg m}^2\ \text{s}^{-1} \quad \text{and} \quad I \approx 0.5 \times 10^{-4}\ \text{A}$$

It is obvious that the frequency of orbital motion of the electron is

$$f = \frac{v}{2\pi r} \approx 3 \times 10^{14}\ \text{Hz}$$

The angular momentum is very small quantity but the current is around 50 μA that can be easily measured. Also we have

$$\frac{q_e}{2m_e} \approx 0.88 \times 10^{11}\ \text{C kg}^{-1}, \quad \frac{q_e}{2m_p} \approx 0.48 \times 10^{8}\ \text{C kg}^{-1} \tag{6.11}$$

and both factors are very large. Here the unit of measurements of a charge is Coulomb (C). It may be appropriate to recall that a concept of the angular momentum was introduced from the classical mechanics. In particular, this implies that $\boldsymbol{L}$ can change arbitrarily. Further, proceeding from the quantum mechanics we will describe a completely different behavior of the angular momentum of an electron and nucleus as well as their magnetic moments.

6.3. MOTION OF ATOMIC MAGNETIC DIPOLE IN AN EXTERNAL MAGNETIC FIELD

Until now we have assumed that the external magnetic field $\boldsymbol{B}$ was absent. Next consider an influence of the constant ambient field $\boldsymbol{B}$ on a motion of atom's particles. For comparison it is useful to recall a motion of an elementary current circuit caused by uniform magnetic field (Chapter 3). In this case, the system does not rotate and therefore the angular momentum is zero. As concerns the magnetic moment of the current loop, the former experiences a rotation in the plane formed by this moment $\boldsymbol{p}$ and the field $\boldsymbol{B}$ until both vectors become oriented along the same line; that is, the torque tries to line up the moment of the system with the direction of the magnetic field. Completely different picture takes place when a charge with a mass, for instance, an electron, rotates and therefore it has an angular momentum $\boldsymbol{L}$. In this case, the magnetic moment $\boldsymbol{p}$ will not line up with the field $\boldsymbol{B}$, but instead of it we observe a precession of the angular momentum $\boldsymbol{L}$, and in accordance with Equation (6.8) the magnetic moment is involved in the same motion. In other words, the small orbit with a particle will behave as a gyroscope and the vector $\boldsymbol{p}$ moves on the conical surface around the field $\boldsymbol{B}$. To explain this phenomenon we will use two approaches.

6.3.1. *The first approach*

As before let us start from the case of the orbital motion of an electron. Suppose that the external magnetic field is absent and the angular momentum of an electron is equal to $\boldsymbol{L}_1$. We already know that a motion of an electron forms a magnetic dipole in the vicinity of point 0 (Fig. 6.1(b)), and its magnetic moment as well as the angular momentum is oriented along the line perpendicular to the electron's orbit. Now we place this system into an external field $\boldsymbol{B}$ which forms some angle with the dipole moment $\boldsymbol{p}$ and remains constant with time. This means that we do not consider an electron's motion at the initial time interval when the field $\boldsymbol{B}$ changes from zero to a constant value. By definition, due to this field the magnetic dipole is subjected to an action of a torque:

$$\tau = \boldsymbol{p} \times \boldsymbol{B} \tag{6.12}$$

It is vitally important that the torque is proportional to the magnetic field, because this relation will allow us to establish the linkage between this field and the frequency of precession of magnetic dipoles of atoms. In order to understand the effect caused by this torque, we will proceed from the second Newton's law which can be written as

$$\frac{d\boldsymbol{L}}{dt} = \tau \quad \text{or} \quad \frac{d\boldsymbol{L}}{dt} = \boldsymbol{p} \times \boldsymbol{B} \tag{6.13}$$

that is, the torque is equal to a rate of a change of the angular momentum. In particular, this means that both vectors $d\boldsymbol{L}/dt$ and $\boldsymbol{\tau}$ have the same direction. Note that if the angle between vectors $\boldsymbol{B}$ and $\boldsymbol{p}$ is zero, the angular momentum remains the same. In general, as follows from Equation (6.13), the angular momentum changes with time. Since vectors $\boldsymbol{p}$ and $\boldsymbol{L}$ are oriented along the same line, the direction of the magnetic moment changes, too. Let us take the time interval, Δt, so small that $\boldsymbol{p}$ does not practically change. Then Equations (6.12) and (6.13) give

$$\Delta\boldsymbol{L} = \boldsymbol{L}_2 - \boldsymbol{L}_1 \approx \boldsymbol{\tau}\,\Delta t \approx (\boldsymbol{p} \times \boldsymbol{B})\Delta t \tag{6.14}$$

Bearing in mind that the dipole moment and the angular momentum must be oriented along the same line, we conclude that a change of the angular momentum, $\Delta\boldsymbol{L}$, is perpendicular to the external magnetic field and the angular momentum. This is a very important fact and it defines a behavior of the magnetic dipole. From Equation (6.14), we have

$$\boldsymbol{L}_2 \approx \boldsymbol{L}_1 + (\boldsymbol{p} \times \boldsymbol{B})\Delta t \tag{6.15}$$

and these three vectors form a rectangle (Fig. 6.1(b)). Since the second vector is small, the difference between magnitudes of the vectors $\boldsymbol{L}_1$ and $\boldsymbol{L}_2$ is much smaller and we have

$$L_1 = L_2 \tag{6.16}$$

Thus, the torque $\boldsymbol{\tau}$ does not change the magnitude of the angular momentum. Multiplying both sides of Equation (6.15) by the vector $\boldsymbol{B}$, we obtain

$$\boldsymbol{L}_2 \cdot \boldsymbol{B} = \boldsymbol{L}_1 \cdot \boldsymbol{B} + (\boldsymbol{p} \times \boldsymbol{B}) \cdot \boldsymbol{B}\,\Delta t$$

Taking into account that the last term at the right-hand side is zero, we have

$$\boldsymbol{L}_2 \cdot \boldsymbol{B} = \boldsymbol{L}_1 \cdot \boldsymbol{B}$$

that is, during time interval Δt the angle between the magnetic field and angular momentum remains the same, and it is obvious that this behavior is observed at any time. If the moving vector preserves both its magnitude and the angle with the field direction, this means that it moves on the conical surface and its arrow describes the circle, as is shown in Fig. 6.1(b). Taking into account that the magnetic moment behaves similarly, the vector $\boldsymbol{p}$ also moves on the conical surface. This behavior of the magnetic moment is called a precession. We can imagine that a small current loop which describes the magnetic dipole periodically changes its orientation in such a way that the angle between its normal and the field $\mathbf{B}$ causing this motion remains constant. As follows from the vector algebra, the constant angle θ between the field and the magnetic dipole is defined as

$$\cos\theta = \frac{\mathbf{p}\cdot\mathbf{B}}{pB} \tag{6.17}$$

As was pointed out earlier, if the magnetic field and the magnetic moment are oriented along the same line, the cone is transformed into line and a precession is naturally absent. On the contrary, when these vectors are perpendicular to each other, the conical surface coincides with the plane where the magnetic moment rotates around the field $\mathbf{B}$. In the same manner, we can speak about the precession of magnetic moment, associated with an electron spin and a motion of a nucleus. As follows from the quantum mechanics, it is impossible to determine a position of both vectors $\mathbf{L}$ and $\mathbf{P}$ on the conical surface; that is, all possible positions are equivalent to each other.

6.3.2. Frequency of precession

Because of the external magnetic field, an orientation of the magnetic dipole changes and the end of the vector $\mathbf{L}$, rotating on the conical surface, periodically returns to the same position. Now we determine the frequency of this precession ω_p. From Fig. 6.1(b), we can see that a change of the angular momentum during the time interval Δt is equal to

$$\Delta L = L\sin\theta\cdot\phi \tag{6.18}$$

Here ϕ is the angle between two positions of the circle's radius and it is given by

$$\phi = \frac{2\pi}{T}\Delta t = \omega_p\,\Delta t$$

Thus, the rate of a change of the angular momentum is

$$\frac{\Delta L}{\Delta t} = \omega_p L\sin\theta$$

and it is equal to the torque's magnitude:

$$\tau = pB\sin\theta$$

From the last two equations we obtain an expression for the frequency of precession in terms of the magnetic moment p, angular momentum L, and the field B:

$$\omega_p L = pB \quad \text{and} \quad \omega_p = \frac{pB}{L} \tag{6.19}$$

Thus, this frequency is independent of the angle θ and it is defined by the ratio p/L and directly proportional to the magnitude of the external magnetic field. Let us represent Equation (6.19) in a slightly different form. Taking into account Equations (6.11) and (6.19), we have in the case of electron

$$f_p = \frac{\omega_p}{2\pi} = \frac{g}{2\pi}\left(\frac{q_e}{2m_e}\right)B \approx g \times 1.4 \times 10^{10}\ [\mathrm{Hz\,T^{-1}}]B$$

or

$$f_p \approx g \times 1.4 \times 10^{6}\ [\mathrm{Hz\,G^{-1}}] \cdot B$$

and for proton

$$f_p = \frac{g}{2\pi}\left(\frac{q_e}{2m_p}\right)B \approx g \times 0.76 \times 10^{7}\ [\mathrm{Hz\,T^{-1}}]B \tag{6.20}$$

or

$$f_p \approx g \times 0.76\ [\mathrm{kHz\,G^{-1}}]B$$

and, as we already know, the factor g depends on an atom. Note that the ratios

$$\gamma_e = \frac{p}{L} = g\left(\frac{q_e}{2m_e}\right), \quad \gamma_p = \frac{p}{L} = g\left(\frac{q_e}{2m_p}\right) \tag{6.21}$$

are called the gyromagnetic ratio for the electron and proton, respectively. Thus, we have

$$\omega_p = \gamma B \tag{6.22}$$

For illustration, assume that $B = 0.5\,\mathrm{G}$, that approximately corresponds to the earth's magnetic field and $g = 5.58$ (hydrogen). Then the frequencies of precession of the electron and proton are

$$f_p \approx 3.91 \times 10^{6}\ \mathrm{Hz} \quad \text{and} \quad f_p \approx 2120\ \mathrm{Hz}$$

respectively. Since the magnetic dipoles, caused by such a motion of electrons and nucleus, periodically change their direction, they generate sinusoidal fields with the frequency of their precession.

6.3.3. *The second approach*

It is useful to derive the same result solving the system of the differential equations (6.13). We choose the Cartesian system of coordinates and assume that the field $\boldsymbol{B}$ is directed along the z-axis. Then we have

$$\boldsymbol{p} = (p_x, p_y, p_z) \quad \text{and} \quad \boldsymbol{B} = (0, 0, B) \tag{6.23}$$

where B is the z-component of the field. Correspondingly, Equation (6.13) can be written as

$$\frac{d\mathbf{L}}{dt} = \begin{vmatrix} \mathbf{i} & \mathbf{j} & \mathbf{k} \\ p_x & p_y & p_z \\ 0 & 0 & B \end{vmatrix}$$

Taking into account Equation (6.21), we obtain the system of the ordinary and linear equations with respect to components of the magnetic moment for a particle:

$$\frac{dp_x}{dt} = \gamma p_y B, \quad \frac{dp_y}{dt} = -\gamma p_x B, \quad \frac{dp_z}{dt} = 0 \tag{6.24}$$

First of all, from the last equation it follows that $p_z =$ constant, that it does not change with time. In other words, it is equal to its initial value. In order to find the horizontal components of the moment, we perform a differentiation of the first two equations and arrive at the same homogeneous differential equations of the second order with constant coefficient for each component:

$$\frac{d^2p_x}{dt^2} + \gamma^2 B^2 p_x = 0 \quad \text{and} \quad \frac{d^2p_y}{dt^2} + \gamma^2 B^2 p_y = 0 \tag{6.25}$$

Their solution is well known and it has a form of a sinusoidal function

$$A \cos(\omega_\mathrm{p} t + \varphi)$$

Letting

$$p_x = p_{xy} \cos(\omega_\mathrm{p} t + \varphi) \tag{6.26}$$

and substituting it into the first equation of the set (6.24), we obtain:

$$p_y = -p_{xy} \sin(\omega_\mathrm{s} t + \varphi) \quad \text{or} \quad p_y = p_{xy} \sin(\omega_\mathrm{s} t + \varphi + \pi) \tag{6.27}$$

Here

$$p_{xy} = (p_x^2 + p_y^2)^{1/2} \quad \text{and} \quad \omega_\mathrm{p} = \gamma B$$

The above equations give us functions which describe a behavior of the magnetic moment as a function of time, but the amplitude and phase remain unknown since we do not know the initial conditions, that is, a function $B(t)$ when it changes from zero to the constant value B. As is seen from Equations (6.26) and (6.27) and Fig. 6.2, the projection of the magnetic moment on the

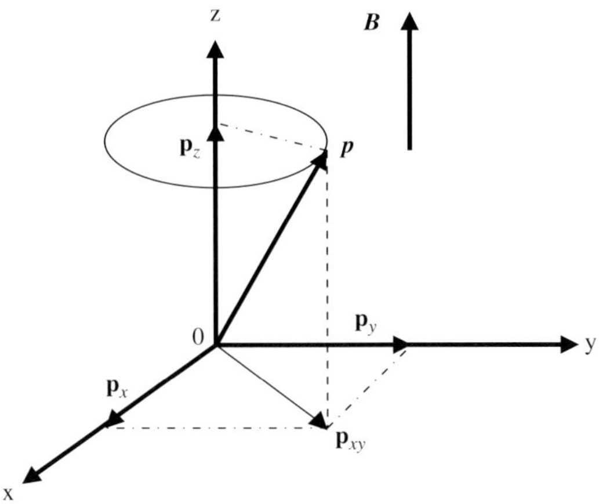

Fig. 6.2. Components of the magnetic moment.

horizontal plane is

$$\boldsymbol{p}_{xy}(t) = p_x(t)\boldsymbol{i} + p_y(t)\boldsymbol{j}$$

and it rotates in this plane with the precession frequency ω_{p} but its magnitude p_{xy} remains constant, while the component p_z does not vary. Therefore, the total vector $\boldsymbol{p}$ is located on the conical surface and performs the periodical motion with the precession frequency, and the ratio p_{xy}/p_z characterizes the cone with the apex at the point 0 and an orientation of the magnetic moment with respect to the field $\boldsymbol{B}$.

6.4. MAGNETIC MOMENT, ANGULAR MOMENTUM, SPIN, AND ENERGY STATES OF ATOMIC SYSTEM

In order to explain some important features of magnetism, we have to proceed from the quantum mechanics and discuss again such concepts as a magnetic moment and angular momentum. At the same time, it is assumed that two equations derived in classical physics which describe energy of the atomic system in the presence of an external magnetic field and a relation between the magnetic moment and angular momentum (Equation (6.8)) remain valid. As before, at the beginning we focus on a motion of electrons.

6.4.1. Magnetic moment

Suppose that in absence of the external field, energy of an atom is U_0. As was shown in Chapter 3, an additional energy of the magnetic dipole placed in the ambient magnetic field is

$$\Delta U = -(\boldsymbol{p} \cdot \boldsymbol{B}) \quad \text{or} \quad \Delta U = -p_B B$$

and, correspondingly, the total energy becomes

$$U = U_0 - p_B B$$

Thus, in terms of energy, the magnetic moment can be defined in the following way. Its component along the field p_B is the coefficient of proportionality between the energy and this field. Also we can still treat the magnetic moment in the classical sense, namely, as a characteristic of the magnetic dipole which creates the secondary magnetic field. It is natural that this field contains information about magnetic properties of substances and this is the main reason why we pay attention to this parameter of an atom.

6.4.2. Angular momentum

In accordance with the quantum mechanics, the angular momentums of an electron and nucleus possess some features which fundamentally differ from those in the classical mechanics. Inasmuch as the world of atoms greatly differs from the world which we directly observe, these features seem very strange and simply shocking but they describe the physical word in a small scale of atoms and we have to accept them as we accept, for example, Newton's law as well as other fundamental laws of the classical physics. For instance, in a macroscopic scale the angular momentum $\mathbf{L}$ is a vector and it has a single scalar component on any axis of coordinate; its value may continuously change from $-L$ to L. In particular, if its value is L, this means that the angular momentum is directed along the axis ("up"), and on the contrary when the component is equal to $-L$, the vector is antiparallel to this axis ("down"). In quantum mechanics we can still use the name "angular momentum" but this concept has different properties and some of them are:

1. Along any direction the angular momentum has a finite number of its components.
2. The difference between values of two neighbor projections is the same, since the value of an angular momentum along any direction is always an integer or half integer of $\hbar$, where $\hbar = h/2\pi$ and h is Planck's constant ($h = 6.626 \times 10^{-34}$ J s). The coefficient of proportionality between the angular momentum and the constant $\hbar$ is called the spin of a system, j, and it varies for different atoms.
3. This set of projections is the same for any axis of coordinates:

$$L = j\hbar, (j-1)\hbar, (j-2)\hbar, \ldots, -(j-2)\hbar, -(j-1)\hbar, -j\hbar \tag{6.28}$$

4. The spin j can be either integer or fractional number, but twice of j has to be integer.

   As follows from Equation (6.28), there are $2j+1$ energy states of atomic system with the same energy U_0.
5. With an increase of the system the number of possible values of projections of an angular momentum increases and then it becomes possible to apply the classical mechanics.

Thus, in the absence of an external magnetic field an atomic system has several "states" with the same energy and their number is defined by a spin j. Each of them is characterized by a certain value of a projection of the angular momentum (Equation (6.28)), and the probability of each "state" is the same.

6.4.3. Magnetic energy of atomic particle

Taking into account this behavior of the angular momentum, consider again an expression for the magnetic energy of an atom placed in the external magnetic field. Assuming that the field $\boldsymbol{B}$ is directed along the z-axis, we can say that this field produces a change of energy by amount

$$\Delta U = -p_z B \tag{6.29}$$

where p_z is the projection of the magnetic moment on the z-axis and this component defines the vector of magnetization. In accordance with Equation (6.8), we have for electron

$$p_z = -g\left(\frac{q_e}{2m_e}\right)L_z = -\gamma_e L_z \tag{6.30}$$

Here the z-component of an angular momentum is given by Equation (6.28) and γ_e is gyromagnetic ratio for electrons. The last two equations are of great importance since they define magnetic energy related to electrons. As follows from Equation (6.29), this energy can be written in the form:

$$U_{\text{mag}} = g\left(\frac{q_e}{2m_e}\right)L_z B = \gamma_e L_z B \tag{6.31}$$

Therefore, the magnetic energy of electrons can have only certain values; for instance, its maximal and minimal values are

$$U_{\max} = g\left(\frac{q_e}{2m_e}\right)\hbar jB \quad \text{and} \quad U_{\min} = -g\left(\frac{q_e}{2m_e}\right)\hbar jB \tag{6.32}$$

In the case $j = 1/2$, there are only two states (Fig. 6.3(a)). For systems with larger spins, we observe more states (Fig. 6.3(b)), with the positive and negative values of the component of an angular momentum. Each state of an atom corresponds to a certain value of the energy, and the larger the value of a spin, the more the states. It may be proper to notice the following. When the external field is absent these possible states have the same energy and their number is defined by a value of a spin, but in the presence of the field $\boldsymbol{B}$ each state obtains slightly different values of energy. In this sense, the energy of the atomic system is "split into $2j+1$ levels". It is essential that a difference between two neighbor states is the same and

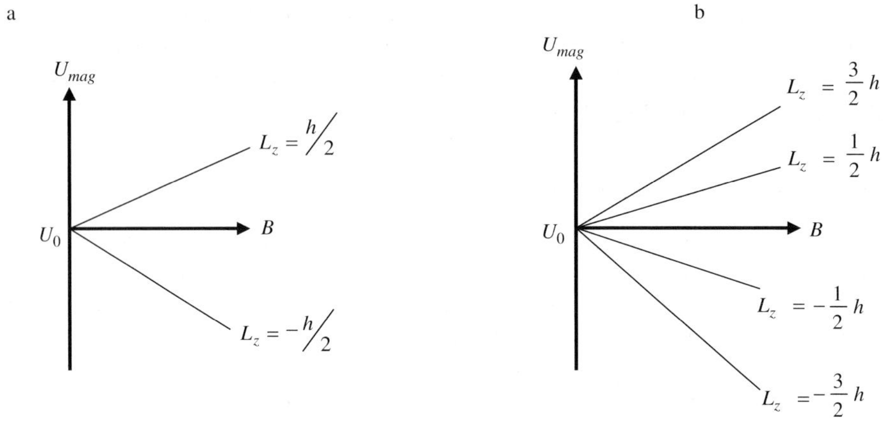

Fig. 6.3. The magnetic energy of electrons as a function of the external field B.

it is equal to

$$\Delta U_{\mathrm{mag}} = g\left(\frac{q_{\mathrm{e}}}{2m_{\mathrm{e}}}\right)\hbar B = \gamma_{\mathrm{e}}\hbar B \tag{6.33}$$

that is defined by parameters of the electrons and the magnetic field. It is important to notice for the given magnetic field "the energy's spacing" (Equation (6.33)) can be expressed in terms of precession frequency. In fact, as follows from Equations (6.8) and (6.20), we have

$$\Delta U_{\mathrm{mag}} = g\frac{q_{\mathrm{e}}}{2m_{\mathrm{e}}}\hbar B = \hbar\omega_{\mathrm{p}} \tag{6.34}$$

Note that Equation (6.34) is used for measuring magnetic moment of an atom. Thus, the difference of energy of two neighbor levels is equal to product of the constant $\hbar$ and precession frequency.

Our study is based on an assumption that the external magnetic field is relatively small, so that the atomic system does not change. It is clear that ΔU_{mag} is very small and energy of each level (state) differs from each other only slightly. Of course, if an atom has several electrons placed in the field $\boldsymbol{B}$, each of them has only one value of energy corresponding to a certain energy level. Until now we discussed a magnetic energy of electrons at different levels, but similar behavior is observed in the case of protons. For instance, in the place of Equation (6.34), we have

$$\Delta U_{\mathrm{mag}} = \hbar\omega_{\mathrm{p}} \quad \text{and} \quad \gamma_{\mathrm{p}} = g\left(\frac{q_{\mathrm{e}}}{2m_{\mathrm{p}}}\right) \tag{6.35}$$

Here γ_{p} is the gyromagnetic ratio for protons and ω_{p} its precession frequency. In essence, we are ready to discuss the magnetic properties of some substances as well as some methods of measuring the magnetic field, but before it is useful to describe the first experiments which proved this amazing behavior of an angular momentum and, as a result, an existence of different energy levels of an atomic system too; also they allowed one to determine the magnetic moment of an atomic system that is of great importance for study of magnetism.

6.4.4. The Stern–Gerlach experiment

This experiment was performed in 1922 and can be considered as an experimental proof of quantization of an angular momentum. The device consists of four main parts shown in Fig. 6.4, and they are: oven (1), screen with a hole (2), couple of magnets (3), and glass plate (4). The hot oven evaporates silver atoms which move away in all directions. With a help of a system of screens with very small holes (one of them is shown), it is possible to form a sufficiently narrow beam of atoms (dotted line), traveling between two magnets which create a magnetic field directed along the z-axis. One of the magnets has a very sharp edge in order to increase the rate of a change of the field in the z-direction. It happens because its field behaves as that of a thin line of fictitious magnetic charges that provides a relatively rapid change. At the beginning when atoms approach the space between magnets, the beam of silver atoms is almost parallel to this edge, that is, perpendicular to the z-axis. As we know, in accordance with the classical physics the additional energy of each particle, for instance, electron in the presence of the external field $\boldsymbol{B}$, is equal to

$$\Delta U_{\mathrm{mag}} = -\boldsymbol{p} \cdot \boldsymbol{B} = -pB \cos\theta \tag{6.36}$$

Here p is the magnitude of the magnetic moment of an electron which is constant and θ the angle between the orientation of magnetic moment of some atom of a beam and the field $\boldsymbol{B}$. Inasmuch as the external field $\boldsymbol{B}$ rapidly varies, the extra energy also changes in the same way along the z-axis. By definition, the force acting

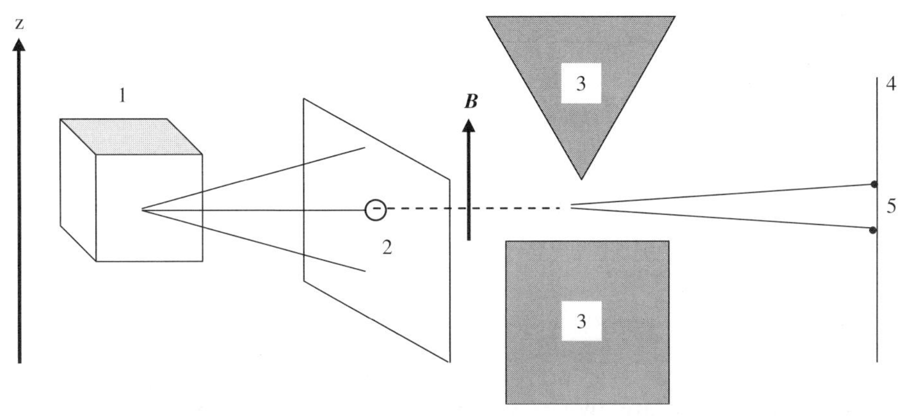

Fig. 6.4. Stern–Gerlach experiment.

on each atom is

$$\boldsymbol{F} = -\mathrm{grad}(\Delta U_{\mathrm{mag}})$$

Correspondingly, for the vertical component of this force, we have

$$F_z = -\frac{\partial(\Delta U_{\mathrm{mag}})}{\partial z} = p \cos\theta \left(\frac{\partial B}{\partial z}\right) \tag{6.37}$$

and its direction depends on the angle θ. Here p and θ are parameters and they do not depend on the z-coordinate. Because of a construction of the magnet, the force is sufficiently large to produce some noticeable displacement of silver atoms of the beam in the z-direction. For illustration consider several cases. In accordance with Equation (6.37), if the magnetic moment is oriented horizontally, $\theta = \pi/2$, the force F_z is zero; such atom continues to move horizontally and reaching the glass plate it appears there as a tiny spot. If the moment is directed along the z-axis, $\theta = 0$, the force pushes an atom upward and correspondingly the path of these atoms forms also a tiny spot on the glass plate above the first one. On the contrary, if the moment of an atom and the direction of the field have opposite directions, $\theta = \pi$, the component F_z is negative and an atom experiences the force which moves it downward. In this case, the tiny spot of atoms appears below the first spot. It is natural to assume that an orientation of magnetic moments of atoms produced by a hot oven is arbitrary; that is, all angles are possible, since the angle θ continuously varies from 0 to π. Then, in accordance with the classical physics the extra energy of the atom in the presence of the field $\boldsymbol{B}$, as well as the force component F_z, varies in the same manner. This means that we should expect on the glass plate a deposition of silver atoms in the form of a very thin strip along the vertical. The top and the bottom of this strip correspond to two values of the angle θ: 0 and π, while the height of this spot is directly proportional to the magnetic moment (Equation (6.37)). Thus, measuring the height it is possible to find the vertical component of the magnetic moment and this was one of the purposes of Stern–Gerlach experiment. Inasmuch as the deflection of atoms was very small, the measurements of the magnetic moment were not very precise. At the same time, an importance of this experiment was related to the fact that instead of an expectable thin strip of silver atoms deposited on the glass plate, there were only two tiny spots at some distance from each other, shown in Fig. 6.4. Thus, this result contradicts the classical physics, but it can be easily explained from the quantum mechanics. In fact, the presence of two spots of silver atoms shows that a silver beam contains electrons with two values of the additional energy; that is, the spin of the silver atom is $j = 1/2$, and in accordance with Equation (6.32), these energies are

$$\Delta U_{\mathrm{mag}} = \frac{g}{2}\left(\frac{q_{\mathrm{e}}}{2m_{\mathrm{e}}}\right)\hbar B \quad \text{and} \quad \Delta U_{\mathrm{mag}} = -\frac{g}{2}\left(\frac{q_{\mathrm{e}}}{2m_{\mathrm{e}}}\right)\hbar B$$

Respectively, there are two possible forces, which split the beam into two smaller separate beams. As pointed out earlier, measuring a deflection of the beams we can determine the magnetic moment. In order to understand better a behavior of electrons and be more prepared to discuss methods of measuring the magnetic field, based on concepts of the quantum mechanics, let us describe one more experiment for measuring magnetic moments and with this purpose in mind consider very briefly the following topic.

6.4.5. *Alternating magnetic field and transition between energy levels of atom*

One of the fundamental features of an atom is its ability to have several possible states in the stationary conditions, for instance, when the ambient magnetic field is absent. In accordance with the quantum mechanics, each state is characterized by certain energy (Fig. 6.5). This energy E_s is plotted along the vertical axis. Here it may be proper to mention that a free electron may move with an arbitrary speed and therefore its energy is positive and can have any value. But we consider an electron bound in atom and its energies are not arbitrary. Let us denote allowed values of energy, for example, as E_0, E_1, E_2, and E_3. One of the features of the atom is the following: it does not remain forever in one of the "excited states", but earlier or later it radiates a light energy. This is associated with a transition of the atom to the state with the lower energy. The frequency of this radiation is determined from the quantum mechanics and the principle of conservation of energy:

$$\hbar\omega = \Delta U \tag{6.38}$$

For instance, in the case of a transition from the state E_3 to E_1, this frequency is

$$\omega_{31} = \frac{E_3 - E_1}{\hbar} \tag{6.39}$$

The latter is called the characteristic frequency of atom and it defines emission line. If there is a transition from state E_3 to E_0, we have

$$\omega_{30} = \frac{E_3 - E_0}{\hbar} \tag{6.40}$$

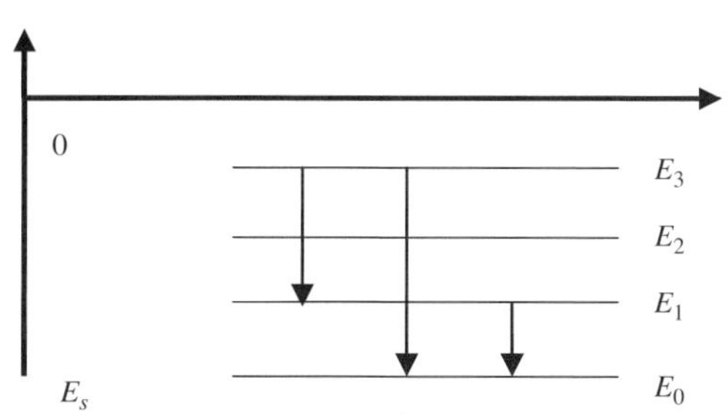

Fig. 6.5. Energy levels of an electron.

Inasmuch as the difference of energies of different states of an atom is relatively large, the frequencies of the electromagnetic waves of the emission correspond to that of a light. Until now we discussed the case when a transition takes place to the states with the lower energy. It turns out that by the absorption of light it is also possible to cause a transition from the lower energy level to the upper energy level of an atom (Chapter 7).

The similar behavior can be observed when atoms, surrounded by a constant field $\boldsymbol{B}$, are placed in the alternating electromagnetic field. For instance, a difference of magnetic energies between two neighbor states ΔU_{mag} (Equation (6.33)) is the same, and for electron and proton we have

$$\Delta U_{\text{mag}}^{\text{e}} = \gamma_{\text{e}} \hbar B = \hbar \omega_{\text{p}}^{\text{e}} \quad \text{and} \quad \Delta U_{\text{mag}}^{\text{p}} = \gamma_{\text{m}} \hbar B = \hbar \omega_{\text{p}}^{\text{p}} \tag{6.41}$$

respectively. First of all, there can be a transition from the levels with higher energy to levels of the lower energy. Since the energy difference is much smaller than in the case shown in Fig. 6.5, this transition is accompanied by a radiation of waves with much smaller frequencies. Also with absorption of these waves, a passage to levels with higher energy is possible. Let us imagine that an atom is placed in an external constant magnetic field $\boldsymbol{B}$. Then, in addition, applying a relatively small sinusoidal magnetic field, an atom may slightly increase its magnetic energy and, correspondingly, there is some probability that it can move to higher energy level. As follows from the quantum mechanics, such probability strongly increases when the quantum of this external energy coincides with the difference of energies between two neighbor levels (Equation (6.41)). In other words, this transition occurs when the frequency ω of an additional field is equal to that of precession, ω_{p}:

$$\omega = \omega_{\text{p}} \tag{6.42}$$

If these frequencies differ from each other, the probability of transition becomes much smaller, that is, from point of a probability of transition, there is a strong resonance at ω_{p}. Therefore, measuring the frequency of this resonance in the known external field $\boldsymbol{B}$ and making use of Equation (6.20), we can calculate the quantity $g(q_{\text{e}}/2m_{\text{e}})$. This allows us to determine the g-factor with a great precision. We discussed this subject considering energy levels of electrons in the presence of the external magnetic field, but the similar phenomenon takes place in the case of nucleus. Here it may be appropriate to make one note. As follows from quantum mechanics, particles are able to absorb only a quantum of energy (Equation (6.32)) or its integer number, and it happens regardless of a frequency of an alternating field. At the same time, we know that the precession frequency for proton in the magnetic field of the earth is around 2000 Hz, and for electrons it is much higher. Inasmuch as except some special cases the constant magnetic field is either equal or stronger than that of the earth the transition from different energy levels takes place, if the frequency of the alternating field exceeds at least 1000 Hz.

6.4.6. The Rabi molecular beam method

In this device, as in the first experiment, there is an oven which produces a stream of atoms moving through a system of three pairs of magnets (Fig. 6.6). We assume that atom's spin j is equal to 1/2. From Equation (6.37), it follows that the force acting on the magnetic dipole and the derivative $\partial \boldsymbol{B}_z/\partial z$ have the same direction if $\theta = 0$ and opposite when $\theta = \pi$:

$$F_z = p\frac{\partial B_z}{\partial z} \quad \text{and} \quad F_z = -p\frac{\partial B_z}{\partial z} \tag{6.43}$$

The first pair of magnets plays the same role as the magnets in the Stern–Gerlach experiment: between them the derivative $(\partial B_z/\partial z)>0$ and the field $\boldsymbol{B}$ is directed upward. We know that the angular momentum and magnetic moment of an electron have opposite directions and electrons with the angular momentum $L_z = \hbar/2$ and $-\hbar/2$ tend to move in the opposite directions along the z-axis. As an example, consider a small beam of atoms a_0 which move in the radial direction toward upper magnet 1. It contains atoms with both values of the angular momentum. For atoms with positive value of an angular momentum, when the magnetic moment is directed downward $(\theta = \pi)$, the force F_z has a negative component. For this reason atoms begin to bend away from the upper magnet and form beam (a_1) which moves through a hole of the screen. As concerns the atoms with $L_z = -\hbar/2$, they bend toward the upper magnet, since $\theta = 0$, and are not able to reach the first hole. In the space between the second magnets, the field $\boldsymbol{B}$ is uniform and, therefore, atoms of this beam are not subjected to the magnetic force. Correspondingly, they move radially toward the lower part of the magnet 3. In such a case, they reach the space between the pair of this magnet where the z-component of the force acting on electrons of this beam is positive

$$F_z = -p\frac{\partial B_z}{\partial z} > 0$$

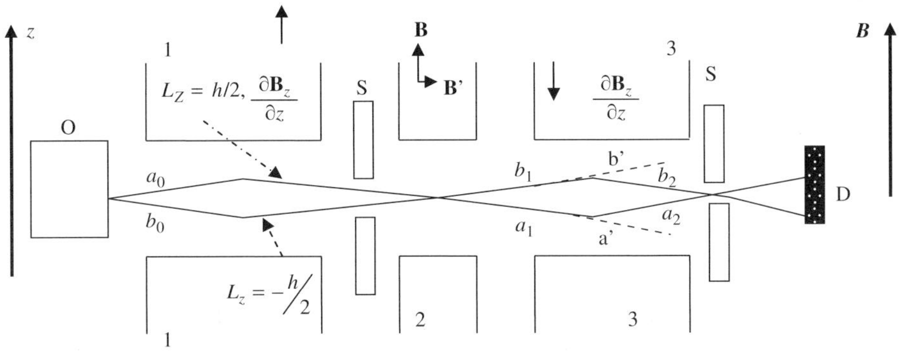

Fig. 6.6. O is oven, and 1, 2, and 3 are magnets. a, a′, b, and b′ are paths of atoms. S is a screen with small holes. D is detector.

since $(\partial B_z/\partial z)<0$. Correspondingly, this force is directed upwards and atoms with $L_z = \hbar/2$, which were pushed down in the magnet 1, experience now a displacement in the opposite direction and move along the path a_2 through the hole of the second screen toward detector. The similar picture is observed for atoms in the beam b_0. Those atoms which have an angular momentum $L_z = -\hbar/2$, that is, $\theta = 0$, experience in the space between the first magnets the force directed upwards:

$$F_z = p\frac{\partial B_z}{\partial z} > 0$$

It happens because their magnetic moment has positive projection on the z-axis. The beam of these atoms goes through the hole of the first screen and moves toward the upper magnet 3 (path b_1). Since the component of the magnetic moment is positive, $\theta = 0$, but $(\partial B_z/\partial z)<0$, we have

$$F_z = p\frac{\partial B_z}{\partial z} < 0$$

and this beam is bent down. After passing the second screen these atoms reach the detector, too. As concerns a detector, it can be realized by different ways, depending on atoms of a beam. For example, in the case of alkali metal like sodium, the detector can be a thin hot wire connected to a sensitive ampere-meter. When these two beams reach the hot wire, Na^+ ions evaporate and electrons remain and form the current. Now we describe the most important feature of this device that makes it essentially different from Stern–Gerlach experiment. Alternating current in coils installed near magnets 2 creates the horizontal magnetic field between them, which is much smaller than the uniform constant field along the z-axis. A frequency of the current in these coils may vary. Suppose that this frequency is equal to that of the precession, ω_p. Then due to the alternating field some atoms will change their value of the angular momentum. This means that an electromagnetic energy is absorbed and electrons move at a different level of the magnetic energy.

For instance, in place of $L_z = \hbar/2$ it becomes $L_z = -\hbar/2$ and correspondingly the z-component of the magnetic moment becomes positive in the path a_1 $(\theta = 0)$. Therefore, for the z-component of the force between magnets 3 acting on these dipoles, we have

$$F_z = p\frac{\partial B_z}{\partial t} < 0$$

Because of this atoms of the beam move down (path a′) and therefore they are not able to pass through the hole of the last screen. The same happens with some atoms of the beam b_1. Thus, at this frequency the current becomes smaller. In such a way of varying a frequency of the alternating field, it is possible to determine the precession frequency and therefore the parameter g (Fig. 6.7). This method is usually called "molecular" beam resonance experiment which allows one to find ω_p with a very high accuracy.

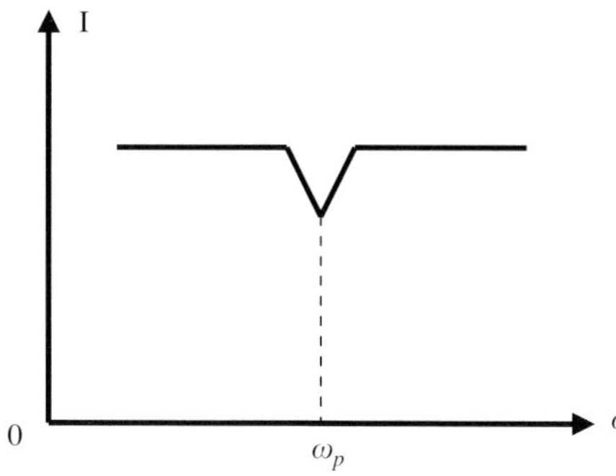

Fig. 6.7. Resonance curve of the current in detector.

Until now we focus on a behavior of the magnetic moments caused by a motion of electrons, but the magnetic dipoles of nuclei have a similar behavior and this subject will be considered in some detail in Chapter 7. For illustration, suppose that the resultant moment of electrons is zero as it takes place in substance like water. In such a case, there is still the magnetic moment of atoms due to the magnetic moment of the hydrogen nuclei. Note that the magnetic moment of nucleus is thousand times smaller than that due to motion of an electron. Let us put a small amount of water into a constant magnetic field $\boldsymbol{B}$ directed along the vertical z-axis. Inasmuch as the spin of a proton of hydrogen is 1/2, it is characterized by two possible energy states, and in each elementary volume of the water there is almost the same amount of atoms, which have magnetic moments directed either along or opposite to this field. As was shown earlier, if the magnetic moment p_z is directed along the field, the proton is in the lower energy state than the proton with the opposite orientation of this moment. It turns out that due to the presence of the constant field $\boldsymbol{B}$, there is slightly more protons with the magnetic moment directed along the field. In other words, the number of protons with the lower energy level exceeds the number of them with higher energy level and this is vitally important to realize the NMR. The difference between them is extremely small but yet it can be detected. Now we imagine that along with the constant vertical field, we create an alternating magnetic field, directed horizontally and caused by coils with a current, as was done in the Rabi's experiment. If the frequency of this field ω is equal to the precession frequency of proton ω_{p}, then the transition between two energy levels may take place. For instance, in the case when a proton moves from the upper level to lower level, it gives up the energy

$$p_z B = \hbar\omega_{\mathrm{p}} \tag{6.44}$$

On the contrary, a transition from the lower to upper levels of energy is accompanied by absorption of energy from the source of the alternating current. It is clear there are both a radiation and an absorption of the energy. Taking into account that there are slightly more protons at the lower energy state, the resultant effect will be absorption of energy from the coil source. As we know, the energy absorption can be detected when

$$\omega = \omega_{\rm p} = g\left(\frac{q_{\rm e}}{2m_{\rm p}}\right)B = \gamma_{\rm p}B \tag{6.45}$$

Here q_e is the proton charge and m_p its mass. Note that this resonance can be achieved by two ways, namely, a change of a frequency of alternating field and a change of a magnitude of the field $\boldsymbol{B}$ while ω remains fixed. This analysis shows that when $\omega = \omega_p$, some energy of the circuit with a coil is absorbed by water and correspondingly the current in the coil becomes smaller. In essence, we described one of the principles of the NMR.

6.5. DIAMAGNETISM

So far we mainly considered a single atom and described a relation between its magnetic moment and angular momentum. Next let us place a diamagnetic material into a magnetic field $\boldsymbol{B}$ and, applying the concepts of the classical physics, explain the mechanism of magnetization. Each of its elementary volume contains many atoms, and in the absence of the external field they are involved in a random motion. By definition of the diamagnetic material, the resultant magnetic moment and angular momentum of each atom related to a motion of electrons are equal to zero. This means that an interaction of atomic currents (dipoles) is absent. Besides, we assume that the density of conduction currents is zero. It seems that such a material does not make influence on the magnetic field but it is not correct. In order to understand what happens, we suppose that at the beginning the external magnetic field $\boldsymbol{B}$ changes with time and then becomes constant. In accordance with Faraday's law, during a time interval when the field $\boldsymbol{B}$ varies, the inductive electric field $\boldsymbol{E}$ arises at each point. Direction of this field depends on different reasons but usually if the magnetic field is uniform within some volume then the field $\boldsymbol{E}$ may also have almost the same magnitude and direction, as is shown in Fig. 6.8. In order to evaluate an influence of this field we will consider a closed path with radius r, located in the plane perpendicular to the field $\boldsymbol{B}$, and apply the Faraday's law of the electromagnetic induction:

$$\oint_l \boldsymbol{E}\cdot d\boldsymbol{l} = -\int_S \frac{d\boldsymbol{B}}{dt}\cdot d\boldsymbol{S} \tag{6.46}$$

Here $d\boldsymbol{l}$ is the element of the path and $d\boldsymbol{S}$ the element of the surface S, bounded by the path l, and directions of vectors $d\boldsymbol{l}$ and $d\boldsymbol{S}$ obey the right-hand rule. We are interested by an action of the component of the field $\boldsymbol{E}$ along the path l. Taking into account the fact that the magnetic field is uniform, we can write

$$\oint_l E\,dl = -\pi r^2\frac{dB}{dt} \quad \text{or} \quad E2\pi r = -\pi r^2\frac{dB}{dt}$$

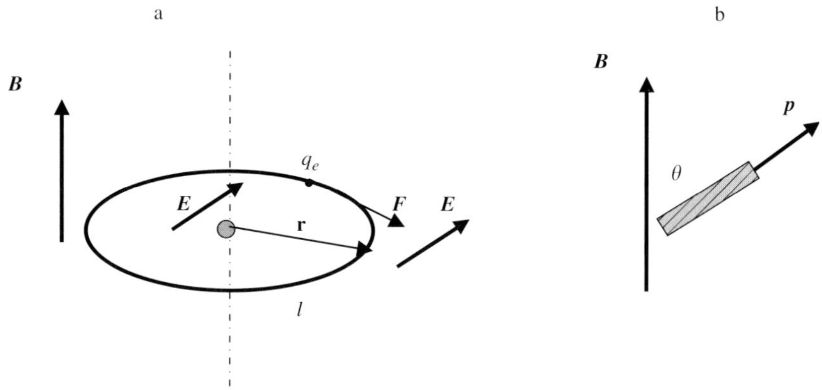

Fig. 6.8. (a) Inductive electric field. (b) Illustration of Equation (6.57).

where E is an average value of the component of electric field tangential to the circle, which has different values and sign at points of this path. Thus, the inductive electric force acting on an electron is

$$F = -q_e E = \frac{rq_e}{2}\frac{dB}{dt} \tag{6.47}$$

This force is perpendicular to the radius-vector $\boldsymbol{r}$ and it is located in the plane of the circle l. Because of this force an electron is subjected to the torque of the inductive origin τ_{in} and its magnitude is equal to

$$\tau_{in} = rF = \frac{q_e r^2}{2}\frac{\partial B}{\partial t} \tag{6.48}$$

and its direction depends on the rate of a change of the magnetic field. The latter suggests that the torque exists only when the field $\boldsymbol{B}$ varies. In accordance with the second Newton's law, we have

$$\frac{dL}{dt} = \tau_{in} = \frac{q_e r^2}{2}\frac{dB}{dt} \tag{6.49}$$

Performing an integration within the time interval during which the field $\boldsymbol{B}$ changes, we obtain

$$\Delta L = \frac{q_e r^2}{2} B \tag{6.50}$$

Here B is the magnitude of the constant magnetic field, while ΔL is the difference between the final value of the angular momentum and its value in the absence of this field. Thus, ΔL appears due to the induction effect during the time of changing of the magnetic field. Inasmuch as the angular momentum and the magnetic moment of the electron are related with each other (Equation (6.5)), we see that a change of the magnetic moment of the electron is

$$\Delta \boldsymbol{p} = -\frac{q_e}{2m_e}\Delta \boldsymbol{L} = -\frac{q_e^2 r^2}{4m_e}\boldsymbol{B} \tag{6.51}$$

The presence of minus sign indicates that the directions of the magnetic moment and the field are opposite (susceptibility is negative). This result follows directly from Lenz's rule according to which induced currents are trying to reduce the magnetic flux due to the field $\boldsymbol{B}$ through an area surrounded by the current. Thus, an appearance of the induced magnetic dipole with the direction opposite to the ambient magnetic field explains the behavior of the diamagnetic. The same effect takes place in any material, in particular, in paramagnetic substance, where an atom has already the magnetic dipole and its moment is involved in precession. In such a case, diamagnetism only slightly reduces the total moment. We obtained Equation (6.51) when there is one electron, but in general, a summation over all electrons gives the total moment. It is more convenient to write Equation (6.51) in the form

$$\Delta \boldsymbol{p} = -\frac{q_e^2}{6m_e}(r^2)_{av}\boldsymbol{B} \tag{6.52}$$

and in accordance with the quantum mechanics, $(r^2)_{av}$ is the average of the square of the distance from the center for the probability distribution of an electron.

As was mentioned earlier, diamagnetism only slightly depends on temperature and, taking into account an inductive origin of this phenomenon, such behavior becomes almost obvious. There are different materials which are diamagnetic, for instance, water, many metals (mercury, gold, and bismuth), and most of organic substances like petroleum and plastics. For example, susceptibility of water is equal to $\chi = -9.05 \times 10^{-6}$, while bismuth has almost the largest value of χ among diamagnetic materials (-166×10^{-6}). Nevertheless, even this value is in order of magnitude smaller than that due to paramagnetism, and for this reason it is very difficult to measure an influence of diamagnetism in paramagnetic materials. Speaking strictly, the field inside the diamagnetic material is a sum of the primary (ambient) and secondary fields. At the same time, the secondary field is extremely small and, as was pointed out above, the magnetic moment is practically defined by the ambient field; that is, interaction between dipoles is negligible. As a result, when the external field is removed, diamagnetic effect vanishes. It is also proper to notice that every electron shell of a diamagnetic material represents pairs of magnetic moments with equal magnitude and opposite directions, and, correspondingly, the resultant magnetic moment due to spin rotation and orbital motion of electrons is equal to zero. At the same time, as we already know, the nucleus has the magnetic moment, and its magnitude and frequency of precession can be thousand times smaller than those for an orbital motion of an electron.

6.6. PARAMAGNETISM

Now we establish a relation between a vector of magnetization of paramagnetic materials and magnetic moments of atoms. The phenomenon of paramagnetism is observed when the dipole moment due to a motion of electrons in each atom is not zero, like atoms of aluminum (Al) and sodium (Na). In general, any atom with odd number of electrons has a magnetic moment. For instance, in the atom of sodium there is one electron in its unfilled shell and it provides the angular momentum and the magnetic moment. It is useful to notice the following. When molecules are formed extra electrons in the outside shells of different atoms (valence electrons) interact with each other. Usually their magnetic moments are equal by magnitude but have opposite directions; that is, they cancel each other. This is a reason why very often the magnetic moment of molecules is zero even though there are exceptions. In most cases, paramagnetism is observed in such substances where inner electron shell of atoms is not filled and there is a resultant angular momentum and magnetic moment. In the absence of the ambient magnetic field $\boldsymbol{B}$, these atomic magnetic dipoles are oriented randomly and it happens because of the thermal agitation. Correspondingly, the resultant magnetic moment of each elementary volume at a room temperature is zero. When the external magnetic field arises, these moments tend to align along the field $\boldsymbol{B}$ and magnetization of a medium takes place. The dipoles line up with the magnetic field due to the torque: $\boldsymbol{\tau} = \boldsymbol{p} \times \boldsymbol{B}$, where $\boldsymbol{p}$ is the dipole moment of an atom. Since some of the dipoles are oriented orderly we observe a magnetization and, unlike diamagnetism, these magnetic moments and the field are oriented in the same direction. The presence of the dipole moments of atoms is the most essential feature of paramagnetic substances. Also it is assumed that the magnetic field caused by these dipoles is several orders smaller than the ambient field which causes magnetization. Now proceeding from the classical theory of paramagnetism and quantum mechanics, we derive an expression for the vector of magnetization $\boldsymbol{P}$ provided that its magnitude is relatively small.

6.6.1. Classical physics approach

By definition, $\boldsymbol{P}$ is the magnetic moment of the unit volume (Chapter 2) and it can be represented as

$$\boldsymbol{P} = N(\boldsymbol{p})_{\text{av}} \tag{6.53}$$

Here N is a number of atoms per unit volume and $(\boldsymbol{p})_{\text{av}}$ their average moment. Our goal is to determine the relationship between the vector of magnetization $\boldsymbol{P}$ and the external field $\boldsymbol{B}$. Each elementary volume of paramagnetic (gas, liquid, or crystal) is full of atoms and we treat them as small permanent magnets. As we know, in the absence of the external magnetic field they are involved in a thermal motion and as a result moments are oriented in all directions. Correspondingly, the average moment $(\boldsymbol{p})_{\text{av}}$ is zero and magnetization is absent. But in the presence of the magnetic field, a part of magnetic dipoles lines up along the field and magnetization occurs. Thus, there are two factors which produce opposite effects, namely, with an increase of magnetic field the number of dipoles which line up along the field becomes larger, while an increase of temperature tends to destroy this ordered arrangement. This dependence of the magnetization was studied by P. Curie and has the following form:

$$P = C\frac{B}{T} \tag{6.54}$$

Here T is absolute temperature measured in Kelvin and C a constant, which depends on a paramagnetic material and our purpose is to find an expression of this constant. In order to solve this task and understand when we can use this law, let us discuss such topic as a probability of a certain orientation of the magnetic dipole of an atom with respect to the field. Earlier we found out an expression for the magnetic energy of a magnetic dipole in the presence of the external field (Equation (6.36)):

$$U(\theta) = -\boldsymbol{p} \cdot \boldsymbol{B} = -pB \cos \theta$$

It may be useful to notice that a minimum energy occurs at the stable point of equilibrium. In fact, suppose that a direction of the dipole is slightly changed, then a torque due to the external field returns the dipole to its original position. On the contrary, when these vectors are opposite to each other, we observe unstable point of equilibrium. Now we are prepared to make next step and find a distribution of the magnetic dipoles as a function of the angle θ. In the absence of the external magnetic field, the magnetic energy of atoms or molecule is zero and there is equal amount of magnetic dipoles with any direction of their moments. Completely different situation takes place when paramagnetic substance is placed into the field $\boldsymbol{B}$, since atoms have a different magnetic energy. As follows from Boltzmann's law the number of magnetic dipoles of atoms, having the magnetic energy U, is proportional to

$$\exp\left(-\frac{U}{kT}\right) \tag{6.55}$$

Here k is Boltzmann's constant and it is equal to

$$k = 1.381 \times 10^{-23}\ \mathrm{J\,K^{-1}} \tag{6.56}$$

At the beginning proceeding from the classical physics assume that for thermal equilibrium an orientation of dipoles can be arbitrary. Then, taking into account Equations (6.54) and (6.55) and letting $n(\theta)$ be the density of dipoles with given orientation θ per unit solid angle, we have

$$n(\theta) = n_0 \exp\left(\frac{pB \cos \theta}{kT}\right) \tag{6.57}$$

The latter depends on the angle θ, the field magnitude B, dipole moment, p, and, of course, the temperature T. Even when the magnetic field B is relatively strong and T is around 300 K, the exponent in Equation (6.55) is much smaller than 1 and, correspondingly, Equation (6.57) can be represented as

$$n(\theta) = n_0\left(1 + \frac{pB\cos\theta}{kT}\right) \tag{6.58}$$

Here n_0 is unknown and it is independent of the angle, and our goal is to determine this constant. It is clear that number of magnetic moments with given θ are distributed uniformly within elementary volume. In order to find n_0, we take into account that by definition the density of magnetic moments remains the same inside an elementary cone with solid angle $d\omega$. Since the solid angle is equal to the elementary area of the sphere with unit radius intersected by a cone, we have

$$d\omega = \sin\theta\, d\varphi\, d\theta$$

Therefore, the number of dipoles inside the elementary cone with the angle θ is

$$dN(\theta) = n_0\left(1 + \frac{pB\cos\theta}{kT}\right)\sin\theta\, d\varphi\, d\theta$$

Taking into account the symmetry with respect to φ and performing integration over all angles θ, we obtain the total number of dipoles N inside the unit volume of paramagnetics

$$n_0 = \frac{N}{4\pi} \quad \text{and} \quad n(\theta) = \frac{N}{4\pi}\left(1 + \frac{pB\cos\theta}{kT}\right) \tag{6.59}$$

Thus, we obtained a formula which describes a distribution of the magnetic dipoles as a function of the angle θ inside a unit volume of the paramagnetic substance. It may be proper to emphasize again that within this volume there are all groups of magnetic dipoles distributed uniformly and their orientation continuously varies from 0 to π. As follows from Equation (6.59), the density of dipoles with a small angle is relatively larger while the number of those dipoles where the moment is oriented opposite to the field is smaller. Inasmuch as these groups have different number of dipoles, their sum is not zero; that is, there is a magnetization. In calculating the vector sum of dipole moments, we take into account the fact that in an isotropic medium the vector of magnetization has the same component as the field $\boldsymbol{B}$. By definition, this component of the vector of magnetization is a sum formed by corresponding components of the magnetic moments inside of the unit volume

$$P = \sum p\cos\theta$$

For instance, the vertical component of the vector magnetization due to dipoles with the angle θ is

$$dP = pn(\theta)\cos\,\theta\,d\omega$$

Therefore, in order to find the resultant component, caused by all dipoles of a unit volume, we have to integrate and it gives

$$P = \int_0^{\pi} n(\theta)p2\pi\,\cos\,\theta\,\sin\,\theta\,d\theta \tag{6.60}$$

Substitution of $n(\theta)$ from Equation (6.59) gives for the vertical component of the magnetization vector

$$P = \frac{N}{2}\int_0^{\pi}\left(1 + \frac{pB}{kT}\cos\,\theta\right)p\,\cos\,\theta\,\sin\,\theta\,d\theta$$

After integration we arrive at Curie's law:

$$P = \frac{Np^2B}{3kT} \tag{6.61}$$

and, as it follows from Equation (6.54), the constant C is equal to

$$C = \frac{Np^2}{3kT} \tag{6.62}$$

The approximate formula (Equation (6.61)) is valid only when the field $\boldsymbol{B}$ is relatively small, and in this case the vector of magnetization is directly proportional to the field. This allows one to introduce susceptibility χ. In fact, by definition we have $P = (\chi/\mu_0)B$ and therefore

$$\chi = \frac{Np^2\mu_0}{3kT} \tag{6.63}$$

It is obvious that with an increase of the number of dipoles and their magnetic moment, the magnetization P also increases, but with an increase of temperature T an influence of a random motion is stronger and P becomes smaller. It is interesting to note that the magnetic moment p increases the magnetization for two reasons; first one follows from definition (Equation (6.53)), and the second is related to the fact that the force acting on the magnetic dipole is proportional to its moment. Correspondingly, the constant C is proportional to square of p. Also note that Equation (6.61) implies an absence of interaction of magnetic dipoles. In other words, $\boldsymbol{B}$ is an ambient field only. It is proper to point out that performing a summation of components of the magnetic moments perpendicular to the field **B**: $p \sin\theta \cos\varphi$ and $p \sin\theta \sin\varphi$, we obtain zero, that is expectable since vectors $\boldsymbol{p}$ and $\boldsymbol{B}$ are parallel. As follows from Equation (6.61), an unlimited increase of the field results in infinitely large value of the magnetization. However, it is incorrect, since there is always such large value of the field $\boldsymbol{B}$, when all dipoles are oriented along this field and saturation occurs. Then, an increase of the field does not affect a value of the vector of magnetization.

6.6.2. Quantum mechanics approach

Until now we described paramagnetism using concept of classical physics; next we take into account that in accordance with quantum mechanics the electron system is characterized by a spin and there are only certain energy levels (states). As usual, for simplicity, consider an electron with the spin $j = 1/2$. In the absence of the magnetic field atoms of paramagnetic material have the same energy, U_0, but the presence of this field gives two different levels of the magnetic energy and, as was shown earlier, the changes of energies related to spin-up and spin-down are

$$\Delta U_1 = g\left(\frac{q_e\hbar}{2m}\right)\cdot\frac{1}{2}B \quad \text{and} \quad \Delta U_2 = -g\left(\frac{q_e\hbar}{2m}\right)\frac{1}{2}B \tag{6.64}$$

since

$$L_z = \pm\frac{1}{2}\hbar$$

Introducing notation

$$p_0 = g\left(\frac{q_e\hbar}{2m}\right)\frac{1}{2} \tag{6.65}$$

Equations (6.64) can be written as

$$\Delta U = \pm p_0 B \tag{6.66}$$

Comparison with Equation (6.36) shows that $-p_0$ is the z-component of the magnetic moment in the case of the spin-up, while p_0 is the z-component when the spin-down. As we know, these cases indicate that the magnetic moment is oriented either along the field or opposite and this is fundamental difference with the classical physics. Applying again the principle of statistical mechanics, we can say that in the presence of a constant magnetic field a probability that an atom in one of this states is proportional to

$$\exp\left(-\frac{\Delta U}{kT}\right)$$

Therefore, in the unit volume a number of atoms with spin-up and spin-down can be represented as

$$N_{\text{up}} = a \exp\left(\frac{-p_0 B}{kT}\right) \quad \text{and} \quad N_{\text{down}} = a \exp\left(\frac{p_0 B}{kT}\right)$$

Here a is unknown and it is determined from the condition that the total number of atoms in the unit volume is N:

$$N_{\text{up}} + N_{\text{down}} = N \tag{6.67}$$

Thus

$$a = \frac{N}{\exp(p_0 B/kT) + \exp(-p_0 B/kT)} \tag{6.68}$$

By definition, the average magnetic moment $(p)_{\text{av}}$ of the atom along the z-axis is defined as

$$(p)_{\text{av}} = \frac{(-p_0)N_{\text{up}} + (p_0)N_{\text{down}}}{N} \tag{6.69}$$

and the magnitude of the vector of magnetization is equal to

$$P = (p)_{\text{av}} N$$

Finally, we obtain

$$P = Np_0 \frac{\exp(p_0 B/kT) - \exp(p_0 B/kT)}{\exp(p_0 B/kT) + \exp(p_0 B/kT)} \quad \text{or} \quad P = Np_0 \tanh \frac{p_0 B}{kT} \tag{6.70}$$

and a behavior of this function is shown in Fig. 6.9. In this light let us make two comments: (1) for ordinary temperatures and relatively large fields, around 1 T,

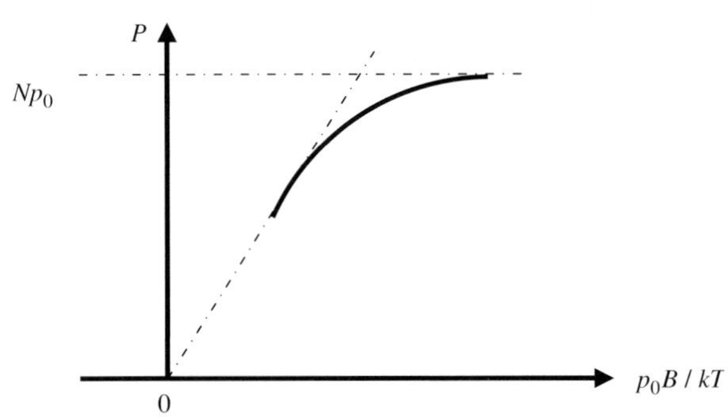

Fig. 6.9. Magnetization as a function of the field and temperature.

the ratio p_0B/kT is much less than 1 and, correspondingly, we observe the linear dependence, as it follows from the classical theory; (2) the difference between number of dipoles directed up and down is extremely small, since $p_0B/kT \ll 1$.

6.7. FERROMAGNETISM

6.7.1. Introduction

In the previous sections we described magnetic properties of diamagnetic and paramagnetic materials and emphasized their different origin, and as a result, the opposite direction of magnetic moments with respect to the external field. At the same time, they have three common features:

1. Alignment of magnetic moments takes place only due to the external magnetic field; that is, an interaction between atomic magnetic dipoles is negligible and it happens because their magnetic fields are extremely small. Correspondingly, the susceptibility χ, characterizing the vector of magnetization, $\boldsymbol{P} = (\chi/\mu_0)\boldsymbol{B}$, has order of 10^{-4} to 10^{-5}.
2. In the absence of the external field $\boldsymbol{B}$ the magnetization disappears in both cases.
3. With an increase of temperature the vector of magnetization decreases even though this dependence is different for diamagnetic and paramagnetic substances.

Now we describe the third much more complicated phenomenon which is called ferromagnetism. Unlike the diamagnetic and paramagnetic substances, in materials with ferromagnetic properties the secondary magnetic field caused by atomic magnetic dipoles is relatively large and very often it greatly exceeds the ambient field. This fundamental feature of ferromagnetism is related to the fact that there is a strong interaction between these dipoles. One can say that a mechanism of such interaction is a foundation of ferromagnetism and it is not yet completely understood. It is natural that in the case of ferromagnetism the vector of magnetization $\boldsymbol{P}$ is not directly proportional to the ambient field $\boldsymbol{B}_0$, but instead of it there is usually a rather complicated relation between them. Inasmuch the theory does not yet allow us to obtain an analytical expression of the function $B(H)$ for different ferromagnetic materials, we will proceed from an experiment when the field $\boldsymbol{H}$ is known at each point of a magnetic medium. Measuring the magnetic field $\boldsymbol{B}$ it is possible to find the magnetic permeability μ and the susceptibility χ, as well as the vector of magnetization as the function of the magnetic field. Thus, at the beginning we describe the main properties of ferromagnetism, which were obtained experimentally and then, using principles of the classical physics and quantum mechanics, make attempt to explain them.

6.7.2. The magnetization curve

With this purpose in mind, we consider a torus of iron with a solenoid at its surface (Fig. 6.10(a)). As was shown in Chapter 2, the conduction current I of the solenoid generates the uniform magnetic field $\boldsymbol{B}_0$ inside the torus. Applying

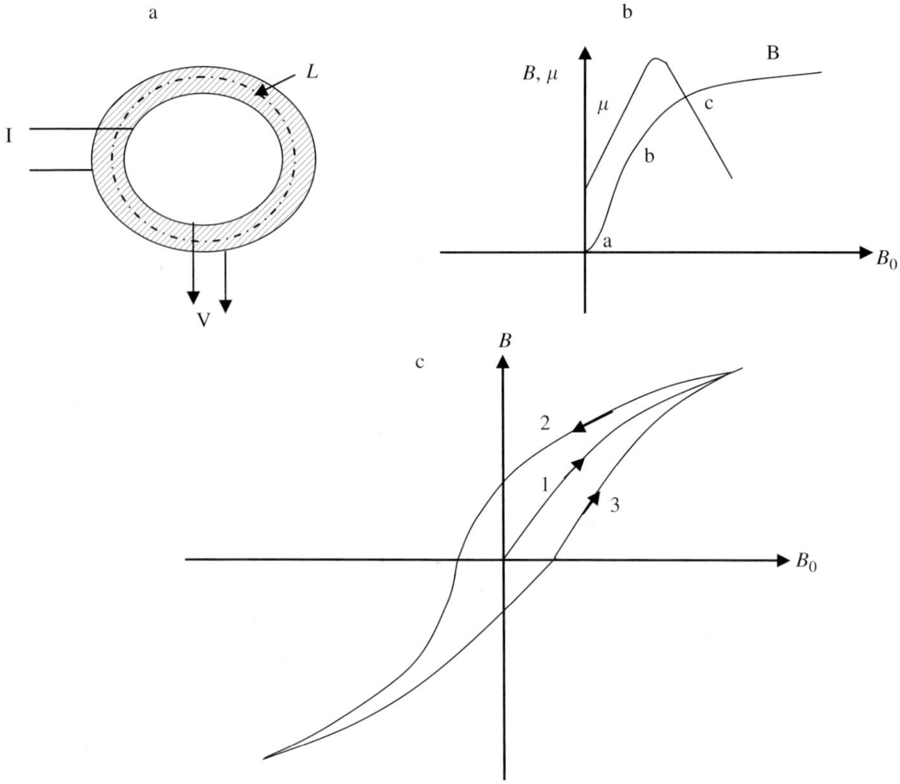

Fig. 6.10. (a) Model of iron torus with current solenoid. (b) Behavior of the functions $B(B_0)$ and $\mu(B_0)$. (c) Hysteresis loop.

the first equation of the magnetic field in the integral form (Fig. 6.10(a)), we have

$$\oint_L \boldsymbol{B}_0 \cdot d\boldsymbol{l} = \mu_0 IN \quad \text{or} \quad B_0 = \mu_0 \frac{IN}{l} = \mu_0 In \tag{6.71}$$

Here N is total number of turns and n their density; l the length of the path. Assuming that any path L inside a torus has the same length, we conclude that the primary magnetic field B_0 is uniform and it has only component tangential to the path L. The secondary magnetic field $\boldsymbol{B}_s$ ($\boldsymbol{B} = \boldsymbol{B}_0 + \boldsymbol{B}_s$) is generated by magnetization and it can be interpreted as the field caused by magnetization currents on the torus surface. In other words, it behaves as a field of a solenoid and also uniform inside the iron. It is essential that the primary field inside the magnet is known (Equation (6.71)), and by definition it is not subjected to an influence of a magnetic medium. At the same time, the total field $\boldsymbol{B}$ is measured by placing a coil on the torus surface

(Fig. 6.10(a)) (voltmeter). In accordance with the law of electromagnetic induction, the electromotive force induced in a measuring coil is

$$\Xi = -\frac{\partial F}{\partial t} \tag{6.72}$$

where F is the flux of the magnetic field through the coil with n_R turns and the cross-section S_R. Since the field is uniform, the flux is

$$F = n_R S_R B$$

and therefore in place of Equation (6.72) we have

$$\Xi = -n_R S_R \frac{\partial B}{dt}$$

Its integration gives

$$B(t) = -\frac{1}{n_R S_R} \int_0^t \Xi(t)\, dt$$

since the field is zero at the first instant. Here $B(t)$ is the field inside the torus but outside it is practically absent. Now we are ready to describe a dependence of the field B on B_0 inside the iron, shown in Fig. 6.10(b). When the primary current in the coil is turned on and it increases, the total field B inside the nonmagnetized iron also increases. First, for very small values of B_0 the field B grows relatively slowly and then its rate of change increases, and the field B becomes much more than the primary field. This means that the magnetic field in the iron is mainly defined by magnetization; that is, the surface magnetization currents play the dominant role, $B_s \gg B_0$. Certainly something special happens with magnetic moments of atoms so that they are able to generate the secondary field greatly exceeding the primary one. We observe here the fundamental difference between the field behavior inside the paramagnetic and ferromagnetic materials. With further increase of B_0 a change of the total field becomes smaller and the path further approaches to the straight line of the unit slope:

$$B = B_0 + \mu_0 P_{max} \tag{6.73}$$

and it represents the saturation stage, because the magnetization vector reaches its maximal value P_{max}. Inasmuch as

$$B = \mu_0(H + P) = \mu H = \frac{\mu}{\mu_0} B_0 \tag{6.74}$$

the magnetic permeability μ changes as a function of the primary field (Fig. 6.10(b)). In particular, it has a maximum and, as follows from Equations (6.73) and (6.74), its right asymptote is

$$\frac{\mu}{\mu_0} = 1 + \mu_0 \frac{P_{max}}{B_0} \tag{6.75}$$

and in the limit the magnetic permeability tends to μ_0.

6.7.3. *Hysteresis loop*

Until now we discussed the behavior of the field, which is represented by the path 1 in Fig. 6.10(c), and assume that a field B_0 is such that saturation takes place. Next we begin to decrease the primary field and observe the second interesting feature, namely, a change of the magnetization does not occur along the same path 1 but instead of it values of the field B follow the path 2. For instance, we see that when the primary field is equal to zero, the magnetization remains and, correspondingly, there is a secondary field which can be significant. Certainly, it is another great difference with paramagnetic material where a small magnetization disappears together with the primary field. If the primary field changes its direction and its magnitude increases, this magnetization becomes smaller and there is such value of B_0, called the coercive force, when the total field and therefore the magnetization vanishes. At this moment the primary and secondary fields have the same magnitudes but opposite directions. With further increase of the primary field magnitude, the total field changes along the path 2 until a magnetization reaches saturation. If we again start to decrease a primary field, the resultant field varies along the path 3. As is seen from Fig. 6.10(c), the total field follows the path 3 with an increase of B_0 and again approaches the asymptote of the path 1. Thus, applying the alternating primary field, so that it varies from large positive to negative values, the field B and magnetization P would change periodically along the closed curve 2–3. This loop caused by oscillations of the primary field is called a hysteresis loop and it is different for different substances. The shape of this loop depends on many factors, such as chemical content of the material and a way of its preparation. Note that if the primary field changes arbitrarily, the hysteresis curve is located somewhere between the loops 2 and 3, shown in Fig. 6.10(c). From this figure we also see that for a given value of the primary field B_0, it is possible to expect one of either two or three values of the total field, and a choice of the correct value depends on the past history of a substance in the primary field. In other words, an action of the primary field at some moment also depends on a magnetization produced by an ambient field at earlier times. This is a reason why in the case of ferromagnetic material we do not have analytical expression of the function $B = f(B_0)$. Later we will attempt to explain some important features of the hysteresis loop but now it is proper to remind that magnetic materials, like iron, found the numerous and important applications, such as electric motors, transformers, and electromagnets. With the help of the ferromagnetic material, it is possible to greatly increase the magnetic field caused by a given conduction current and also control a distribution of this field, for instance, generate this field mainly in a space between poles of magnets. Now we consider one example illustrating an application of a hysteresis loop for measuring magnetic field.

6.8. PRINCIPLE OF THE FLUXGATE MAGNETOMETER

In order to understand this method of measuring the magnetic field, consider, first, a circuit and suppose that the current in a generator is a sinusoidal function of the frequency ω. Correspondingly, the magnetic field caused by this current inside the coil without magnetic core varies in the same way:

$$B_c(t) = p \sin \omega t$$

Besides, there is the magnetic field of the earth B_e and therefore the total field is

$$B_0(t) = B_e + p \sin \omega t \tag{6.76}$$

Next we place inside a coil a material like iron. The behavior of the magnetic field B inside of it is shown in Fig. 6.10(c). Let us assume that the current of the oscillator is chosen in such a way that the field B approaches the saturation stage; that is, the function $B(B_0)$ is not linear one. During the time interval when the external field B_0 decreases, the total field inside the magnetic core follows the path 2 but in the second part of the period it increases along the path 3, as is seen from Fig. 6.10(c). In both cases, the field B is an odd function of the external field and between stages of saturation it can be written for each path as

$$B = B_0(a - bB_0^2) \tag{6.77}$$

We neglected here the third and higher powers of B_0 because their contribution is very small. Substitution of Equation (6.76) into Equation (6.77) gives

$$B(t) = (B_e + p \sin \omega t)[a - b(B_e^2 + 2B_e p \sin \omega t + p^2 \sin^2 \omega t)]$$

or

$$B(t) = \{aB_e - bB_e^3 - 2B_e^2 pb \sin \omega t - B_e bp^2 \sin^2 \omega t + ap \sin \omega t - pbB_e^2 \sin \omega t \\ - 2p^2 bB_e \sin^2 \omega t - bp^3 \sin^3 \omega t\}$$

Thus, for an electromotive force induced in the receiver coil surrounding the coil with a magnet, we have

$$\Xi = -Sn\{p\omega[a - 3bB_e^2] \cos \omega t - 3B_e p^2 \omega b \sin 2\omega t - 3bp^3 \omega \sin^2 \omega t \cos \omega t\}$$

Here S and n are the area and number of coil turns, respectively. Since

$$\sin^2 \omega t = \frac{1 - \cos 2\omega t}{2}$$

we have

$$\sin^2 \omega t \cos \omega t = \frac{1}{2}\cos \omega t - \frac{1}{2}\cos \omega t \cos 2\omega t$$

Taking into account that

$$\cos \omega t \cos 2\omega t = \frac{1}{2}(\cos \omega t + \cos 3\omega t)$$

we have

$$\sin^2 \omega t \cos \omega t = -\frac{1}{4}\cos 3\omega t$$

Thus

$$\Xi = -Sn\left\{ p\omega[a - 3bB_e^2]\cos \omega t - 3B_e p^2 \omega b \sin 2\omega t + \frac{3}{4}bp^3\omega \cos 3\omega t \right\} \quad (6.78)$$

In this approximation the signal measured by a coil contains three harmonics. The first and third exist even when the field B_e is absent. As concerns the second harmonic, it is directly proportional to this constant field of the earth. Removing the first and third harmonics and measuring the second one, we can determine B_e. In essence, this approach allows one to transform the constant magnetic field into a sinusoidal signal which is much simpler to measure. Let us make several comments:

1. Constants a and b are parameters of the hysteresis loop, while p characterizes the magnitude and sign of the magnetic field caused by the oscillating current.
2. Nonlinearity of a hysteresis loop is essential ($b \neq 0$); otherwise the electromotive force would be insensitive to the constant magnetic field. At the same time, there is no need to reach the saturation stage of the magnetization curve.
3. The first and third harmonics of the electromotive force contain odd powers of parameter p but the second harmonic is directly proportional to square of p. This means that a change of the sign of p leads to a change of sign of the first and third harmonics while the sign of the second harmonic remains the same.

Now we are ready to describe the fluxgate device (Fig. 6.11(a)). The principle of the device based on measuring the second harmonic is very simple. There are two identical coils with magnetic bars of the high permeability. The direction of wounding them is opposite to each other. A measuring coil surrounds both magnets, and electromotive force is defined by a rate of a change of fluxes through each ferromagnetic core. At any instant at each magnetic core there is one of the

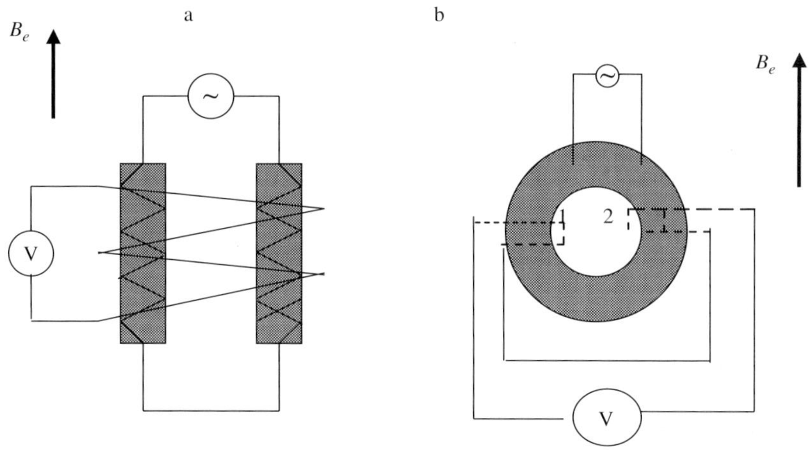

Fig. 6.11. (a) Fluxgate device with two magnetic cores. (b) Fluxgate device with one torus.

two external fields:

$$B_e + p \sin \omega t \quad \text{and} \quad B_e - p \sin \omega t \quad (p>0)$$

This means that the magnetic fields inside ferromagnetic cores are defined by different portions of some path of the hysteresis loop. As follows from Equation (6.78), the first and third harmonics of the field do not affect the resultant electromotive force, but the second harmonic, which is proportional to the constant magnetic field, is doubled. Also it is useful to outline the main idea of the device, which consists of one torus shown in Fig. 6.11(b). The source of the alternating current generates inside the torus the constant flux, but its magnetic field has opposite directions in the vicinity of receiver coils 1 and 2, where the magnetic field B_e has the same direction. Correspondingly, measuring a sum of the electromotive forces induced in these coils, we again measure only the second harmonic proportional to this field.

6.9. MAGNETIZATION AND MAGNETIC FORCES

Now we continue to study the main features of ferromagnetism. Proceeding from the experimental data, we already described dependence of the magnetic field inside a ferromagnetic material on the ambient magnetic field. Earlier it was emphasized that the hysteresis loop cannot be derived analytically, since both functions $B(B_0)$ and $P = P(B_0)$ depend on magnetization in the past. At the same time, it is instructive to make an attempt to describe analytically the magnetization curve 1, shown in Fig. 6.10(c), and then compare results of calculations with experiments. With this purpose in mind, we will use the same approach as in the case of the paramagnetism. Suppose that atoms of the ferromagnetic material are at the state of thermal equilibrium and, as before, there are two factors which affect magnetization. One of them is the magnetic field acting on the dipole which tries to line up it along this field, and the other is a random motion of dipoles that produces an opposite effect and decreases a magnetization. In accordance with Equation (6.70), we can write

$$P = P_s \tanh \frac{p_0 B_a}{kT} \tag{6.79}$$

Here $P_s = N p_0$ is the maximal magnitude of the vector of magnetization that corresponds to the stage of saturation and B_a the field acting on the atomic magnetic dipole. In a paramagnetic substance, an interaction between magnetic dipoles is neglected and the actual field B_a coincides with the ambient magnetic field, $B_a = B_0$, but it is not correct in the case of ferromagnetic material. In fact, the hysteresis loop clearly shows that the magnetic field $\boldsymbol{B}$ inside this substance greatly differs from the primary field $\boldsymbol{B}_0$. In other words, the secondary magnetic field caused by magnetic dipoles of atoms usually play dominant role and we have

$$\boldsymbol{B} = \boldsymbol{B}_0 + \boldsymbol{B}_s \tag{6.80}$$

When we consider the magnetic field $\boldsymbol{B}$ in a macroscopic scale, its behavior inside a magnetic material is very simple; it is a continuous function which gradually changes from point to point as the primary field $\boldsymbol{B}_0$ does. On the contrary, within an atom the behavior of the field magnitude B is very peculiar. For instance, near a nucleus it can be extremely large while between atoms it becomes much smaller; correspondingly, this field cannot describe the field which actually acts on the magnetic dipole of an atom. In order to find this field $\boldsymbol{B}_a$, let us mentally draw a small spherical surface around some point where we would like to find the actual field (Fig. 6.12(a)). Then we can represent the magnitude of field $\boldsymbol{B}$ at the center of the sphere as a sum

$$B = B_1 + B_2 \tag{6.81}$$

where B_1 is the magnetic field at the center of the sphere caused by magnetization currents of this small volume, while B_2 is the field generated by all other currents including also the currents producing the ambient field B_0. In other words, the latter is a part of the field B_2. Inasmuch as the currents causing this field are located at some finite distance from the sphere center, the function B_2 slowly varies in the vicinity of this point and therefore it may serve as the actual field: $B_a = B_2$. As follows from Equation (6.81)

$$B_a = B_2 = B - B_1 \tag{6.82}$$

In order to use Equation (6.79), we have to express the right-hand side of this equation in terms of known quantities and magnetization P. With this purpose in

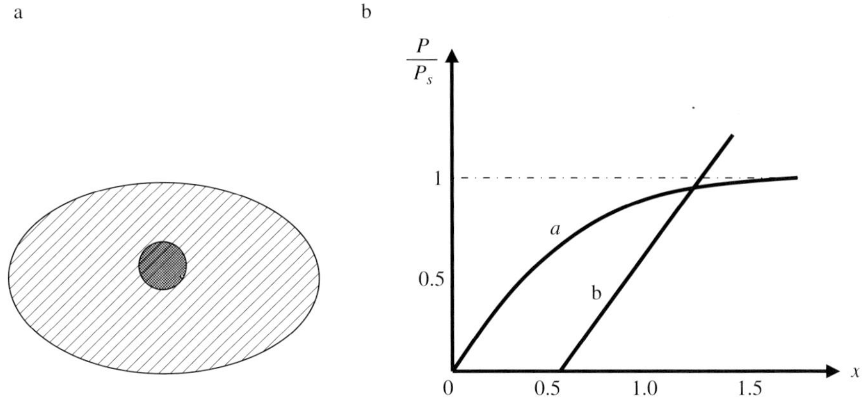

Fig. 6.12. (a) Illustration of Equation (6.81). (b) A graphical solution of Equation (6.88).

mind, it is natural to make use of results derived in Chapter 3. In accordance with Equations (3.141) and (3.143), inside the sphere magnitudes of the field and the vector of magnetization are

$$B_i = \frac{3\mu}{\mu + 2\mu_0} B \quad \text{and} \quad P = \frac{3}{\mu_0} \frac{\mu - \mu_0}{\mu + 2\mu_0} B \tag{6.83}$$

since

$$M = P \frac{4}{3} \pi a^3$$

Thus, the field and magnetization are uniform inside the sphere and they are independent of the radius a. This fact is very important because it allows us to take an arbitrary small sphere and think that the actual field is the field caused by all currents except the field of the dipole at the sphere center. Bearing in mind that the field inside of the sphere includes the external field $\boldsymbol{B}$, we have

$$B_i = B_1 + B$$

Then, Equation (6.83) gives

$$B_1 = \frac{3\mu}{\mu + 2\mu_0} B - B = 2 \frac{\mu - \mu_0}{\mu + 2\mu_0} B = \frac{2}{3} \mu_0 P \tag{6.84}$$

Respectively, for the actual field we have

$$B_2 = B - \frac{2}{3} \mu_0 P \tag{6.85}$$

At the right-hand side both terms are unknown, and for this reason we make use of the equality

$$B = \mu_0(H + P)$$

Its substitution into Equation (6.85) gives

$$B_a = B_2 = \mu_0 H + \frac{1}{3}\mu_0 P \tag{6.86}$$

Of course, in general the field H is also unknown but there are important exceptions and some of them we already considered, for instance, the field inside the very long solenoid and magnetic torus. In such cases, fictitious sources of the field H are absent and for this reason it coincides with the ambient field H_0. This permits us to rewrite the equation for the actual field as

$$B_a = B_0 + \frac{1}{3}\mu_0 P \tag{6.87}$$

where the external field B_0 is known. This makes Equation (6.79) much more convenient for a study of magnetization. Substitution of Equation (6.87) into Equation (6.79) yields

$$\frac{P}{P_s} = \tanh\left(\frac{B_0 + (\mu_0 p_0/3)P}{kT}\right) \tag{6.88}$$

This is a transcendent equation which allows us to find a relationship between the magnetization P and the primary field B_0. In order to illustrate a solution of this equation which contains the unknown P at the left and right sides of this equation, we introduce a variable x:

$$x = \frac{B_a}{kT} = \frac{B_0}{kT} + \frac{\mu_0 p_0}{3kT}P$$

Then we have two functions of x

$$\frac{P}{P_s} = \tanh x \quad \text{and} \quad \frac{P}{P_s} = \frac{3kT}{\mu_0 p_0 P_s}x - \frac{3B_0}{\mu_0 p_0 P_s} \tag{6.89}$$

and they are represented in Fig. 6.12(b). One of them is the hyperbolic tangent (curve a), but the other is a linear one and it is described by the straight line (curve b). Its slope is directly proportional to the temperature and it intersects the x-axis at the point

$$x_0 = \frac{3B_0}{\mu_0 p_0 P_s} \tag{6.90}$$

The coordinates of the point of intersection of these two graphs satisfy both equations of the set (6.89), and therefore they are solutions of Equation (6.88).

6.9.1. Spontaneous magnetization

Consider a special case when the ambient field is absent, but before let us recall that Equation (6.88) is based on an assumption that an interaction between atomic magnetic dipoles is governed by magnetic forces; that is, they obey Ampere's law. Suppose that the external field $B_0 = 0$ and introduce notation:

$$T_c = \frac{\mu_0 p_0 P_s}{3k} \tag{6.91}$$

Then Equations (6.89) can be written as

$$\frac{P}{P_s} = \tanh x \quad \text{and} \quad \frac{P}{P_c} = \frac{T}{T_c} x \tag{6.92}$$

As we know, solutions of the second equation of the set (6.92) are linear functions, and correspondingly it is proper to distinguish two families of straight lines. One system has relatively large slope $((T/T_c) > 1)$, and they are located above the curve a, for instance, curve 1. The second family has smaller slope, $(T/T_c) < 1$ (curve 3), and finally the boundary between them, line 2, which is tangential to the curve a and the ratio $(T/T_c) = 1$. As is seen from Fig. 6.13, the lines of the first family do not intersect the curve a, and all of them have only one common point with this curve when $P = 0$. This means that the temperature is relatively high and a random motion of atoms prevents the atomic dipoles to line up in a certain direction.

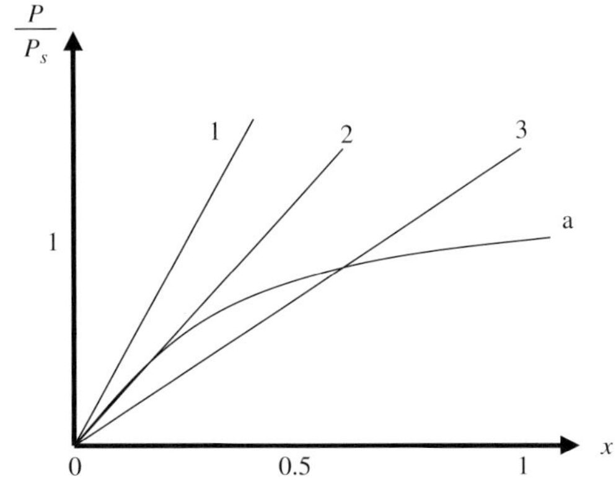

Fig. 6.13. Solution of Equations (6.92) when $B_0 = 0$, and the straight lines 1, 2, and 3 correspond to cases: $(T/T_c) > 1$, $(T/T_c) = 1$, and $(T/T_c) < 1$.

In other words, Equations (6.89) have a solution when a magnetization is zero. Next, consider a solution when temperature of the ferromagnetic material is lower then T_c, that corresponds to the second family of lines $((T/T_c)<1)$. Each straight line intersects the curve a in two points, that is, there are two solutions. One of them is $P=0$, but it is unstable solution since even very small change of temperature leads to sufficiently strong magnetization. At the same time, the second solution is stable, and it shows that with a decrease of temperature the magnetization becomes relatively large. Thus, in accordance with Equations (6.92), in the absence of the external magnetic field the following behavior of a magnetization as a function of temperature takes place. First, there is a range of relatively high temperatures when magnetization is absent. Then, as soon as temperature becomes smaller than T_c, magnetization suddenly arises and we observe a so-called spontaneous magnetization. In this case, the thermal motion is rather weak so that an interaction between atomic magnets is able to line up part of them in the same direction and with a decrease of temperature this effect becomes stronger and the magnetization approaches to its maximal value, P_s. We see that in one range of temperatures the ferromagnetic behaves as a paramagnetic material but with a decrease of T it is magnetized itself. The spontaneous magnetization is a very interesting phenomenon but it seems that it contradicts the results obtained from the hysteresis loop where it was shown that a ferromagnetic substance becomes a permanent magnet only after it was magnetized. In other words, if this material was not placed earlier into a magnetic field the magnetization does not arise regardless of how small the temperature is. Later we will discuss how this contradiction was resolved.

6.9.2. Curie temperature

Now we pay an attention to the case when $T=T_c$ and $B_0=0$, that is, a boundary between two families of curves. As follows from Equations (6.92), we have

$$\frac{P}{P_s}=\tanh x, \quad \frac{P}{P_s}=x \tag{6.93}$$

It is obvious that the straight line 2 is tangential to the curve a since at small values of x: $\tanh x \approx x$, and it has a slope equal to unity. By definition, this line corresponds to temperature when the spontaneous magnetization arises and it is called Curie temperature T_c. As follows from Equation (6.91), it is directly proportional to the atomic magnetic moment and magnetization at the stage of saturation, but is independent of an external magnetic field. For instance, in the presence of the field B_0 we still observe a transition from the paramagnetic to ferromagnetic at the Curie temperature defined by Equation (6.91). Next, consider an influence of this field when its value is relatively small. From Equation (6.91), it follows that due to the presence of B_0 the straight lines with the given temperature are shifted to the right. Inasmuch as at lower values of T the slope of the curve a is very small, this shift causes also a small change of magnetization. At the higher temperatures an intersection with this curve takes place where it varies almost linearly and, correspondingly, an influence of the external field becomes stronger. Finally, let us evaluate the Curie temperature from Equation (6.91) and compare with the experimental data. As an example, consider nickel; in this case, an experiment shows that $T_c = 631$ K. Magnetization of the nickel as that for other ferromagnetic materials is related to magnetic moments of electrons in the inner shell of the atom. In the case when the effect of an orbital motion is absent, for single electron the component of the angular momentum L in any direction is $h/2$ and we have

$$p_0 = \frac{q_e h}{2m} = 0.93 \times 10^{-23} \text{ A m}^2 \tag{6.94}$$

since $g = 2$. Knowing its density and atomic weight, we have for number of atoms N per unit volume

$$N = 9.1 \times 10^{28} \text{ m}^{-3}$$

Since $P_s = p_0 N$, Equation (6.91) becomes

$$T_c = \frac{\mu_0 p_0^2 N}{3k} \tag{6.95}$$

Thus, we have

$$T_c = \frac{4\pi \times 10^{-7} \times 0.86 \times 10^{-46} \times 9.1 \times 10^{28}}{3 \times 1.4 \times 10^{-23}} \approx 0.24 \text{ K}$$

The ratio between the experimental and calculated values is around 2600. This clearly shows that the theory based on an assumption that an interaction of atomic magnetic dipoles is governed by only magnetic forces fails, even though it allowed us to predict the spontaneous magnetization. Such great discrepancy suggests that in addition there is some kind of nonmagnetic interaction between electrons of neighboring atoms. Of course, in order to have an agreement between the experimental and theoretical results, we may artificially change Equation (6.88) and write

$$\frac{P}{P_s} = \tanh\left(\frac{B_0 + \lambda\mu_0 P}{kT}\right) \tag{6.96}$$

where λ is some constant, and the Curie temperature (6.95) is defined as

$$T_c = \lambda\mu_0 p_0^2 N$$

Having taken λ around 900 we arrive at the correct value of T_c. It is interesting to note that the same number gives satisfactory result for other ferromagnetic substances. The term $\lambda\mu_0 P$ characterizes an influence of atomic magnetic dipoles around the place where we calculate magnetization, and certainly magnetic forces are not capable to create such tremendous force of interaction when the value of λ is near 1000. By no means, these estimations can be considered as an explanation of ferromagnetism; they only illustrate that the theory based on the classical physics as well as the concept of a spin and magnetic moment of an atom is not sufficient to describe the behavior of such materials.

6.9.3. Spontaneous magnetization and Weiss domains

Assuming a thermal equilibrium of magnetic dipoles and their interaction in accordance with Ampere's law, we arrived at Equation (6.88) which allowed us to find a relation between the magnetization and the actual magnetic field. To derive this equation we also used concepts of the statistical and quantum mechanics which gave us some insight in the behavior of ferromagnetic materials. First of all, qualitatively, dependence of a magnetization from a temperature, given by Equation (6.88), corresponds to experimental data. At the same time, this equation gives a value of Curie point, which is several orders smaller than the real value of this temperature. Such discrepancy clearly indicates that an interaction which follows from the classical theory has to be much stronger in order to obtain numbers which are close to experiments. Finally, Equation (6.88) shows that there is a temperature when even in an absence of an external field, the interaction of the magnetic dipoles forces them to line up parallel to each other and magnetization arises spontaneously. This phenomenon is not observed in the case of paramagnetic materials where an interaction between magnetic dipoles is negligible, and it contradicts the experimental studies of the hysteresis loop. In order to resolve this difference between the theoretical analysis and experiments, Weiss introduced at the beginning of the 20th century the concept of domains which made a strong impact on the theory of ferromagnetism. He suggested the following explanation. If we consider an extremely small crystal of iron or other ferromagnetic material, which represents a so-called domain, then the spontaneous magnetization arises. As soon as the temperature is below the Curie point, elementary magnetic dipoles of the domain are oriented in the same direction even when the ambient field is absent, and there is a strong interaction between them. Of course, every dipole creates a magnetic field outside the domain. Relatively large piece of a material consists of many domains and in each of them spontaneous magnetization occurs. However, the orientation of the vector of magnetization $\mathbf{P}$ depends on a domain. Besides, in every domain and at relatively low temperatures magnetization is close to that of saturation and for these two reasons an average value of P over all domains is almost zero. Correspondingly, the magnetic field is absent outside the magnet. This clearly demonstrates that magnetic properties of a bulk material and its tiny pieces are very different. Thus, applying the concept of domains we see that Equation (6.88) correctly predicts a spontaneous magnetization for a single domain. Also, it becomes clear why measurements of the hysteresis loop which are carried out for a ferromagnetic material with extremely large number of such domains (the size of each domain is around 10^{-5} m and less) do not discover this phenomenon. Let us continue our discussion of domains and consider two cases.

6.9.4. Case one: Single crystal of ferromagnetic and its domains

Suppose that we a have a single crystal of iron which contains many domains. They may have different shapes and dimensions. Below Curie point at each domain the vector of magnetization may have different direction but almost the same magnitude which is close to a saturation value. The magnetic field caused by magnetization currents in each domain differs from zero, but outside their resultant field is absent. One can say that domains are distributed in such "clever" way that their magnetic fields outside a crystal cancel each other. In order to visualize this distribution of domains let us imagine that there is only one domain, shown in Fig. 6.14(a), where all magnetic dipoles have the same direction. Respectively, there is a field outside that contradicts an experiment. In the next case (Fig. 6.14(b)), there are two domains with opposite directions of the magnetic dipoles that reduce the field outside. It is obvious that an appearance of different domains with different orientation of atomic dipoles (Fig. 6.14(c and d)) results in a decrease of the magnetic field. A dotted line inside the single crystal characterizes the boundary between domains and it is called the "wall" of domain and its existence is associated with the additional energy (wall energy). As follows from the quantum mechanics, it is possible to treat an interaction of neighboring atoms with the help of nonmagnetic forces. In accordance with the exclusion principle formulated by Pauli, if two electrons are located at the same point, they have to have opposite spins and therefore the opposite magnetic moments.

This means that such pair of dipoles does not create the magnetic field. It turns out that as a rule each shell contains pair of electrons and their magnetic moments are opposite to each other. In other words, the exclusion principle explains why most of the materials do not have magnetic properties. Paramagnetic and ferromagnetic substances are exceptions. In the case of paramagnetism, an interaction is negligible and spontaneous magnetization is absent, while the magnetic moments are oriented under an action of only the external field. As concerns the ferromagnetic materials,

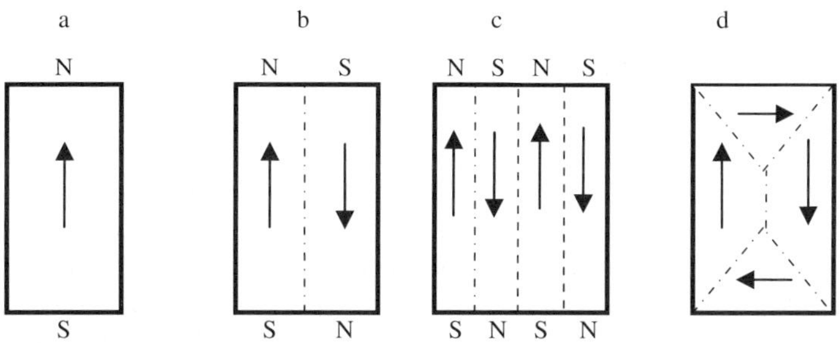

Fig. 6.14. Formation of domains.

the situation is completely different, an interaction between nearby magnetic dipoles is very strong and it forces neighboring dipole to line up parallel to each other. It seems that such an orientation contradicts Pauli principle; however, in reality this indicates that there is a special mechanism of interaction of nonmagnetic origin in the ferromagnetic material. There are several models which are developed to explain this orientation of dipoles when the external field is absent. In one of them free electrons play the main role. As we know, the electron in the internal shell causes the magnetic field of an atom. It may be possible that it is also capable to provide a spin with opposite sign to the free electrons moving around. When these electrons reach next atom, the electron inside of the internal shell of this atom acquires spin with opposite sign. In other words, magnetic moments of electrons inside both internal shells become parallel to each other and the conduction electrons play role of an intermediary. There is another model of an interaction where electrostatic forces caused by electron play the dominant role and their influence also helps to explain the parallel orientation of magnetic dipoles inside each domain.

Next we describe magnetization under an action of the external magnetic field oriented, for example, along the central wall and assume that the magnet is a single crystal shown in Fig. 6.14(d). Since magnetic dipoles tend to be oriented along the external field, the middle domain wall moves to the right so that the region where dipoles are oriented "up" becomes bigger than that with dipoles directed "down". As a result, the magnetic energy of dipoles in the presence of the external field decreases. With further increase of the field, the whole crystal is gradually transformed into a single domain where most of the dipoles are parallel to the external field and their magnetic energy reaches minimum. If the external field is arbitrarily directed with respect to the axes of the crystal, the process of magnetization usually consists of several steps, since it is relatively easy to magnetize along the crystal axis than in any other direction. For this reason at the beginning, when an external field is rather weak, the domains, where an orientation of dipoles is close to that of crystal axes and the external field, begin to grow until the magnetization is directed along one of these directions. Then with an increase of the external field all dipoles become oriented along this field and we again deal with one domain with a strong magnetization.

6.9.5. Case two: Polycrystalline material of ferromagnetic

Suppose that a material consists of many crystals (polycrystalline material); for instance, it can be an ordinary piece of iron. Inside of it there are very many crystals with their crystalline axes having different directions and each of them may have some domains (Fig. 6.15). Now let us apply an external field and, as before, we observe the process of magnetization, shown by curve 1 in Fig. 6.10(b). When this field is weak the domain walls start to move and those domains, where the direction of the vector of magnetization is close to that of the primary field, begin to grow. This part of magnetization curve (a) of Fig. 6.10(b) has one important feature, namely, if the field decreases then the magnetization will follow this part of the curve and finally magnetization vanishes; that is, this process is reversible. Of

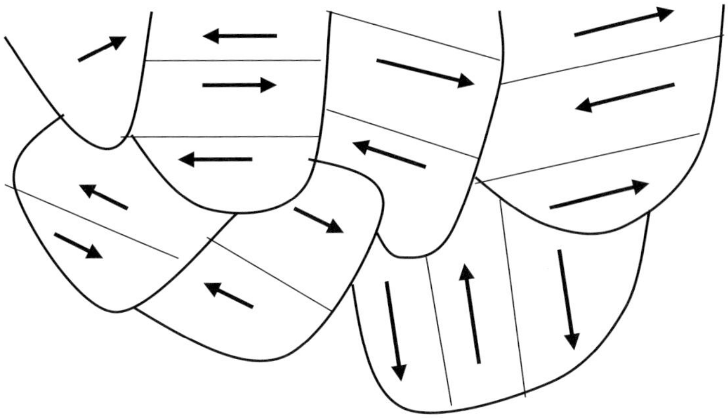

Fig. 6.15. Domains of polycrystalline ferromagnetic material.

course, such a behavior takes place if the external field is very small. With an increase of the field we observe the second part of the curve (b), and in order to describe its behavior it is necessary to take into account the presence in each crystal of such factors as impurities, dislocations, and strains. With an increase of the field the domain wall reaches these imperfections and within some range of the field's change it does not move. Then, an external field becomes sufficiently large and the domain continues to extend its dimensions; that is, it overcomes this obstacle. This interaction between the domain wall and dislocations is accompanied by some losses of the magnetic energy that causes a hysteresis curve. Within this range an increase of magnetization can be represented by a system of small step functions and such behavior is detected by measurements of the electromotive force. Finally when a field is strong enough and most of domain walls are moved and magnetization took place almost in each crystal, there are still some exceptions where the vector of magnetization has a direction different from that of the external field. To force them to be oriented along this field we have to increase it more, and this process is represented by the last part of the curve (c). In essence, we described the hysteresis and the behavior of the magnetization curve, which shows that properties of a magnetic material depend on ability of a wall domain to move under an action of the external field. For instance, the so-called stainless steel which is a mixture of atoms of iron, chromium, and nickel is almost nonmagnetic, even though there are some exceptions with some special composition of the alloy. Perhaps, such behavior of the stainless steel is related to the fact that the domain walls are not able to move easily even when the external field is very large. The other example is a permanent magnet. In this case, a wall domain remains at the same place within a broad range of the external field, and, correspondingly, it has a very wide hysteresis loop. In other words, when an external field has a direction opposite to that of the original magnetization, the magnitude of this field has to be very large in order to change the vector $\boldsymbol{P}$. On the contrary, when we use materials with a few

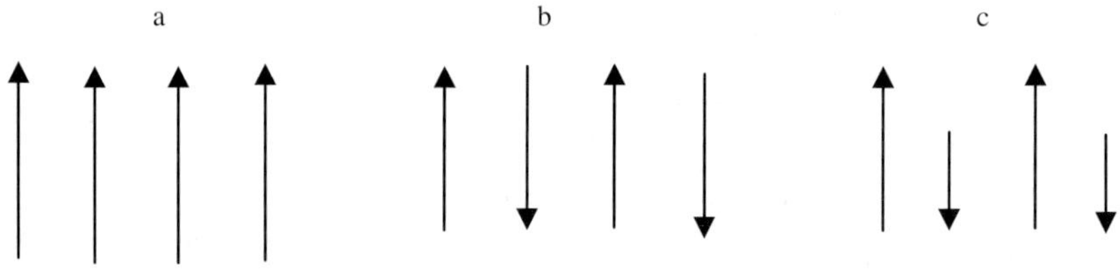

Fig. 6.16. Mutual orientation of magnetic moments: (a) ferromagnetic, (b) antiferromagnetic, and (c) ferrite.

dislocations and impurities, the domain walls move easily under an action of the external field, that is, this substance is sensitive to a change of this field and this is a reason why we say that such a material is magnetically "soft", and it has narrow hysteresis loop (Fig. 6.10).

In addition, let us notice that there are several different groups of elements of the periodic table which have one common feature, namely, their inner electron shell is not complete and, correspondingly, the atoms have the magnetic moment. At the same time, each group is characterized by a certain mutual orientation of these dipoles and, as a result, they cause different magnetic fields. In the first group (ferromagnetic), the moments of neighbor dipoles have the same direction (Fig. 6.16(a)), and magnetization currents may create very strong magnetic fields. In the second group, called antiferromagnetic, the magnetic moments of the neighbor dipoles have opposite direction (Fig. 6.16(b)), and, therefore, they do not create the magnetic field in spite of the fact that each atom has the magnetic moment. For instance, chromium and manganese are antiferromagnetic. In the third group of materials, a mutual position of dipoles is more peculiar and it is shown in Fig. 6.16. In such a case, both the direction and magnitude of the dipole moment periodically change. This group of materials is called ferrites and their influence on the magnetic field is relatively small. Besides, they are insulators and this is the reason why they are useful when the high-frequency electromagnetic fields are used. Also, there are other groups of magnetic materials.

Chapter 7

Nuclear Magnetism Resonance and Measurements of Magnetic Field

7.1. INTRODUCTION

Earlier we pointed out that a study of an interaction between a magnetization caused by nuclear magnetic dipoles and the constant and high-frequency electromagnetic fields allows us to understand different features of a structure of solids, fluids, and gases and it found a broad application in physics, chemistry, medicine and etc. Taking into account the fact that this phenomenon is also used for measuring magnetic fields, for evaluation of physical properties of sediments in the borehole geophysics, and even there are attempts to apply NMR in the surface exploration geophysics, we will describe the principles of this method in some detail. In Section 6.4 we already illustrated an action of the constant and alternating fields on the behavior of the nuclei, as well as an interaction between these particles and a surrounding medium. First, let us review some results obtained in the previous chapter. Suppose that the single nucleus is located in the constant magnetic field $\boldsymbol{B}_0$ directed along the z-axis. As is shown in Fig. 7.1(a), the magnetic moment of nucleus $\boldsymbol{p}$ rotates on the cone surface around this field and the frequency of precession is related to the field as

$$\boldsymbol{\omega}_0 = -\gamma \boldsymbol{B}_0 \tag{7.1}$$

where γ is gyromagnetic constant which is different for different nuclei, and $\boldsymbol{\omega}_0$, the vector directed opposite to magnetic field, if $\gamma > 0$. This fact directly follows from the Newton's second law

$$\frac{d\boldsymbol{L}}{dt} = (\boldsymbol{p} \times \boldsymbol{B}_0)$$

In accordance with the classical physics, the magnetic energy of the magnetic dipole is equal to

$$\Delta U_m = -pB_0 \cos\theta \tag{7.2}$$

Here θ is the angle of precession which may vary from 0 to π. Thus, the energy of the nuclear dipole is minimal when its moment is directed along the field and it

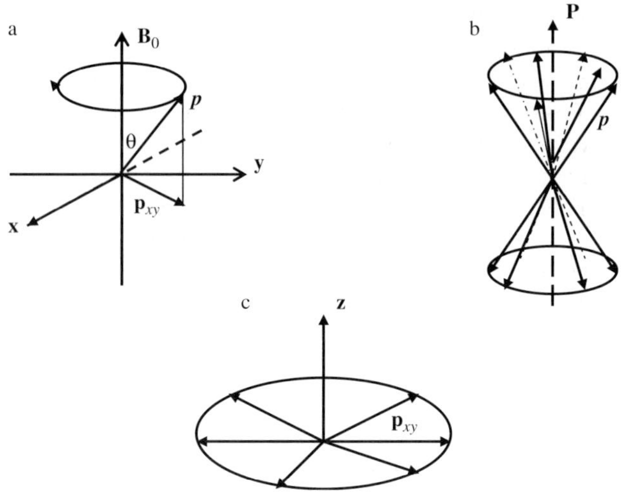

Fig. 7.1. (a) Precession of the nuclear magnetic dipole; (b) orientation of magnetic dipoles on the cone surface; (c) arbitrary distribution of phases of horizontal components $P_\perp$.

reaches a maximum if these vectors have opposite directions. At the same time, as was described in Chapter 6

$$\Delta U_m = \gamma \hbar m B_0 \tag{7.3}$$

where $m = j, j{-}1, \ldots, -j$. For instance, in the case $j = 1/2$, we have

$$(\Delta U_m)_{\max} = \frac{\hbar\omega_0}{2} \quad \text{and} \quad (\Delta U_m)_{\min} = -\frac{\hbar\omega_0}{2} \tag{7.4}$$

As was shown in Section 6.3 components of the magnetic moment are described by equations

$$p_x = p_{xy}\cos(\omega_0 t + \varphi),\ p_y = -p_{xy}\sin(\omega_0 t + \varphi),\ p_z = \text{constant} \tag{7.5}$$

From these equations it follows that the horizontal component of the magnetic moment rotates and describes the circle with the radius $p_{xy} = (p_x^2 + p_y^2)^{1/2}$, but the vertical component p_z remains constant. It is clear that the angle of precession θ, characterizing a cone, is defined from the ratio:

$$\theta = \tan^{-1}\frac{(p_x^2 + p_y^2)^{1/2}}{p_z}$$

Note that in accordance with the quantum mechanics it is possible to speak only about a probability of a location of the moment on the cone's surface. In other words, the magnetic field defines a frequency of precession and a position of the cone with respect to this field but it does not determine a location of the magnetic dipole at each instant. This means that the initial phase φ in Equation (7.5) can have an arbitrary value. Besides, as was discussed in Section 6.2, the component p_z may have only discrete values defined by the spin j. For instance, in the case of $j = 1/2$ the vector of magnetic moment can be located either on the upper or lower parts of the cone. For an arbitrary value of the spin j there are $(2j+1)$ energy levels and, correspondingly, the magnetic moment can be located at lateral surfaces of cones with different values of θ. Taking into account Equation (7.4) we see that a difference between two neighbor levels is the same:

$$\Delta E = E_m - E_{m-1} = \gamma \hbar B_0 = \hbar \omega_0 \tag{7.6}$$

Thus, the frequency of precession coincides with that of transition between neighbor levels of energy and can be also represented as

$$f_0 = \frac{\Delta E}{h} = \frac{\gamma}{2\pi} B_0 \tag{7.7}$$

that is, the frequency of transition is directly proportional to the constant ambient field and the coefficient of proportionality, which is defined by type of the nuclei. In this light it is proper to make several comments:

1. The constant magnetic field generates magnetization which in own turn depends on a distribution of nuclei with different energy levels. For instance, with an increase of difference between number of nuclei with the low and higher energy levels it is natural to expect an increase of magnetization.
2. If the frequency of an external electromagnetic field coincides with that of transition it is possible to observe an absorption and radiation of energy. In the first case there is a transition to the higher energy level, but in the second the energy of a nucleus becomes smaller.
3. As follows from the classical physics with an increase of the solid angle of the cone this energy becomes higher, in particular, when it is equal 2π the magnetic energy is equal to zero, since the angle between the field and magnetic dipole is $\pi/2$.
4. In the classical physics the magnetic dipole represents a relative small current loop and its motion takes place due to the force caused by the magnetic field (Ampere's law). Respectively, its magnetic moment continuously changes its direction until it coincides with that of the magnetic field. At this moment the dipole has a minimum of magnetic energy and equilibrium is stable. In contrary, when the dipole is directed opposite the field equilibrium is unstable. Such motion is observed if we assume that the angular momentum of the dipole is neglected. However, when it is present and has the same or opposite direction as the magnetic moment, the latter will be involved in precession with some frequency around the magnetic field. In accordance with the quantum mechanics, unlike the classical one, instead of a continuous motion of the magnetic moment there is a discontinuous transition from one position on the

conical surface to another, for example, from the lower to upper part of the cone. At the same time in both cases it is convenient to think that the magnetic field causes the real force acting on the magnetic dipole.

7.2. THE VECTOR OF NUCLEAR MAGNETIZATION

Next, proceeding from the concept of a magnetic moment of a nucleus, we again arrive at the macroscopic quantity of the vector the magnetization. With this purpose in mind consider a system of magnetic moments of nuclei and, by analogy with the diamagnetic and paramagnetic materials, introduce the vector of magnetization as

$$\boldsymbol{P} = \sum \boldsymbol{p}_i \tag{7.8}$$

which characterizes the dipole moment of the unit volume. Here $\boldsymbol{p}_i$ is the magnetic moment of a nucleus. To illustrate Equation (7.8) suppose that the spin $j = 1/2$, and all dipoles are located either at the upper or lower surfaces of the same cone (Fig. 7.1(b)). Summation of these moments gives the vector of magnetization $\boldsymbol{P}_0$ directed along the ambient field $\boldsymbol{B}_0$. It is essential that at the state of equilibrium this vector is time independent and, unlike the magnetic moments of particles, it is not involved in precession. Inasmuch as the ambient field defines only the precession frequency of each magnetic moment and an orientation of the cone of precession, the motion of these moments usually occur with different phases. In other words, magnetic moments of different dipoles are located at different places of the conical surface. Then, performing a summation of the transversal components $\boldsymbol{p}_\perp$ located in the plane perpendicular to the field, we observe the destructive interference and their sum is equal to zero $P_\perp = 0$ (Fig. 7.1(c)). In the case of the thermal equilibrium a distribution of these dipoles obeys the Boltzmann's law and, therefore, a sum of the components along the field is not equal to zero. This happens because magnetic dipoles are located at different energy levels, and the number of dipoles located at the lower level slightly exceeds that on the higher level. In other words, a number of vectors $\boldsymbol{p}_i$ in the upper cone are more than that in its opposite part. Thus, the static nuclear magnetization for the state of equilibrium is not equal to zero:

$$\boldsymbol{P}_{0z} = \boldsymbol{P}_0 = \sum \boldsymbol{p}_{zi} \tag{7.9}$$

As in the case of the diamagnetic and paramagnetic substances the static magnetization is directly proportional to the ambient field, if $\hbar\omega_0 \ll kT$:

$$\boldsymbol{P}_0 = \frac{\chi_n}{\mu_0} \boldsymbol{B}_0 \tag{7.10}$$

where χ_n is the nuclear susceptibility and its value is very small ($\approx 10^{-6}$), while for the electronic paramagnetic and diamagnetic materials it is much stronger

(10^{-4}–10^{-6}). As we already know, sedimentary formations consist of mainly the diamagnetic as well as paramagnetic materials. Of course, if there are ferromagnetic particles it is necessary to take into account their presence. Under an action of the ambient magnetic field $\boldsymbol{B}_0$, the magnetization arises and the resultant moment of atomic dipoles differs from zero. These dipoles have different origin: they may appear due to a motion of electrons of an atom and induced currents of a diamagnetic origin, as well as a motion of nuclei. Our goal is to study a magnetization caused by magnetic moments of nuclei. Note that a resultant magnetic moment of dipoles due to an orbital motion of electrons and their rotation (spin) in a water and oil molecule is equal to zero. Our attention is mainly paid to the magnetic moments of protons of the hydrogen in the presence of the constant magnetic field which is relatively small and, correspondingly, produces rather small magnetization. This means that the number of dipoles N_{up} with positive component along the field is only slightly exceeds those, N_{down} with negative component. Applying the same approach as in the case of paramagnetic materials (Equation (6.65)), we find that in the state of the thermal equilibrium:

$$\frac{N_{\text{up}}}{N_{\text{down}}} \approx 1 + 2\frac{pB_0}{kT}$$

Assuming, for example, that $p \approx 10^{-26}$, $k \approx 1.4 \times 10^{-23}$, $T = 300°$, $B_0 = 10^{-1}$ T, we see that the second term is extremely small

$$2\frac{10^{-26} \times 10^{-1}}{1.4 \times 10^{-23} \times 3 \times 10^2} \approx 0.5 \times 10^{-6}$$

Inasmuch as the magnetization is very small, the secondary magnetic field caused by these dipoles is almost negligible with respect to $\boldsymbol{B}_0$. As was pointed out earlier there is one more feature of this magnetization. Every magnetic moment rotates around the ambient magnetic field (the z-axis) with the precession frequency ω_0, and a rotation of its horizontal component is described by the sinusoidal function

$$A\cos(\omega_0 t + \varphi)$$

Because of a thermal motion, the initial phases φ for each magnetic moment can be different. Also, the neighbor dipoles have a slightly different frequency of precession, since the field $\boldsymbol{B}_0$ changes from point to point. Sometimes this factor is called dephasing (Fig. 7.1(c)). Correspondingly, due to the destructive interference the horizontal component of the resultant magnetic moment of all dipoles in an elementary volume becomes equal to zero. At the same time the sum of vertical components of magnetic dipoles gives the magnitude of the vector of magnetization (Equation (7.9)). Therefore, in order to use a magnetization as a tool for studying parameters of a medium we have to create, first of all, an additional field, $\boldsymbol{B}_1(t)$, which in general changes a magnitude and direction of the vector of magnetization, $\boldsymbol{P}_0$. Then, measuring behavior of the total magnetization caused by the ambient and additional fields with time, it is possible to obtain information about some important features of a medium. There are several methods of measuring magnetization caused by nuclei; that is NMR, which allows one to obtain information about a structure of atoms and molecules as well as some physical properties of a medium. It may be proper to emphasize that in all of them we observe radiation and absorption of an electromagnetic energy at the frequency of precession and, correspondingly, a transition from one to another energy level. In the first approach, briefly described in Chapter 6, a change of the frequency of the additional field and measurements of a loss of its energy allowed us to determine the precession frequency ω_0. The same result is obtained when the frequency of the additional field remains constant but the magnitude of the ambient constant field B_0 changes. The third method is based on the use of the constant additional field which usually greatly exceeds the ambient field ($B_1 \gg B_0$). For instance, such case is shown in Fig. 7.2(a), when these fields are perpendicular to each other. Due to the field $\boldsymbol{B}_1$ we move the initial vector of magnetization away from the original position of equilibrium and create magnetization which has a different direction and magnitude. Because of an influence of a surrounding medium the magnitude of the vector of magnetization does not instantly reach its static value and requires some time before it becomes constant (Fig. 7.3(b)). This behavior already contains information about a medium. Note that during this time the magnetic moments $\boldsymbol{p}_i$ perform a rotation around the resultant field $\boldsymbol{B}$. Then, the additional field is turned off and we begin to observe a precession of the magnetic moments of nuclei around the original ambient field $\boldsymbol{B}_0$. For this reason the vector of magnetization $\boldsymbol{P}$ also rotates around this field with the frequency of precession; that is, its tip moves along a spiral, and an evolution depends on the ratio between time during which the vector of magnetization approaches to its static value and time when the horizontal component vanishes. It is obvious that with time this vector becomes equal to $\boldsymbol{P}_0$ that corresponds to the state of equilibrium when the ambient field is $\boldsymbol{B}_0$. Thus, this transient behavior of the vector of magnetization $\boldsymbol{P}(t)$ results a decrease of its magnitude ($B_1 > B_0$) and the angle between the ambient field and magnetization; this motion also includes a precession around $\boldsymbol{B}_0$. Certainly, this is a complicated motion and an influence of a surrounding medium on its behavior will be briefly discussed later. As illustration, a function $P(t)$ is given in Fig. 7.3(b) when the

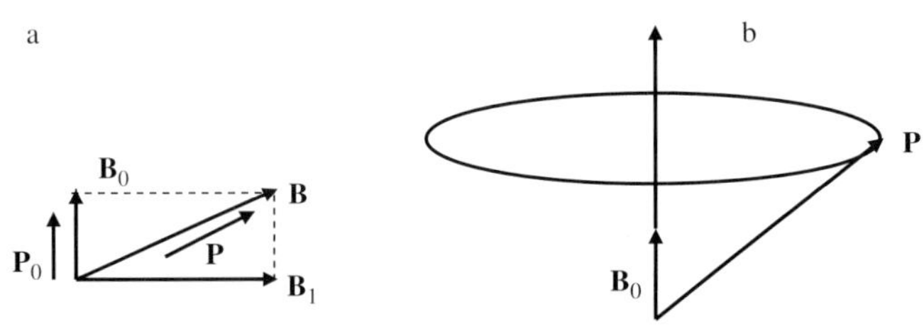

Fig. 7.2. (a) Resultant field; (b) precession of the vector of magnetization.

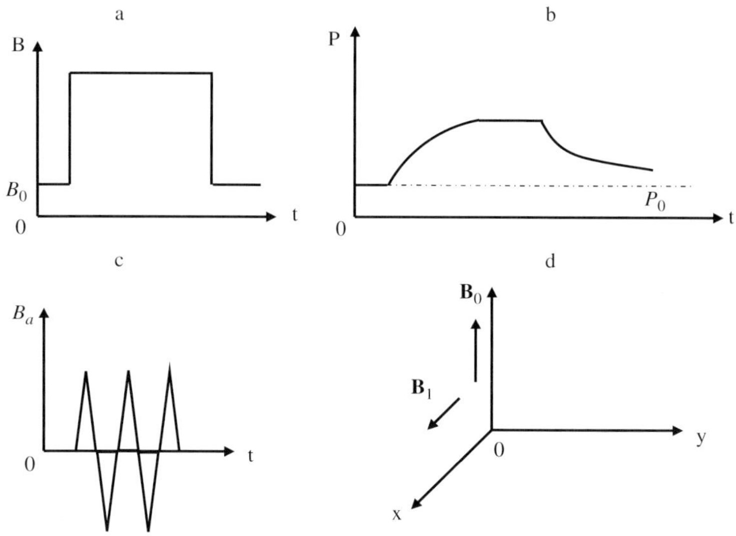

Fig. 7.3. (a, b) Behavior of the magnetic field and magnitude of magnetization; (c) sinusoidal impulse; (d) Orientation of the ambient and additional fields.

additional magnetic field behaves as an impulse with the constant magnitude. The vector of magnetization is obtained by measuring an electromotive force in the receiver coil perpendicular to the horizontal plane where the vector $\boldsymbol{P}$ is located at the initial moment. From the point of energy this process represents a quasi-static reorientation of the magnetic dipole system to the new magnetic axis, followed by a change in the Boltzmann population of energy levels, if $\boldsymbol{B}_1 \neq \boldsymbol{B}_0$. Adjustment of level populations occurs due to relaxation processes shown in Fig. 7.3(b).

Suppose that the additional field is many times greater than the ambient field. Then, at the beginning of the motion (after the field B_1 is turned off) the vector $\boldsymbol{P}$ rotates almost in the horizontal plane, Fig. 7.2(b and a) coil in the plane parallel to the ambient field will measure an electromotive force. This signal behaves as a sinusoidal function with the frequency of precession ω_0, and its amplitude is directly proportional to the horizontal projection of the vector of magnetization; with time the amplitude of this electromotive force decreases.

Next consider one more approach when an additional field is described by the sinusoidal impulse of finite duration and begin to study its influence on the vector of magnetization. Also, it is assumed that the fields $\boldsymbol{B}_0$ and $\boldsymbol{B}_1$ are perpendicular to each other. Inasmuch as in this case it is not simple to predict an effect produced in such field, it is important to derive an equation of motion for the vector $\boldsymbol{P}$.

7.3. EQUATIONS OF THE VECTOR OF MAGNETIZATION

In deriving in this section a relation between the vector of magnetization and the magnetic field we consider only the time interval when both the ambient and additional fields are present. This means that is an influence of the magnetic and electrical fields caused by other particles of a medium is neglected. It seems that such approximation is hardly possible since in real conditions every atom is surrounded by a complex system of other atomic particles and their magnetic and electric fields may have a strong contribution. For instance, if magnetic moments of electrons do not completely compensate each other, then their magnetic field in the vicinity of a nucleus can be very large, specially, in solids and sometimes it is in many times larger than the earth's field. At the same time, due to a relatively strong influence of the random motion of micro-particles in fluids and gases, an influence of the magnetic fields caused by electrons is usually greatly reduced. As usual, there are exceptions; for instance, in paramagnetic molecules like O_2, NO_2, ClO_2 the magnetic field caused by electrons can be significant. Inasmuch as most of molecules of fluids are diamagnetic, it is possible to neglect a magnetic interaction between nuclei and electrons, provided that there is a rather intensive thermal motion. It is also true for the electrical interaction between micro-particles in fluids and gases which is also small. Thus, as the first approximation, we can think that a magnetic dipole of a nucleus of a substance placed in a uniform field is

$$\boldsymbol{B} = \boldsymbol{B}_0 + \boldsymbol{B}_1 \tag{7.11}$$

and it behaves as a single particle with the given magnetic moment $\boldsymbol{p}$ and angular momentum $\boldsymbol{L}$. We will proceed from the known equation for the single particle derived in Chapter 6

$$\frac{d\boldsymbol{p}}{dt} = \gamma(\boldsymbol{p} \times \boldsymbol{B}) \tag{7.12}$$

Let us mentally imagine an elementary volume where the field $\boldsymbol{B}$ is the same. This equation can be written for each nuclear dipole of this volume. Then, performing a summation of these equations we obtain the equation for the macroscopic quantity of magnetization:

$$\frac{d\boldsymbol{P}}{dt} - \gamma(\boldsymbol{P} \times \boldsymbol{B}) = 0 \tag{7.13}$$

which allows us to study an establishment of magnetization caused by the resultant field $\boldsymbol{B}$ (Equation (7.11)). It is essential that the vector of magnetization $\boldsymbol{P}$ is a function of the magnetic field but not vice versa since an influence of the secondary field, caused by the nuclear magnetic dipoles, on the field $\boldsymbol{B}$ is neglected. As follows from Equation (7.13)

$$\frac{d\boldsymbol{P}}{dt} = \gamma(\boldsymbol{P} \times \boldsymbol{B}_0) + \gamma(\boldsymbol{P} \times \boldsymbol{B}_1) \tag{7.14}$$

and this means that the vector $\boldsymbol{P}$ can be represented as a result of action of the ambient and additional fields. The relation between the vectors $\boldsymbol{P}$ and $\boldsymbol{B}$ is the linear and homogeneous differential equation of the first order in the vector form with the zero right-hand side. In the Cartesian system of coordinates this equation is

$$\frac{d\boldsymbol{P}}{dt} = \gamma \begin{vmatrix} \boldsymbol{i} & \boldsymbol{j} & \boldsymbol{k} \\ P_x & P_y & P_z \\ B_x & B_y & B_z \end{vmatrix} \tag{7.15}$$

We restrict ourselves to the cases when the field B_0 is directed along the z-axis and the additional field B_1 is located in the horizontal plane ($B_{1z} = 0$). Then Equation (7.15) can be written as the system of three differential equations of the first order with three unknowns:

$$\frac{dP_x}{dt} = \gamma(B_{0z}P_y - B_{1y}P_z), \quad \frac{dP_y}{dt} = \gamma(B_{1x}P_z - B_{0z}P_x)$$

and

$$\frac{dP_z}{dt} = \gamma(B_{1y}P_x - B_{1x}P_y) \tag{7.16}$$

To illustrate a solution of this system consider two examples and start from the simplest one.

7.3.1. Case 1: Additional field is absent

In this case the vector of magnetization $\boldsymbol{P}_0$ is at the state of equilibrium and it is directed along the ambient field. Correspondingly, the left- and right-hand sides of Equation (7.14) are equal to zero

$$\frac{d\boldsymbol{P}_0}{dt} = \gamma(\boldsymbol{P}_0 \times \boldsymbol{B}_0) = 0 \tag{7.17}$$

that is, one of solutions of Equation (7.13) corresponds to the static case. Of course, in this case vectors $\boldsymbol{P}_0$ and $\boldsymbol{B}_0$ are related as $\boldsymbol{P}_0 = \chi_n \boldsymbol{B}_0/\mu_0$.

7.3.2. Case 2: The additional field is horizontal

Earlier we described qualitatively an influence of this type of the additional field. Now consider this case in some detail proceeding from Equation (7.15). Assume that until the moment $t = 0$ the vector of magnetization $\boldsymbol{P}_0$ is at the state of equilibrium and has only the vertical component. Then the additional field $\boldsymbol{B}_1(t)$ arises and it is directed along the x-axis. Inasmuch as the resultant magnetic field is located in the plane $x0z$, we have

$$\frac{dP_x}{dt} = \gamma B_0 P_y, \quad \frac{dP_y}{dt} = \gamma(B_{1x}P_z - P_x B_0), \quad \frac{dP_z}{dt} = -\gamma P_y B_{1x} \tag{7.18}$$

Here, $P_x(t)$, $P_y(t)$, and $P_z(t)$ are functions of time. As follows from the set (7.18) in order to find the vector of magnetization we have to solve simultaneously three differential equations of the first order.

7.4. ROTATING SYSTEM OF COORDINATES

It turns out that it is more convenient to study a motion of the vector of magnetization in the system of coordinates rotating around the static field $\boldsymbol{B}_0$, when it moves in the same direction as the precession of nuclear moments. This advantage becomes specially obvious if the additional field is horizontal and rotates with the same velocity and has the same direction as the rotation of magnetic moments. Importance of this case is related to the fact that the sinusoidal magnetic field acting in some direction can be expressed in terms of two constant and rotating fields. This is the reason why we are going to represent Equation (7.13) in such system of coordinates. Transition from the static (laboratory) to rotating system of coordinates is very conventional approach and used in many situations. Perhaps, the most common and well-known example is the case of the classical mechanics when we study a motion of a body under an action of some forces. In principle, it is possible to solve this problem either in the inertial system of coordinates or in the system of coordinates of the rotating earth where any particle of the earth's surface at the equator moves with the linear velocity around 1700 km/h. In spite of this earth's motion it is more convenient to use this rotating system. This means that in such system a motion is much simpler as well as equations describing this behavior. However, the use of the system rotating together with the earth requires a change in the Newton's second law and new terms appear which have the dimensions of the force. These terms are interpreted as fictitious forces such as centrifugal and Coriolis forces. Thus, it is not surprising that in the rotating system Equation (7.13) may have different form, and new terms can be considered as some fictitious magnetic forces acting on the nuclear magnetic dipoles. The static and rotating systems are shown in Fig. 7.4. Thus, our goal is to represent Equation (7.13) in the system x', y', z', where $z = z'$, and we start from the derivative $d\boldsymbol{P}/dt$. By definition in both systems we have

$$\boldsymbol{P}(t) = P_x\boldsymbol{i} + P_y\boldsymbol{j} + P_z\boldsymbol{k} = P_{x'}\boldsymbol{i}_1 + P_{y'}\boldsymbol{j}_1 + P_{z'}\boldsymbol{k}_1 \tag{7.19}$$

Here $\boldsymbol{i}$, $\boldsymbol{j}$, $\boldsymbol{k}$ and $\boldsymbol{i}_1$, $\boldsymbol{j}_1$, $\boldsymbol{k}_1$ are unit vectors in the static and rotating systems of coordinates, respectively. Differentiation of Equation (7.19) in the static frame gives

$$\frac{d\boldsymbol{P}}{dt} = \frac{dP_x}{dt}\boldsymbol{i} + \frac{dP_y}{dt}\boldsymbol{j} + \frac{dP_z}{dt}\boldsymbol{k} = \frac{dP_{x'}}{dt}\boldsymbol{i}_1 + \frac{dP_{y'}}{dt}\boldsymbol{j}_1 + \frac{dP_{z'}}{dt}\boldsymbol{k}_1 + P_{x'}\frac{d\boldsymbol{i}_1}{dt} + P_{y'}\frac{d\boldsymbol{j}_1}{dt} + P_{z'}\frac{d\boldsymbol{k}_1}{dt} \tag{7.20}$$

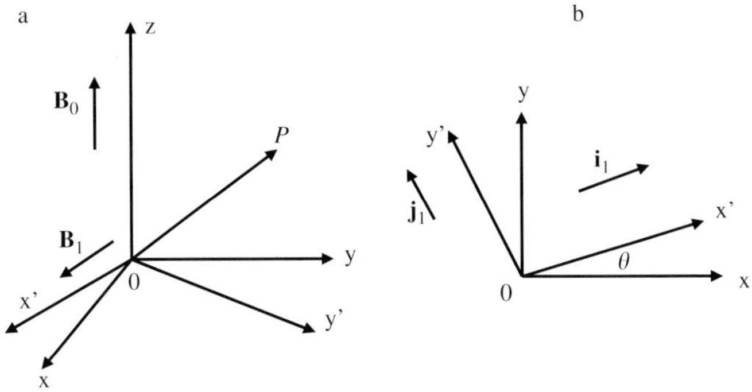

Fig. 7.4. Static and rotating system of coordinates.

The sum

$$\left(\frac{d\boldsymbol{P}}{dt}\right)_r = \frac{dP_{x'}}{dt}\boldsymbol{i}_1 + \frac{dP_{y'}}{dt}\boldsymbol{j}_1 + \frac{dP_{z'}}{dt}\boldsymbol{k}_1 \tag{7.21}$$

is the derivative of the vector $\boldsymbol{P}$ in the rotating system of coordinates, since for the observer rotating together with this system the unit vectors $\boldsymbol{i}_1$, $\boldsymbol{j}_1$ and $\boldsymbol{k}_1$ do not change their directions. At the same time, for observer in the static system these vectors change their orientation and, correspondingly, the last three terms at the right-hand side of Equation (7.20) differ from zero. As is seen from Fig. 7.4(b)

$$\boldsymbol{i}_1 = \boldsymbol{i}\cos\theta + \boldsymbol{j}\sin\theta, \quad \boldsymbol{j}_1 = -\boldsymbol{i}\sin\theta + \boldsymbol{j}\cos\theta$$

Thus,

$$\frac{d\boldsymbol{i}_1}{dt} = (-\boldsymbol{i}\sin\theta + \boldsymbol{j}\cos\theta)\frac{d\theta}{dt} = \boldsymbol{j}_1\omega = \omega(\boldsymbol{k}_1 \times \boldsymbol{i}_1)$$

or

$$\frac{d\boldsymbol{i}_1}{dt} = \boldsymbol{\omega} \times \boldsymbol{i}_1 \tag{7.22}$$

since $\boldsymbol{\omega} = \omega\boldsymbol{k} = \omega\boldsymbol{k}_1$. In the same manner, we obtain

$$\frac{d\boldsymbol{j}_1}{dt} = \boldsymbol{\omega} \times \boldsymbol{j}_1, \quad \frac{d\boldsymbol{k}_1}{dt} = \boldsymbol{\omega} \times \boldsymbol{k}_1 \tag{7.23}$$

Substitution of Equations (7.22) and (7.23) into Equation (7.20) gives a relation between the derivatives of the vector of magnetization in the static and rotational systems of coordinates:

$$\left(\frac{d\boldsymbol{P}}{dt}\right)_s = \left(\frac{d\boldsymbol{P}}{dt}\right)_r + \boldsymbol{\omega} \times \boldsymbol{P} \tag{7.24}$$

Thus, replacing the derivative in Equation (7.13) we obtain the equation of motion of the vector of magnetization in the rotating system of coordinates

$$\left(\frac{d\boldsymbol{P}}{dt}\right)_{\mathrm{r}} = \gamma(\boldsymbol{P} \times \boldsymbol{B}) - (\boldsymbol{\omega} \times \boldsymbol{P})$$

and a change of an order of vectors in the last term yields

$$\left(\frac{d\boldsymbol{P}}{dt}\right)_{\mathrm{r}} = \gamma\left[\boldsymbol{P} \times \left(\boldsymbol{B} + \frac{\boldsymbol{\omega}}{\gamma}\right)\right] \tag{7.25}$$

Here all terms are considered in the rotating system of coordinates, that is, each term is calculated by an observer who rotates together with this system. The term ω/γ has dimensions of the magnetic field and it can be treated as a fictitious magnetic field which arises due to a rotation. Analogy with forces like Coriolis or centrifugal ones is obvious. Introducing a notation

$$\boldsymbol{B}_{\mathrm{eff}} = \boldsymbol{B} + \frac{\boldsymbol{\omega}}{\gamma} \tag{7.26}$$

Equation (7.25) becomes

$$\left(\frac{d\boldsymbol{P}}{dt}\right)_{\mathrm{r}} = \gamma(\boldsymbol{P} \times \boldsymbol{B}_{\mathrm{eff}}) \tag{7.27}$$

Comparison with Equation (7.13) allows us to conclude that for an observer in the rotating system of coordinates the vector of magnetization performs only a precession around the effective field, and this fact greatly simplifies a study of a motion of the vector of magnetization.

7.5. BEHAVIOR OF THE VECTOR *P* IN THE ROTATING SYSTEM OF COORDINATES

Now we consider several examples of the magnetic field behavior which are important for application of NMR and some of them demonstrate advantages of the use of the rotating system of coordinates.

7.5.1. Example 1

Suppose that there is the constant ambient field only

$$\boldsymbol{B} = \boldsymbol{B}_0 \tag{7.28}$$

directed along the z-axis, and the system of coordinates rotates around the z-axis with the frequency of precession $\boldsymbol{\omega}$:

$$\boldsymbol{\omega} = \boldsymbol{\omega}_0 = -\gamma \boldsymbol{B}_0 \tag{7.29}$$

It is essential that a direction of this rotation coincides with that of precession. Then, substituting Equation (7.29) into Equation (7.25) we see that the right-hand side is equal to zero and

$$\frac{d\boldsymbol{P}}{dt} = 0 \tag{7.30}$$

As we already know, this means that a superposition of nuclear magnetic dipoles gives the vector of magnetization oriented along the constant ambient field, and it is not involved in precession.

7.5.2. Example 2: The additional field rotates in the horizontal plane

Now we assume that in the static system of coordinates x, y, z, there are two forces: the constant ambient field $\boldsymbol{B}_0$ directed along the z-axis, and the additional field $\boldsymbol{B}_1$ which is located in the horizontal plane and it rotates around the z-axis with the frequency ω. The magnitude of this field is constant. Again, as in the first example, we introduce the system of coordinates x', y', z' which rotates around the z'-axis with the same frequency as the field $\boldsymbol{B}_1$. In accordance with Equation (7.26) the effective field is

$$\boldsymbol{B}_{\text{eff}} = \boldsymbol{B}_0 + \frac{\boldsymbol{\omega}_0}{\gamma} + \boldsymbol{B}_1 \tag{7.31}$$

At the beginning it is convenient to consider the special case when the frequency ω is equal to that of precession.

7.5.3. The case of resonance ($\omega = \omega_0 = \gamma B_0$)

Inasmuch as the magnitude and direction of the vectors $\boldsymbol{\omega}$ and $\boldsymbol{\omega}_0$ are equal to each other, substitution of Equation (7.29) into Equation (7.31) gives

$$\boldsymbol{B}_{\text{eff}} = \boldsymbol{B}_1 \tag{7.32}$$

Thus, Equation (7.13) becomes

$$\frac{d\boldsymbol{P}}{dt} = \gamma(\boldsymbol{P} \times \boldsymbol{B}_1) \tag{7.33}$$

We see that in the case of the resonance ($\omega = \omega_0$) the vertical field $\boldsymbol{B}_0$ is completely compensated by the fictitious field, while the horizontal additional field

$\boldsymbol{B}_1$ is constant. Since its direction is not important suppose that it is directed along the x'-axis. Then, from Equation (7.33) it follows that in the rotating system of coordinates the vector of magnetization is involved in precession in the vertical plane around the horizontal axis x'; that is, the lateral surface of the cone is transformed into the y', z'-plane. Certainly, this simple behavior of the vector $\boldsymbol{P}$ was difficult to predict, and this result shows again the advantage of the use of the rotating frame. Observations in this system suggests that the motion of $\boldsymbol{P}$ is caused by the additional field and the frequency of precession is equal to

$$\omega_1 = -\gamma \boldsymbol{B}_1 \tag{7.34}$$

We demonstrated that the constant additional field, which rotates in the horizontal plane of the static system, produces a motion of the magnetization vector in the vertical plane of the rotating system around the field $\boldsymbol{B}_1$ (Fig. 7.5(a)). This means that the field $\boldsymbol{B}_1$ allows us to change an orientation of the vector $\boldsymbol{P}$ in the vertical plane which at the beginning was directed along the z'-axis. It is obvious that its motion has a periodical character, and examples of different positions of this vector caused by this additional field are shown in Fig. 7.5. Changing the time of action of the additional field it is possible to change arbitrarily an orientation of the vector of magnetization. It is instructive to evaluate the precession frequency around the additional field. Assuming that the gyroscopic ratio is approximately equal to

$$\gamma \approx 10^8 \text{ s}^{-1}\text{T}^{-1}$$

and the additional field varies within a range:

$$10^{-4}\text{ T} < B_1 < 10^{-2}\text{ T}$$

we see that in the case of the pulsed NMR, a precession frequency range is

$$10^4\text{s}^{-1} < \omega_1 < 10^6\text{s}^{-1}$$

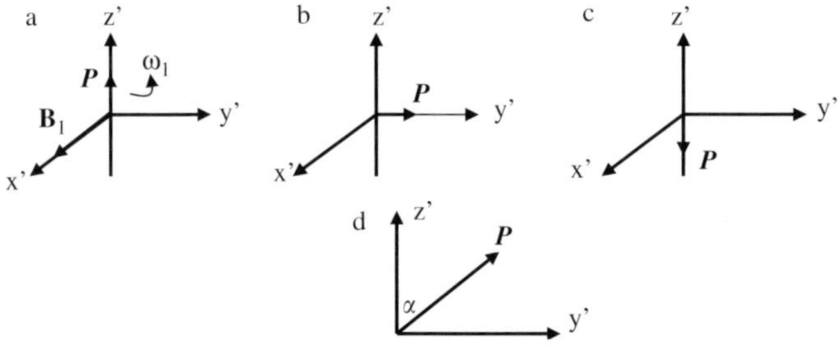

Fig. 7.5. A change of an orientation due to the additional field: (a) initial position; (b) rotation by 90°; (c) rotation by 180°; (d) orientation of the vector ***P***.

Summarizing, we may say that if the additional field is a rotating vector in the static frame with a constant magnitude and the frequency ω_0, then the vector of magnetization rotates clockwise in the $y'0z'$-plane toward the y'-axis, and this motion has a periodical character. It may be proper to notice again that this plane is the lateral surface of the cone of precession. At each instant t the angle α between the z'-axis and the vector $\boldsymbol{P}$ is equal to

$$\alpha = \omega_1 t = \gamma B_1 t \tag{7.35}$$

and this relation is very useful in the impulse methods of NMR.

We found out that in our case a motion of the vector of magnetization in the rotating system is extremely simple: it is only precession around the additional field $\boldsymbol{B}_1$. At the same time, as was demonstrated earlier, in the static system this motion is much more complicated because it is a combination of a precession around the field $\boldsymbol{B}_1$ and a rotation of the frame $x'y'z'$ with respect to the field $\boldsymbol{B}_0$ with the frequency ω_0. Thus, motions in these systems greatly differ from each other. In this light let us note that practical measurements are made in the rotating frame using special devices, like "phase-sensitive" detectors.

7.5.4. General case ($\omega \neq \omega_0$)

Until now we assumed that the frequency of rotation of the additional field is equal to that of precession one ω_0. This condition allowed us to eliminate an influence of precession around the ambient field, provided that observations are performed in the rotating frame. Next, consider a more general case when these frequencies are not equal to each other. In particular, they may have the same magnitude but opposite directions. In such general case the effective field is defined by all three terms at the right-hand side of Equation (7.31). It is essential that vectors: $\boldsymbol{B}_0$ and $\boldsymbol{\omega}/\gamma$ are located at the z'-axis but have opposite directions, while the vector $\boldsymbol{B}_1$ is perpendicular to both of them (Fig. 7.6). As is seen from Fig. 7.6 the magnitude of the effective field is equal to

$$B_{\text{eff}} = \left[\left(B_0 - \frac{\omega}{\gamma}\right)^2 + B_1^2\right]^{1/2} \quad \text{or} \quad B_{\text{eff}} = \frac{1}{\gamma}[(\omega_0 - \omega)^2 + \omega_1^2]^{1/2} \tag{7.36}$$

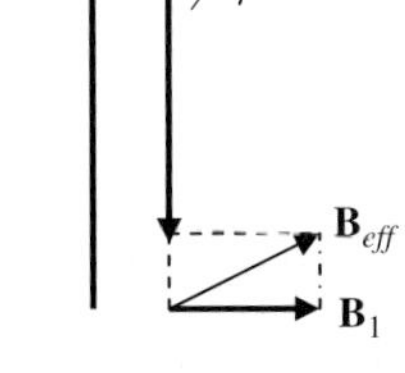

Fig. 7.6. Mutual positions of magnetic fields.

Here $\omega_0 = \gamma B_0$, $\omega_1 = \gamma B_1$ and by analogy with Equation (7.29) the last equation can be written as

$$\boldsymbol{\omega}_e = -\gamma \boldsymbol{B}_{eff} \quad \text{and} \quad \omega_e = \gamma B_{eff} \tag{7.37}$$

Thus, in the rotating system of coordinates the vector of magnetization precesses around the effective field with the frequency

$$\omega_e = [(\omega_0 - \omega)^2 + \omega_1^2]^{1/2} \tag{7.38}$$

From Equation (7.19):

$$\frac{d\boldsymbol{P}}{dt} = \gamma(\boldsymbol{P} \times \boldsymbol{B}_{eff})$$

it follows that the vector $\boldsymbol{P}$ is involved in precession around the effective field which is located in the plane $x'0z'$. In other words, the vector of magnetization is located on the conical surface and its tip rotates with the frequency ω_e along the circle in the plane perpendicular to the field $\boldsymbol{B}_{eff}$ (Fig. 7.7). To characterize a motion of the vector $\boldsymbol{P}$ we introduce the angle α between the ambient field $\boldsymbol{B}_0$ and the vector $\boldsymbol{P}$ which varies with time. Our goal is to find an expression of this angle in terms of

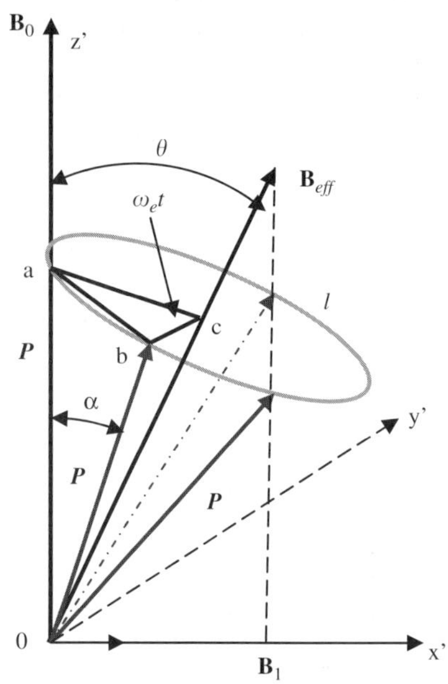

Fig. 7.7. Motion of the vector of magnetization.

time, the frequency of precession ω_e around the effective field, and the angle θ between the ambient and effective fields. It is obvious that the angle α characterizes a deviation of the vector of magnetization from the z'-axis with time, since at the moment $t = 0$ this angle is equal to zero. This is a reason why there is a point where this axis touches the circle of rotation on the conical surface. Taking into account Equation (7.36) we have for the angle θ:

$$\tan\theta = \frac{B_1}{B_0 - (\omega/\gamma)} = \frac{\omega_1}{\omega_0 - \omega} \tag{7.39}$$

or

$$\sin\theta = \frac{B_1}{B_{\text{eff}}} = \frac{\omega_1}{\omega_e}, \quad \cos\theta = \frac{B_0 - (\omega/\gamma)}{B_{\text{eff}}} = \frac{\omega_0 - \omega}{\omega_e} \tag{7.40}$$

Now we are ready to determine the angle α. First of all, let us express the radius r of the circle of rotation l in terms of the magnitude of $\boldsymbol{P}$. As is seen from the triangle $0ac$ (Fig. 7.7):

$$ac = bc = r = P\sin\theta$$

From the triangle abc located in the horizontal plane we can calculate the length of the chord ab:

$$ab = 2r\sin\frac{\omega_e t}{2} = 2P\sin\theta\sin\frac{\omega_e t}{2}$$

The same length can be determined from the triangle $a0b$. This gives

$$ab = 2P\sin\frac{\alpha}{2}$$

Hence

$$\sin\frac{\alpha}{2} = \sin\theta\sin\frac{\omega_e t}{2} \tag{7.41}$$

Thus, we expressed the angle $\alpha(t)$ in terms of the angle θ, the precession frequency ω_e, and time t and this allows us to find a position of the vector of magnetization at any instant. It is proper to emphasize that the frequency ω_e depends on the frequency of rotation of the field B_1, as well as two frequencies of precession: ω_0 and ω_1. Since

$$2\sin^2\frac{\alpha}{2} = 1 - \cos\alpha$$

and taking a square from both sides of Equation (7.41), we obtain a different form of this relation

$$\cos\alpha = 1 - 2\sin^2\theta\sin^2\frac{\omega_e t}{2} \tag{7.42}$$

As seen from Fig. 7.7 the tip of the vector $\boldsymbol{P}$ rotates along the circle which is located at one side with respect to the z'-axis. For illustration, consider again the case of resonance when $\omega = \omega_0$. As follow from Equations (7.39) and (7.40) the angle θ is equal to $\pi/2$ and, therefore, the circle of rotation is located in the plane $y'0z'$. At the same time, as it should be, Equation (7.38) gives $\omega_e = \omega_1$. Until now we did not make any assumptions about relative magnitudes of the ambient and additional fields. Taking into account the fact that in most cases of NMR

$$B_1 \ll B_0 \tag{7.43}$$

let us consider the effect caused by such small additional field in some details. As before the vectors $\boldsymbol{B}_0$ and $\boldsymbol{B}_1$ are perpendicular to each other. First, assume that an inequality

$$|\omega - \omega_0| \gg \omega_1 \tag{7.44}$$

takes place. This may happen if the frequency of rotation of the additional field $\boldsymbol{B}_1$ strongly differs from that of precession around the ambient field. Then, as follows from Equation (7.39), the angle θ is very small; that is, the effective field is almost directed along the z'-axis. Correspondingly, Equation (7.42) gives $\cos\alpha \approx 1$. This means that a circle of rotation of the vector $\boldsymbol{P}$ has very small radius and its deviation from the z'-axis is practically negligible. In other words, in such case the small additional field is not able to change an orientation of the magnetization vector. Next, consider a range of frequencies which are close to that of precession ω_0 and they obey the condition

$$|\omega - \omega_0| \leq \omega_1 \tag{7.45}$$

For instance, at the boundary of this range we have $|\omega - \omega_0| = \omega_1$ and this gives $\theta = \pi/4$.

Then Equation (7.42) becomes

$$\cos\alpha = \cos^2\frac{\omega_e t}{2} \tag{7.46}$$

This shows that the resonance occurs not only at the frequency ω_0 but also at its vicinity and the width of resonance range: $|\omega - \omega_0|$ is of the order of ω_1. Note that at frequencies close to ω_0, the vector of magnetization rotates on the conical surface which is almost parallel to the plane $y'0z'$.

7.5.5. *Additional field B_1 is a sinusoidal function*

Now we will demonstrate that the sinusoidal additional field acting along the x-axis of the static frame produces practically the same effect as the constant field rotating in the horizontal plane with the same frequency as that of precession. In fact, suppose that the additional field is directed along the x-axis of the static system and its sinusoidal function of time:

$$B_1(t) = B_x \cos \omega_0 t \tag{7.47}$$

Here, ω_0 is the frequency of precession around the ambient field directed along the z-axis.

Let us introduce two vectors located in the horizontal plane:

$$\boldsymbol{B}_1^{(1)}(t) = \boldsymbol{i}\frac{B_x}{2}\cos \omega_0 t + \boldsymbol{j}\frac{B_x}{2}\sin \omega_0 t, \ \boldsymbol{B}_1^{(2)}(t) = \boldsymbol{i}\frac{B_x}{2}\cos \omega_0 t - \boldsymbol{j}\frac{B_x}{2}\sin \omega_0 t \tag{7.48}$$

Both vectors have the same magnitude and they rotate with the frequency of precession ω_0 in the opposite directions around the ambient field. Comparison of Equation (7.47) and (7.48) gives

$$\boldsymbol{B}_1(t) = \boldsymbol{B}_1^{(1)}(t) + \boldsymbol{B}_1^{(2)}(t) \tag{7.49}$$

Thus, we represented the sinusoidal field acting only along the x-axis as a sum of two vector fields with equal and constant magnitudes rotating in the horizontal plane with the frequency of precession ω_0 but in the opposite directions. The field $\boldsymbol{B}_1^{(2)}(t)$ rotates with the frequency of precession $\boldsymbol{\omega}_0$ and therefore it causes a reorientation of the vector of magnetization. As concerns the field $\boldsymbol{B}_1^{(1)}(t)$ its frequency of rotation is $\boldsymbol{\omega} = -\boldsymbol{\omega}_0$. Therefore in place of Equation (7.39) we have

$$\tan \theta = \frac{B_1^{(1)}}{B_0 + (\omega_0/\gamma)} = \frac{\omega_1}{2\omega_0} \tag{7.50}$$

Taking into account Equation (7.43) we conclude that the angle θ, as well as α are small, that is, the rotating field $B_1^{(1)}$ does not produce noticeable change in orientation of the vector $\boldsymbol{P}$, if $B_1 \ll B_0$. This means that the sinusoidal field acting along some line in the horizontal plane may produce the same effect as the additional field $B_1^{(2)}$ rotating around the ambient field with the frequency $\boldsymbol{\omega}_0$. Thus, applying sinusoidal impulses of the field $\boldsymbol{B}_1$ with a different duration along, for example, the x-axis, it is possible to change the orientation of the vector of magnetization in plane perpendicular to this field. In other words, the angle between the ambient field $\boldsymbol{B}_0$ and the vector of magnetization $\boldsymbol{P}$ may vary from 0 to π. We found that a relatively small horizontal and sinusoidal field $\boldsymbol{B}_1$ is able to rotate the vector of magnetization in the plane perpendicular to this field.

7.6. MAGNETIZATION CAUSED BY THE ADDITIONAL FIELD

First, suppose that at some instant $t = 0$ we generated a constant magnetic field $\boldsymbol{B}_1$, which, for instance, is much stronger than the vertical field $\boldsymbol{B}_0$, and it is located in the horizontal plane. It is clear, the total field is almost horizontal and it begins to produce the magnetization, since at the initial instant $t = 0$ the sum of components of magnetic moments along the total field is nearly equal to zero; that is, there are almost an equal number of these components with the same magnitude but opposite signs. With an increase of time the number with positive projection along the field becomes more than with negative components. This process of magnetization takes some time and it occurs when the nuclear dipole system loses energy to a heat reservoir – the "lattice." We finally arrive at equilibrium when the magnetization vector $\boldsymbol{P}$ is defined by Equation (7.10). Next we assume that the horizontal field $\boldsymbol{B}_1$ is turned off. Then the magnetic dipoles of protons remain under an action of only the field $\boldsymbol{B}_0$. At this instant they begin to precess around this field with the Larmor frequency. At the beginning this motion occurs almost in the horizontal plane but then the angle θ between the vector of magnetization $\boldsymbol{P}$ and the field $\boldsymbol{B}_0$ becomes smaller and it approaches zero. The time interval during which this transition takes place depends on a surrounding medium. Since both the angle θ and the magnitude P decrease $(B_1 > B_0)$, the horizontal component $\boldsymbol{P}_h$ also decreases with time. In order to describe the behavior of magnetization caused by an additional field it is convenient to distinguish two processes. One of them is a growth of magnetization along the ambient field, and its duration is characterized by the time constant T_1. The second process describes a decay of magnetization in the direction perpendicular to the ambient field and its rate of change is described by time constant T_2. It is usual that T_1 is called the longitudinal time constant or longitudinal relaxation time, and T_2 is transversal time constant or transversal relaxation time. Also, different notations are used for them: $T_{\parallel}$ and $T_{\perp}$. In other words, they are time constants, characterizing decay of components of the vector of magnetization which are parallel and perpendicular to the ambient field B_0, respectively.

As was pointed out in real conditions, specially, in solids, each magnetic dipole of a particle is subjected to an influence of the surrounding medium, and among different types of such interactions we distinguished the dipole–dipole and dipole–lattice ones. In this light let us notice that in calculating the vector of magnetization for the diamagnetic and paramagnetic materials, caused by the constant magnetic field, we neglected by interaction between elementary magnetic dipoles. However, now we study such parameters as the frequency of precession, the cone of precession, the transition between energy levels, time during which the magnetic dipole of nucleus has a certain level of magnetic energy, and in such cases it is necessary to take into account the effect of the surrounding medium.

A change of the nuclear magnetization $\boldsymbol{P}(t)$ may happen due to either the constant or alternating radio-frequency field $\boldsymbol{B}_1$. In both cases a change of magnetization is related with absorption of the external energy and transition between energy levels. Correspondingly, it is useful to have an equation which characterizes a rate of the change of the vector $\boldsymbol{P}$. Earlier we wrote such equation

$$\frac{d\boldsymbol{P}}{dt} = \gamma(\boldsymbol{P} \times \boldsymbol{B}) \tag{7.51}$$

In particular, it is valid for the system of isolated magnetic dipoles when we can neglect by an interaction with each other. In such case it is possible to neglect by local magnetic fields and assume that all elementary dipoles are located in the same field $\boldsymbol{B}$. In real situations there is an interaction with the surrounding medium and for each elementary volume it is proper to write equations where $\boldsymbol{B}$ is the sum:

$$\boldsymbol{B}_0 + \boldsymbol{B}_1(t) + \boldsymbol{B}_{\text{loc}}$$

Here $\boldsymbol{B}_{\text{loc}}$ is the local field created by dipoles of the surrounding medium and it is unknown. Correspondingly, substitution of the last sum into Equation (7.51) gives an equation with two unknown vectors $\boldsymbol{P}$ and $\boldsymbol{B}$ since a relation between them is also unknown. For this reason creating a macroscopic theory of NMR, F. Bloch proceeded from the same Equation (7.51) but used a different approach. He assumed that the field $\boldsymbol{B}$ at the right-hand side of this equation is only sum of the constant ambient and additional fields:

$$\boldsymbol{B} = \boldsymbol{B}_0 + \boldsymbol{B}_1(t)$$

but the absence of the local fields at the cross-product of Equation (7.51) was replaced by introducing additional terms of relaxation which characterize an interaction of magnetic dipoles with the surrounding medium.

7.7. BLOCH EQUATIONS

F. Bloch suggested that the process of magnetization can be described by the following system of equations for the horizontal and vertical components of the vector of magnetization in Cartesian system of coordinates:

$$\frac{dP_x}{dt} = \gamma(P_y B_z - P_z B_y) - \frac{P_x}{T_2}, \quad \frac{dP_y}{dt} = \gamma(P_z B_x - P_x B_z) - \frac{P_y}{T_2}$$

and

$$\frac{dP_z}{dt} = \gamma(P_x B_y - P_y B_x) - \frac{P_z - P_0}{T_1} \tag{7.52}$$

or

$$\frac{d\boldsymbol{P}_{\text{h}}}{dt} = \gamma(\boldsymbol{P} \times \boldsymbol{B})_{\text{h}} - \frac{\boldsymbol{P}_{\text{h}}}{T_2}, \quad \frac{d\boldsymbol{P}_z}{dt} = \gamma(\boldsymbol{P} \times \boldsymbol{B})_z - \frac{\boldsymbol{P}_z - \boldsymbol{P}_0}{T_1} \tag{7.53}$$

where

$$(\boldsymbol{P} \times \boldsymbol{B})_{\rm h} = (\boldsymbol{P} \times \boldsymbol{B})_x \boldsymbol{i} + (\boldsymbol{P} \times \boldsymbol{B})_y \boldsymbol{j}$$

and $\boldsymbol{P}_0$ is the vector of magnetization at equilibrium, and $\boldsymbol{P}_{\rm h}$ is the component of the vector of magnetization perpendicular to the ambient field. Certainly, these equations can be considered as a generalization of Equation (7.51) and, in particular, they describe the behavior of magnetization after we turned off the horizontal field $\boldsymbol{B}_1$. As we already know, this includes precession of magnetic dipoles around the ambient magnetic field and a change of the magnitude of the vector of magnetization when it approaches to the value $\boldsymbol{P}_0$. As was already mentioned, the longitudinal $\boldsymbol{P}_{||}$ and transversal $\boldsymbol{P}_{\perp}$ components vary with time differently: the first one changes its magnitude from the initial value to P_0, while the second rotates in the horizontal plane with the precession frequency and its magnitude goes to zero. For illustration, consider one example.

7.7.1. Solution of Bloch's equations when the additional field is absent

Suppose that with the help of the additional field $\boldsymbol{B}_1$, the vector of magnetization is taken from the state of equilibrium when $\boldsymbol{P} = (0, 0, P_0)$ and at some instant $t = 0$ the field $\boldsymbol{B}_1$ is turned off. Our goal is to study the behavior of the vector $\boldsymbol{P}$ when it returns to the initial position; that is, we have to find a solution of Bloch's equations. Letting $\boldsymbol{B} = B_0 \kappa$ Equation (7.52) becomes

$$\frac{dP_x}{dt} = \gamma P_y B_0 - \frac{P_x}{T_2}, \quad \frac{dP_y}{dt} = -\gamma P_x B_0 - \frac{P_y}{T_2}, \quad \frac{dP_z}{dt} = -\frac{P_z - P_0}{T_1} \tag{7.54}$$

First, consider the behavior of the vertical component of magnetization which obeys the last equation of this set:

$$\frac{dP_z}{dt} + \frac{P_z}{T_1} = \frac{P_0}{T_1} \tag{7.55}$$

This is the linear inhomogeneous differential equation of the first order and its solution is a sum:

$$P_z = P_z^{(1)} + P_z^{(2)}$$

where $P_z^{(1)}$ is a solution of the homogeneous equation:

$$\frac{dP_z^{(1)}}{dz} + \frac{P_z^{(1)}}{T_1} = 0$$

and $P_z^{(2)}$ is a partial solution of Equation (7.55). It is obvious that

$$P_z^{(1)} = C\exp\left(-\frac{t}{T_1}\right) \quad \text{and} \quad P_z^{(2)} = P_0$$

Thus, for the vertical component of magnetization we have

$$P_z(t) = C\exp\left(-\frac{t}{T_1}\right) + P_0 \tag{7.56}$$

where C is unknown constant. In order to determine this constant we assume that at the instant when the horizontal field is turned off the initial value of $P_z(0)$ is known. Then Equation (7.56) gives

$$C = P_z(0) - P_0$$

Finally, we have

$$P_z(t) = P_0 - [P_0 - P_z(0)]\exp\left(-\frac{t}{T_1}\right) \tag{7.57}$$

From this equation it follows that the longitudinal component of magnetization gradually increases with time if $P_0 > P_z(0)$ or becomes smaller when $P_0 < P_z(0)$. In order to determine the horizontal components P_x and P_y we have to solve the first two equations of the set (7.54):

$$\frac{dP_x}{dt} = \omega_p P_y - \frac{P_x}{T_2} \quad \text{and} \quad \frac{dP_y}{dt} = -\omega_p P_x - \frac{P_y}{T_2} \tag{7.58}$$

As was mentioned earlier we expect that these components rotate around the z-axis and decay with time. This indicates that these functions may be represented as

$$P_x(t) = \mathrm{Re}A^* \exp\left(-\frac{t}{T_2}\right)\exp(-i\omega_p t)$$

and

$$P_y(t) = \mathrm{Re}B^* \exp\left(-\frac{t}{T_2}\right)\exp(-i\omega_p t) \tag{7.59}$$

Substituting the latter into Equation (7.59), we obtain the system of equations with respect to unknown complex amplitudes A^* and B^*:

$$iA^* + B^* = 0 \quad \text{and} \quad -iB^* + A^* = 0 \tag{7.60}$$

The determinant of this homogeneous system is equal to zero and the system has nonzero solution if

$$B^* = -iA^* = e^{-i\pi/2} A^* \tag{7.61}$$

Thus, our assumption was correct and the horizontal components of the vector $\boldsymbol{P}$ are described by Equation (7.59), provided that phase of $P_x(t)$ and $P_y(t)$ differ by $\pi/2$ and

$$P_x(t) = |A^*| \exp\left(-\frac{t}{T_2}\right) \cos(\omega_p t + \varphi)$$

and

$$P_y(t) = |A^*| \exp\left(-\frac{t}{T_2}\right) \sin(\omega_p t + \varphi) \tag{7.62}$$

As concerns unknown magnitude and phase, they are determined from the initial condition when the field $\boldsymbol{B}_h$ is turned off. For the magnitude of the transversal component of the vector $\boldsymbol{P}$ we have

$$P_{\perp}(t) = |A^*| \exp\left(-\frac{t}{T_2}\right) \tag{7.63}$$

Thus, Equations (7.57), (7.62), and (7.63) describe an influence of the surrounding medium on the vector of magnetization when the additional field is turned off.

7.8. MEASUREMENTS OF RELAXATION PROCESSES

7.8.1. Introduction

An observation of a decay of the induced signal, caused by magnetic field of nuclear magnetic dipoles, is the main approach which allows us to determine the vector of magnetization $\boldsymbol{P}$. For instance, in the case of the borehole geophysics the signal is caused by the magnetic field of hydrogen proton in fluid molecules. Because of their presence we observe relaxation processes of the longitudinal and transversal components of the vector of magnetization, which depend on petrophysical properties of formations such as a movable fluid porosity, pore size distribution, and permeability. In order to create the large and strong magnetic field, $\boldsymbol{B}_0$, the NMR device has a permanent magnet elongated in the direction of the tool motion; that is, along the borehole. In the vicinity of the borehole the magnet generates field B_0 of around 500 Gauss; it is almost in 1000 times larger than the magnetic field of the earth, and this process of magnetization lasts a few seconds. Further we assume that measurements are performed after an impulse is turned off and a duration time of sinusoidal impulses t_p is small with respect to time constants T_1 and T_2. This means that we can neglect by relaxation during an action of the sinusoidal impulse, $t_p \ll T_1$ and $t_p \ll T_2$. Also, as we know, the frequency of precession is defined as

$$f_0 = \frac{\gamma}{2\pi} B_0 \tag{7.64}$$

In the case of hydrogen nuclei $\gamma/2\pi = 4258$ Hz/Gauss and if $B_0 = 500$ Gauss the field B_1 must have a frequency just above 2.1 MHz. Otherwise, as was above demonstrated, the vector of magnetization would not rotate. This frequency makes NMR a "resonance" technique. In accordance with Equation (7.35), the angle through which the vector of magnetization is turned is equal to

$$\alpha = 360^\circ \frac{\gamma}{2\pi} B_1 t_p \tag{7.65}$$

Here α is the angle of rotation or tip angle in degrees, B_1, one-half of the oscillating field strength in the static (laboratory) frame, and t_p, the time during which the B_1 field acts. For instance, in order to turn the vector of magnetization by 90° we need 1.5 μs if $B_1 = 4$ Gauss.

7.8.2. Measurements of a decay of the longitudinal component of the vector P

First, suppose that the sinusoidal impulse B_1 turned the vector $\boldsymbol{P}$ by 90°, as is shown in Fig. 7.8(a). In accordance with Equation (7.57), we have

$$P_z(t) = P_0 \left[1 - \exp\left(-\frac{t}{T_1}\right)\right] \tag{7.66}$$

since $P_z(0) = 0$, and P_0 is the value of magnetization at the state of equilibrium. If we place the coil in the horizontal plane, then a change of the function $P_z(t)$ causes a transient magnetic field, and the electromotive force appears at the coil. However, the signal is usually insignificant to measure since this process of decaying is very slow. This indicates that, due to an exchange of the energy between nuclear

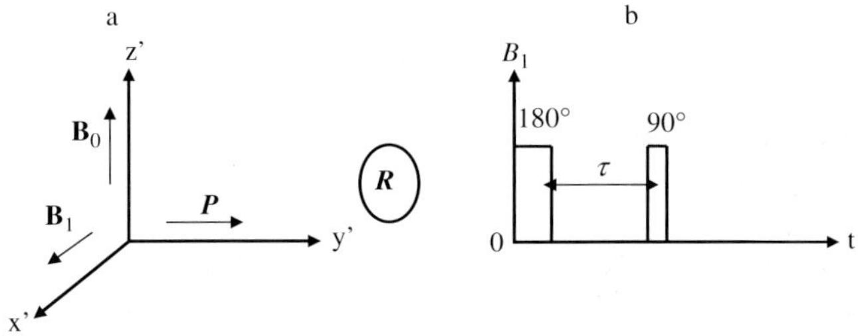

Fig. 7.8. (a) Orientation of vectors of magnetic fields and magnetization; (b) sequence of impulses.

moments and surroundings (the spin–lattice interaction), these moments gradually return to the original position along the z'-axis. This is the main reason why measurements of the longitudinal relaxation require a special approach, and it can be done in different ways, for instance, applying the sequences of two so-called 180° and 90° impulses with different time intervals τ between them (Fig. 7.8(b)). By definition, the first impulse rotates the vector $\boldsymbol{P}$ by 180°, while the second impulse turns $\boldsymbol{P}$ by 90°. Let us consider an action of these two impulses. The first impulse turns the vector of magnetization $\boldsymbol{P}_0$ by 180° and it is directed opposite the z'-axis (Fig. 7.9(a)). At the initial instant the vector of magnetization has the maximal negative component. This means that the magnetic energy of nuclei in an elementary volume is positive, and the difference between number of magnetic dipoles with the positive and negative energy is maximal. With time due to action of the ambient field and a surrounding medium this difference becomes smaller and the magnitude of the vector $\boldsymbol{P}$ changes. For instance, at some moment τ it is equal to $P_{z'}(\tau)$. This vector is not involved in a rotation since it is directed either along or opposite the field $\boldsymbol{B}_0$ and therefore it cannot be detected by using the coil receiver. To overcome this problem, the next even shorter impulse turns at the instant τ the vector of magnetization by 90° and it becomes directed along the y'-axis (Fig. 7.9(a)). As soon as this impulse stops to act, the vector $\boldsymbol{P}(\tau)$ begins to precess around the z'-axis, and it causes the alternating magnetic field which can be detected by the coil (R) located in the vertical plane (Fig. 7.8(a)). Then, measuring the initial electromotive force caused by a rotation of the vector $P_{z'}(\tau)$ in the horizontal plane, we can calculate the value of this component. One can say that 90°-impulse allows us to place the vector of magnetization in the horizontal plane where it performs precession around the ambient field. Next, we wait some time ($t \approx 5T_1$) until the vector of magnetization reaches almost its maximal value along the ambient field. Then the sequences of two impulses with different time intervals τ are repeated, and

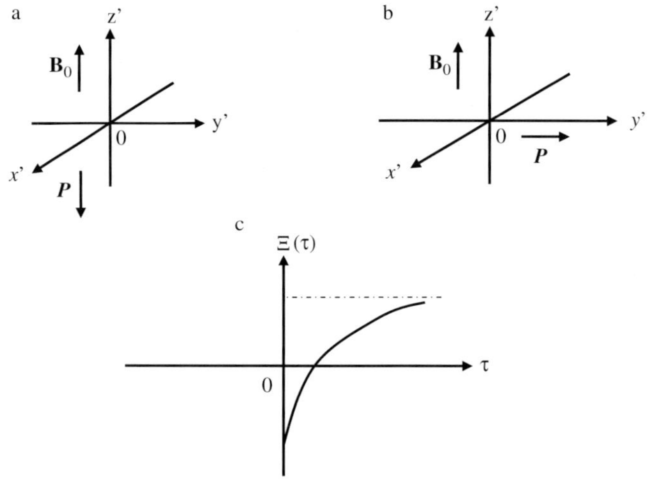

Fig. 7.9. (a) Rotation of the vector $\boldsymbol{P}$ by 180°; (b) rotation of the vector $P_{z'}(\tau)$ by 90°; (c) function $P_{z'}(\tau)$.

the function $P_{z'}(\tau)$ is determined (Fig. 7.9(c)). Let us emphasize that the signal $\Xi(\tau)$ is measured at the initial moment after the second impulse, when magnetic dipoles rotate synchronously in the horizontal plane and create a measurable signal in the receiver coil. As follows from Equation (7.57)

$$\Xi(\tau) = \Xi_0 \left[1 - 2\exp\left(-\frac{\tau}{T_1} \right) \right] \tag{7.67}$$

since $P_{z'}(0) = -P_0$. Thus, we have

$$\ln[\Xi_0 - \Xi(\tau)] = \ln 2\Xi_0 - \frac{\tau}{T_1} \tag{7.68}$$

and the slope of the graph describing this function allows us to find T_1. Perhaps, it is useful to notice that at the instant $\tau = \tau_0$ when $\Xi(\tau_0) = 0$ Equation (7.67) gives

$$\tau_0 = T_1 \ln 2 \approx 0.69 T_1 \tag{7.69}$$

Let us briefly discuss a behavior of the function $\Xi(\tau)$. As was pointed out at the first instant the difference between the number of nuclear dipoles oriented downward and number of dipoles with the opposite direction of the magnetic dipoles is maximal. Under an action of the ambient field this difference becomes smaller with an increase of time, and it is accompanied by interaction with the surrounding medium. There is an instant $t = \tau_0$ when number of magnetic dipoles directed along and opposite the field $\boldsymbol{B}_0$ is equal to each other and magnetization is absent. If $t > \tau_0$ then the number of dipoles with the positive vertical component of magnetic moment exceeds the number of dipoles with the negative component. At greater times this difference increases and finally the vector of magnetization approaches the stage of saturation. For instance, if $\tau = 5T_1$ then we have $P_{z'} \approx 0.993 P_0$. This process of magnetization depends on a substance, for instance, in the water and light oil it takes 3 s or less. For other types of oil the relaxation time T_1 is even smaller but in solids we have to wait minutes and even hours. It may be proper to notice that during this change of the vector $\boldsymbol{P}$ the magnetic moments of some dipoles, $\mathbf{p}_i$, move from the lower part of the cone of precession to its upper part, where they form the angle θ which is smaller than 90°.

7.8.3. Measurements of a decay of the transversal component of the vector P

Now our goal is to study a relaxation of the transversal component of the vector of magnetization using as before the same receiver. Suppose the 90° impulse of the field $\boldsymbol{B}_1$ moves the vector of magnetization on the horizontal plane perpendicular to the ambient field $\boldsymbol{B}_0$. As soon as the impulse stops to act, the magnetic dipoles of nuclei begin to precess around $\boldsymbol{B}_0$. At first, all dipoles rotate in unison; that is, they occupy practically the same position at the horizontal plane, and correspondingly generate a small magnetic field at frequency f_0 that can be detected by the receiver coil. It is natural that the signal is directly proportional to the horizontal component of magnetization, $P_{x'y'}$. Then dipoles gradually lose synchronization and it happens mainly because the constant magnet never provides uniform field $\boldsymbol{B}_0$. Inasmuch as this field changes from point to point, the precession frequency of dipole also varies (Equation (7.64)). In other words, these dipoles have different positions on the horizontal plane, and we observe so-called dephasing. When the dipole directions are uniformly distributed in this surface, the resultant magnetic field produced by them is equal to zero and no further signal is detected by receiver. The decay, measured in the absence of the sinusoidal impulse, is usually exponential and called "free induction decay." In reality its decay time is T_2^* and sign (*) indicates that a decay is not only a property of formation but of the imperfection of the device. If we assume that the field $\boldsymbol{B}_0$ is uniform, then a decrease of the magnetic field caused by intrinsic relaxation processes will be much slower and it is characterized by the relaxation time T_2. Thus, in order to study this relaxation process related to formation it is necessary to remove an influence of dephasing, caused by a change of the field B_0; otherwise NMR method cannot be used. This is one of the steps of measurements.

7.8.4. Spin echoes or refocusing

Now we describe the spin-echo method developed by Hahn which allows one to remove an influence of dephasing. With this purpose in mind he suggested to use the system of two impulses: $90^\circ - \tau - 180^\circ$ when measurements are performed at the instant 2τ. Later, other more efficient modifications of this approach were introduced. The principle of Hahn's approach is the following. First, the system of magnetic dipoles is subjected to an influence of the sinusoidal impulse of the field $\boldsymbol{B}_1$ directed along the x'-axis. As we know this impulse, applied at the instant $t = 0$, turns the vector of magnetization $\boldsymbol{P}$ by 90° and it becomes directed along the y'-axis (Fig. 7.10(a)). By definition, the total vector $\boldsymbol{P}$ is a sum of magnetic moments $\boldsymbol{p}_i$ caused by nuclei located at different places of an elementary volume. Correspondingly, they have slightly different frequency of precession. This is the reason why they begin to diverge, since some of nuclei rotate more rapidly than the system of coordinates but others are slower (Fig. 7.10(b)). For observer at the z'-axis the first group moves clockwise, but the second rotates in the opposite direction. Thus, the observer sees two groups of magnetic dipoles moving in the opposite directions. At some instant τ after the 90° impulse, the second impulse is applied and it also directed along the x'-axis. Under action of this 180° impulse each of vectors $\boldsymbol{p}_i$ is turned by 180° around the x'-axis (Fig. 7.10(c)). For an observer they continue to move in the same direction. Those vectors which move more rapidly than the system of coordinates, that is, clockwise, approach to the y'-axis. At the same time, vectors which move slower (counterclockwise) approach the same axis but from the opposite side. Therefore, at the instant $t = 2\tau$ they have the same phase and directed opposite to y'-axis (Fig. 7.10(d)). Thus, at such moment we observe refocusing of the dipoles and the effect dephasing is removed. Dipoles continue their movement and dephasing again takes place (Fig. 7.10(e)). Performing measuring of the

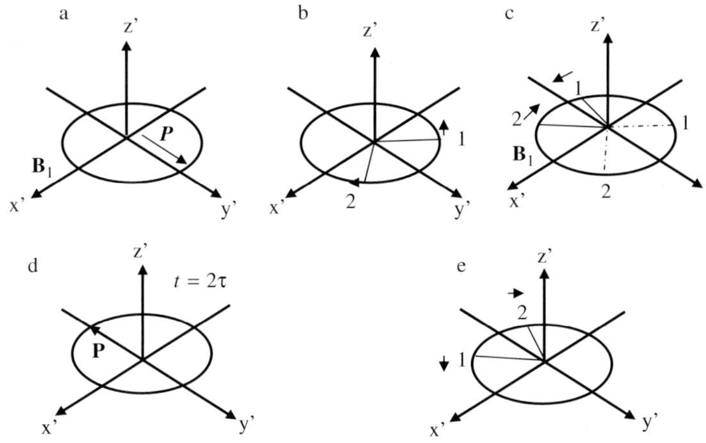

Fig. 7.10. Illustration of Hahn method.

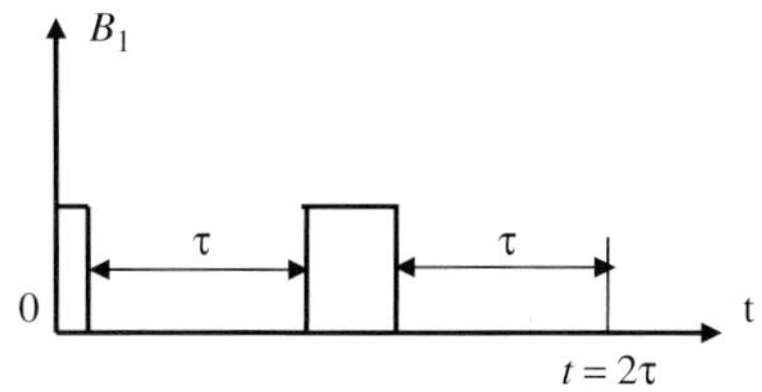

Fig. 7.11. The system of impulses in Hahn method.

electromotive force at the instant $t = 2\tau$, we determine magnetization of a medium $P(2\tau)$. This sequence of impulses is shown in Fig. 7.11. After some time ($\approx 5T_1$) the vector magnetization almost reaches saturation and the same cycle of measurements is performed but with other value of τ. Repeating these measurements with different values of time we obtain the function $P_\perp(t)$, which describes the relaxation processes in a medium and allows us to determine the time constant T_2.

7.9. TWO METHODS OF MEASURING MAGNETIC FIELD

As one more illustration of the magnetic resonance we will describe the basic features of two magnetometers. One of them is the proton precession device based on the nuclear magnetic resonance, while the other is the Cesium vapor magnetometer in which the effect of the electron magnetic resonance is used.

7.9.1. Proton precession magnetometer

The main part of this magnetometer is a small volume of liquid, such as kerosene or water with relatively high density of hydrogen placed in a small cylinder. Under an action of the magnetic field of the earth $\boldsymbol{B}_e$ all nuclear magnetic moments are involved in the precession around this field and, as we know, its frequency f_p is defined as

$$f_p = \frac{\gamma}{2\pi} B_e \tag{7.70}$$

where

$$\gamma = 2.67515418 \times 10^8 \ \mathrm{T/s}$$

For instance, if $B_e = 5.0 \times 10^{-5}$ T we have

$$f_p \approx 2130 \ \mathrm{Hz}$$

Inasmuch as the spin of hydrogen is equal to 1/2, there are two groups of nuclear dipoles.

In one of them the projection of the magnetic moment on the field direction is positive, while the other gives the negative component. Each nuclear dipole is subjected to the thermal motion, and at the state of equilibrium there is slightly more magnetic moments with the positive component than with the negative one. Respectively, the vector of nuclear magnetization $\boldsymbol{P}$ differs from zero even though it is extremely small. Unlike the nuclear magnetic moments this vector does not precess and it is directed along the ambient field $\boldsymbol{B}_e$. This is one reason why it cannot be used to measure the earth's field; the other is its small value. Our goal is to measure the frequency of precession and, then using Equation (7.70), to determine the magnetic field of the earth. With this purpose in mind the coil with constant current creates the constant magnetic field $\boldsymbol{B}_1$, which is practically normal to that of the earth and it is almost in two hundred times greater than the field of the earth ($B_1 \approx 200 B_0$); that is, it is around 100 Gauss. Because of this field the vector of magnetization becomes much stronger and it is practically directed along B_1. At some instant the current is turned off and the field $\boldsymbol{B}_1$ vanishes. Correspondingly, all nuclear magnetic moments and the vector of magnetization begin to precess around the magnetic field of the earth. Earlier we shown that a rotation of the vector $\boldsymbol{P}$ causes an alternating electromagnetic field and its frequency coincides with that of precession. Applying, for example, the coil which generates the field $\boldsymbol{B}_1$ the magnetometer measures an electromotive force which periodically changes a sign that gives the value of f_p and, therefore a value of the magnetic field of the earth. In essence, this procedure is similar to measurements of the longitudinal and transversal relaxation times T_1 and T_2, described earlier. In addition, let us make several comments:

1. In the proton precession magnetometer a relatively strong magnetization is caused by a constant magnetic field normal to the earth's field.
2. There are two steps during each measurement of the field: one of them produces a magnetization along the additional field $\boldsymbol{B}_1$, the other measures the

frequency of the electromotive force induced in the coil. This force decays during several seconds and it happens because the magnetic moments of nuclei have different phases. In average the device allows one to perform 2–3 readings per second and its sensitivity is a fraction of gamma.

3. The fluid which is either water or kerosene is diamagnetic and therefore possesses a magnetic moment, exceeding in several orders the magnetic moment of nuclei. However, its influence on the vector of magnetization is neglected, perhaps because the diamagnetic moment is much more tightly coupled to the thermal reservoir (lattice) than the nuclear dipole system.

7.9.2. Optically pumped magnetometers

Physical principles of this type of magnetometers fundamentally differ from those of the proton magnetometer. Imagine that a small cell, placed in the magnetic field of the earth, contains a gaseous metal (alkali metal like potassium or cesium; Fig. 7.12(a)). As before the device has to perform two functions, namely

1. Relatively strong increase of magnetization of a cell substance.
2. Measurements of a resonance frequency that allows one to determine the magnetic field.

To illustrate the process of magnetization we make a simplification and assume that a spin of atoms is equal to 1/2. As we know this means that in the presence of the ambient magnetic field there are two energy levels, A and B, which are called ground levels (Fig. 7.12(b)). One group of atoms is located at the level A, where the projection of magnetic moments on the field direction is positive, and this means that energy is minimal. The other group has a negative component of the magnetic moment and, correspondingly, their energy is slightly higher and their position is characterized by the level B. Earlier we demonstrated that the number of atoms in both ground levels is practically the same; more precisely amount of them at the

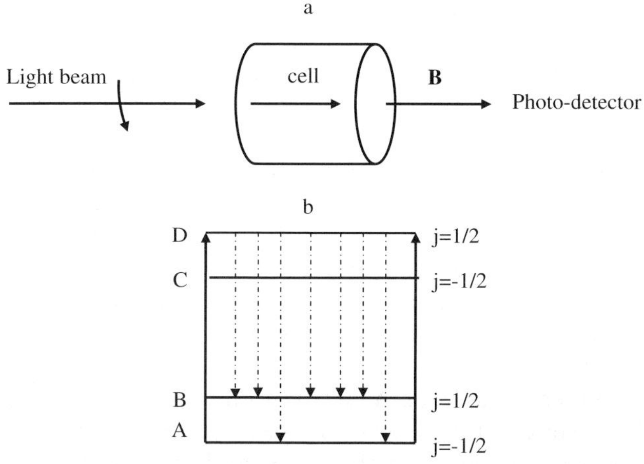

Fig. 7.12. (a) Orientation of the magnetic field and light beam in M_z magnetometer; (b) optical pumping.

level A only slightly exceeds that at the level B, and this difference is extremely small. This is the reason why the vector of magnetization, which is the sum of magnetic moments per unit volume, is also very small. Until now we have a complete analogy with the case of proton magnetometer. In that device we created a relatively strong vector of magnetization which was involved in precession around the earth's field, and it was done with the help of the additional constant field normal to the ambient magnetic field. In the optically pumped magnetometers the vector of magnetization is also created but in a completely different way which is called Optical Pumping. This approach is based on such concepts as an interaction between an electromagnetic energy and a substance, as well as an angular momentum of photon and a selection rule which governs a transition of atoms between energy levels. First of all, imagine that a beam of light moves through a cell along the ambient magnetic field (Fig. 7.12(a)). Note that the wavelength of the light is chosen in such way that photon can be absorbed by atom of the substance of a cell. In other words, there is an interaction between the light and a material. Each photon of the light is characterized by a magnitude and direction of the angular momentum. It is essential that an angular momentum of photon depends on polarization of the light; for instance, in the case of the linear polarization it is equal to zero, but the right- and left-circular polarization is associated with either $+1\hbar$ or $-1\hbar$ angular momentum. We consider the beam with the right-circular polarization and therefore when an atom absorbs a photon its spin is increased by 1. This absorption causes an increase of energy and we can expect that atoms are transformed at excited levels with higher energy (C and D). In fact this happens, but there is one very important exception, namely sublevel B, where atoms cannot absorb photons of the beam with the right-circular polarization, and they remain at this state regardless of the action of light. Such behavior follows from the selection rule. For instance, if we imagine that atom of the sublevel B absorbs a photon then it would move at the excited sublevel $j = 3/2$ which does not exist. Thus, atoms of this ground sublevel are not subjected to an influence of the light. This remarkable feature of the state B is in essence a key point of the process of magnetization.

Next consider a behavior of atoms at the level A. As a result of this absorption they move at the excited level D where the spin becomes equal $1/2$, $((-1/2) + 1)$. As soon as atom is at the state D it very quickly begins to emit a photon, and a spontaneous emission is observed. Unlike the process of absorption the spontaneous emission is accompanied by photons which may increase or decrease a spin of atom by one or preserve the same and, correspondingly, the probabilities of each transition from D to B and from D to A are nearly equal. As we already know atoms which fall at sublevel B cannot move to other states, and they remain there for sufficiently long time greatly exceeding lifetime at the excited states. As concerns atoms at the state A, the situation is completely different, since the selection rule allows them to absorb photons and jumps to the state D. Then, as before, they fall to either states B or A. If atom falls to B it gets stuck and stays there, but if it falls at state A this process is repeated. Finally, the most atoms are located at the same state B, and therefore the total sum of magnetic moments becomes relatively large. In other words, this process of optical pumping creates magnetization of a medium inside a light beam and this is the first important action of the optically pumped magnetometer. In more general case, when there are more sublevels, we observe a similar picture, since transitions caused by absorption of photons are possible from all ground levels except one sublevel. At the same time the inverse transition (spontaneous emission) is permitted to all ground sublevels. Since lifetime of electrons at the excited sublevels is in several orders smaller than that at ground levels, it is possible to observe the optical pumping.

Now we outline the principle of measurements of the magnetic field with the optically pumped magnetometer and, as an example, consider M_z-type of these devices. In this case the light beam is parallel to the ambient magnetic field (Fig. 7.12(a)). As was shown in Chapter 6, a difference of magnetic energies at states A and B is

$$\Delta U_{\mathrm{mag}} = \hbar\omega_{\mathrm{p}}$$

The frequency of precession corresponds to radio-frequency range and it is related with the measured magnetic field as

$$\omega_{\mathrm{p}} = \gamma B_{\mathrm{e}}$$

and γ is gyroscopic constant for electron. First of all, let us notice that when atoms are located at the level B the cell is transparent to the light because they are not able to absorb photons. Now suppose that the cell with a substance is placed into the additional field which is perpendicular to the ambient field and its frequency coincides with ω_{p}. In previous sections we demonstrated that this sinusoidal magnetic field is able to change a direction of the vector of magnetization at any angle with respect to the ambient field, in particular, at angles exceeding $\pi/2$. This means that atoms move from the state B to A where they absorb photons. Because of this energy of light becomes smaller and photo element can detect a minimum of its intensity. Thus, changing a frequency of the additional field we can determine ω_{p} and, correspondingly, calculate the ambient magnetic field B_{e}. Note that the light beam does not have to be completely parallel to the ambient magnetic field. However, the signal is the strongest in that case, and it vanishes when the field and light beam direction form the right angle.

Bibliography

Abragam, A., 1961. Principles of Nuclear Magnetism. Oxford University Press, New York, USA.

Alpin, L.M., 1966. The Theory of Fields. Nedra, Moscow.

Campbell, W.H., 1997. Introduction to Geomagnetic Fields. Cambridge University Press, Cambridge.

Chapman, S. and Bartels, J., 1940. Geomagnetism. Oxford University Press, Oxford, England.

Farrar, T.C., 1971. Pulse and Fourier Transform NMR Introduction to Theory and Methods. Academic Press, New York, USA.

Feynman, R.P., Leighton, R.B. and Sands, M.L., 1965. Feynman Lectures on Physics, Vol. 2. Addison-Wesley Company, Reading, MA, USA.

Ianovski, V.M., 1944. Earth Magnetism. Leningrad, S-Petersburg State University, S-Petersburg, Russia.

Jakosky, J.J., 1950. Exploration Geophysics. Trija, Newport Beach, CA, USA.

Kaufman, A.A., 1992. Geophysical Field Theory and Method, Part A. Academic Press, London, England.

Kaufman, A.A. and Hansen, R.O., 2007. Principles of the Gravitational Method. Elsevier, Amsterdam.

Kleinberg, R.L., 1999. Nuclear Magnetic Resonance, Experimental Methods in the Physical Science, Vol. 35, Academic Press, San-Diego, CA, USA.

Kleinberg, R.L. and Flaum, C., 1998. Review: NMR Detection and Characterization of Hydrocarbons in Subsurface Earth Formations, Methods and Applications in Material Science. Wiley, Weinhelm, Germany.

Merill, R.T. and McElhinny, M.W., 1998. The Magnetic Field of the Earth. Academic Press, San-Diego, CA, USA.

Nettleton, L.L., 1940. Geophysical Prospecting for Oil. McGraw-Hill, New York.

Oreskes, N. (Ed.), 2001. Plate Tectonics. Westview, Boulder, Colorado, USA.

Parasnis, D.S., 1986. Principles of Applied Geophysics. Chapman and Hall, New York, USA.

Parkinson, W.D., 1982. Introduction to Geomagnetism. Elsevier, Amsterdam.

Smythe, W.R., 1968. Static and Dynamic Electricity. McGraw-Hill, London, England.

Stratton, J.A., 1941. Theory of Electromagnetism. McGraw-Hill, New-York.

Whittaker, E.T., 1960. A History of the Theories of Aether and Electricity. Harper, London, England.

Zilberman, G.E., 1970. Electricity and Magnetism. Nauka, Moscow.

Appendix

PALEOMAGNETISM AND PLATE TECTONICS

As is well known, measurements of the magnetic field have found broad applications in exploration, engineering, and global geophysics. These include studies of the surface, airborne, marine, and borehole magnetic methods, the investigation of the present magnetic field of the earth and surrounding space and its behavior in the past, as well as a role of magnetic fields for understanding of dynamic processes in the earth core. In this appendix we will describe the important role of magnetic methods in making one of the greatest discoveries, which fundamentally changed the former theory of tectonics. This historical event happened in the middle of the last century; but before we describe it, let us return to magnetism and review some known facts. Magnetization of rocks occurs mainly due to the presence of different types of ferromagnetic materials. Numerous measurements at various parts of the world with samples of rocks of different geological ages have shown that in general magnetization consists of two parts. These components are the inductive and remanent magnetizations. Each of them is characterized by vectors of magnetization, which may have different magnitude and direction. In particular, the remanent magnetization can have a direction opposite to the direction of the present magnetic field. By definition, the inductive magnetization is caused by the existing magnetic field, while the remanent magnetization arose in past usually during the formation of rocks. Earlier (Chapter 2) we pointed out that from physical point of view both magnetizations have the same inductive origin. At the same time the study of the permanent magnetization is of a great importance because it allows one to reconstruct the history of the geomagnetic field of the earth. This branch of geomagnetism is called paleomagnetism. There are different types of the remanent magnetization. One example is thermo-permanent magnetization (TPM). It plays a dominant role in the magnetization of the igneous rocks. Liquid magma moves through pre-existing formations or appears as lava on the surface. With time the temperature of the lava decreases and it is solidified into igneous rocks. At the beginning due to high kinetic energy the magnetic dipoles of ferromagnetic particles are distributed randomly and magnetization is absent. Then, after the temperature crosses Curie point, dipoles align along the geomagnetic field at that time and thermo-remanent magnetization takes place. This magnetization remains constant and preserves the magnetic field at the time of cooling. Magnetization of sedimentary rocks, containing small particles of magnetic material like magnetite, may occur differently and usually it happens in shallow rivers, lakes, and sea. During the process of deposition these particles, having various magnetic moments, are oriented along the magnetic field. As we know, in such instances the torque is equal to zero. Later, when a formation is consolidated under pressure, the ferromagnetic particles lose their ability to move. This means that the vector of magnetization in every elementary volume is directed along the geomagnetic field of the time when a sedimentary rock was formed. This process is termed as detrital remanent magnetization (DRM).

It is obvious that in general, the total vector of magnetization is a sum of the remanent and inductive magnetization. Their separation and measurement is a very important element of the paleomagnetism. The ratio of these quantities is called Koenigsberger ratio and some examples of its value are given below only for illustration.

Rock	$Q_n = P_r/P_{in}$
Basalt	110
Gabbro	30
Andesite	5
Granite	03–10
Diabase (Dolerite)	0.5

Paleomagnetism studies allowed one to reconstruct the magnetic field of the earth over the last 100 million years to discover many interesting features of its behavior. One of them is the reversal of the polarity that has happened numerous times in geologic history. Reversal episodes appear with increasing frequency during the more recent Miocene, Pliocene, and Pleistocene Eras. It is essential that due to the modern methods of geochronology, the time, when different polarities of the magnetic field took place, is known with sufficient accuracy. This information was vitally important to support the main concept of the plate tectonics. Now we are ready to describe the role of magnetic methods in developing the plate tectonic theory.

In the past, tectonic theory was mainly based on a study of geological structures of continents while the largest part of the earth covered by oceans was sparsely investigated. After the World War II it became possible to obtain detailed information about topography of the ocean bottom. Prior tectonic theory was strongly based on the concept that formation of geological structures was a result of the vertical motions only. The thought of the horizontal motion was not considered. Several scientists, such as Abraham Ortelius (1596), Francis Bacon(1620), Benjamin Franklin, Antonio Snider-Pellegrini (1858) and others noted that a shape of continents on either side the Atlantic Ocean fit together. This fact led to the idea, developed by Wegener, that continents once formed a single landmass which was broken up and the continents moved to their present positions. He formulated the concept of the Continental Drift, which implied the presence of horizontal motion. Wegener gave many examples to support his hypothesis, but was not able to provide an explanation of physical processes which can cause this motion of continents. His suggestion that a motion of continents was caused by a rotation of the earth and fictitious centrifugal force was rejected by the scientific community. At the same time, the idea of Continental Drift received support from several geologists, one of them being Alexander Du Toit, who brought more evidence that the continents drift. Arthur Holmes also supported the idea of Continental Drift and suggested that continents do not float on the mantle, but their motion is caused by convection currents driven by the heat of the interior of the earth. In the mid-1950s, groups of geophysicists led by Irving and Blackett performed intensive measurements of the remanent magnetization of rocks. To illustrate the importance of these data, let us recall (Chapter 1) that there is a relation between angles θ and α, characterizing a position of an observation point and a direction of the vector of magnetization $\boldsymbol{P}$ (Fig. 1)

$$\tan\,\alpha = \frac{1}{2}\tan\,\theta \qquad (1)$$

Measuring the angle α, we calculate the angle θ between the radius R and orientation of the dipole moment $\boldsymbol{M}$. When the line, along which this moment is directed, intersects the earth's surface we obtain a point which represents the magnetic pole corresponding to the dipole part of the magnetic field of the earth. These data together with the knowledge of the age of rock samples allowed one to find a position of the magnetic pole as a function of time at different continents and plot paths of a movement of a pole at each continent (polar wander). It turned out that paths obtained for Australia, India, North America, and Europe differed from each other. If they coincided that meant that these continents did not move with respect to each other, but this result indicated that there was the continental drift. The leading geologists and geophysicists, like Runcorn and Bullard accepted this phenomenon. American geologist Harry Hess, inspired by these paleomagnetic studies, returned to his research which he started before World War II where he

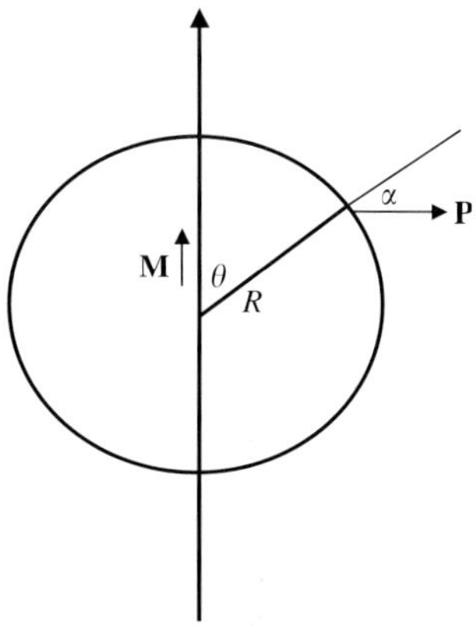

Fig. 1. Illustration of Equation (1).

discussed a role of convection currents in continental drift. Following Holmes, he supposed that the mantle convection might be driving the crust apart at mid-ocean ridges and downward at ocean trenches and causing the continental drift. In the paper, published in 1962, he wrote "…but the general picture on paleomagnetism is sufficiently compelling that is more reasonable to accept than to disregard it." In accordance with Hess's theory, the seafloor itself moves and carries the continents with it and the cause of this phenomenon is the convection currents in the upper mantle. He suggested that mid-ocean ridges are associated with weak zones of the lithosphere through which new magma moves upward and erupts along the crest of the ridges, causing new ocean crust. During many millions of years this motion of magma created a system of mid-ocean ridges extended over 50,000 km. This process of the seafloor spreading implied that the earth is expanding but observations could not explain this phenomenon. In order to overcome this problem Hess also suggested that it must be shrinking elsewhere and new ocean crust continuously moves away from ridges in a conveyor – similar to very deep, narrow canyons (trenches). This continuous process of creating new crust and destroying old oceanic lithosphere takes place simultaneously, but at different places. As a result, in accordance with Hess, the Atlantic Ocean, where ridges are located, was expanding, but Pacific Ocean was shrinking because of presence of trenches. Such mechanism explained many important facts such as a Continental Drift, preservation of the earth's dimensions in spite of the seafloor spreading, the small accumulation of sediments on the ocean floor, and why ocean rocks are much younger than continental rocks. It became clear that the motion of land masses is the result of moving plates – large fragments of the upper part of the earth where continents are located. The thickness of these plates is around 80–100 km and they move at rate of 3–10 cm per year. At the boundary of two plates, there are volcanoes, earthquakes, and mountains, and such coincidences take place because at these locations the plates either collide, move away from each other, or slide past one another. The central idea of this revolutionary theory developed by Hess and others was that the seafloor was spreading and magnetic methods again played an extremely important role to prove this phenomenon. In 1955, a yearlong, the bathymetric survey was performed off the west coast of the United States along with measurements of the magnetic field. Observations resulted in a map which showed a system of strips of positive and negative anomalies with respect to the present magnetic field. The strips were oriented along a north–south direction. These linear anomalies were about 1% or less of the earth's normal field and a few tens of kilometers in width. They were present throughout the length of 2000 km profiles with a few exceptions. This result was very unusual and had not been previously observed on land or at sea. At the beginning, there were attempts to explain them in the old classical way as a result of a change of magnetization of rocks forming ocean bottom. There were several different explanations for the origin of the system of magnetic linear anomalies, but it was difficult to find reasonable geological structures which support this field behavior. Then after 7 years two geologists Vine and Matthew, and independently Morley, offered very elegant explanation which became one of the most important arguments in favor of the seafloor spreading. First of all they were familiar with two ideas, which were not universally accepted at that time: the seafloor spreading and reversals of the earth's magnetic field. Their interpretation of these anomalies was the following. When magma rose through a weak zone of the lithosphere, it spread in both directions parallel to the axis of the mid-ocean ridge. The basalt magma filled this space and began to cool off. As the temperature cooled below Curie point the rock became magnetized and the direction of the vector of magnetization corresponded to that of the magnetic field which existed at that time. As is well known, this magnetization remains unchanged with time but moves away from the ridge as new magma arrives. After several millions of years, the polarity of the magnetic field changes and new portion of magma acquires opposite direction of magnetization. As a result we obtain strips of magnetization moving away from the place where they were created. Since plates move in the opposite directions, magnetic strips are located symmetrically to the place where magma arrives. In the same manner after a change of polarity more pairs of strips appear with opposite directions of magnetization, and so on. Each of these zones create the magnetic field which is either directed as the current field or has opposite direction and, correspondingly, the measurements across this system produce the positive and negative anomalies. The width of each anomaly over some zone of basalt is defined by the extension of the period during which the magnetic field does not change its polarity. In order to prove that this explanation is correct, Vine and Matthews as well as Morley, made use of the known information about polarity of the magnetic field at different times. Then, knowing the distance of each strip from the origin and assuming that the velocity of the lithosphere is about several centimeters per year, they were able to find time during which each of these zones moved, and correspondingly time when reversal of the magnetic field took place. To their great satisfaction they found that these data were in agreement with the known data derived from paleomagnetism and geochronology. This result is considered as one of the most important proofs of the seafloor spreading and this phenomenon was observed at several locations on the earth.

Subject Index